SEMICONDUCTORS AND SEMIMETALS

VOLUME 6

Injection Phenomena

SEMICONDUCTORS AND SEMIMETALS

Edited by R. K. WILLARDSON
BELL AND HOWELL ELECTRONIC MATERIALS DIVISION
PASADENA, CALIFORNIA

ALBERT C. BEER
BATTELLE MEMORIAL INSTITUTE
COLUMBUS LABORATORIES
COLUMBUS, OHIO

VOLUME 6

Injection Phenomena

1970

ACADEMIC PRESS New York and London

ACADEMIC PRESS, INC.
111 Fifth Avenue, New York, New York 10003

United Kingdom Edition published by
ACADEMIC PRESS, INC. (LONDON) LTD.
Berkeley Square House, London W1X 6BA

LIBRARY OF CONGRESS CATALOG CARD NUMBER: 65-26048

PRINTED IN THE UNITED STATES OF AMERICA

Contents

Chapter 4 **Double Injection in Semiconductors**

R. Baron and J. W. Mayer

Chapter 5 **The Photoconductor–Metal Contact**

W. Ruppel

List of Contributors

Numbers in parentheses indicate the pages on which the authors' contributions begin.

ALLEN M. BARNETT, *General Electric Company, Research and Development Center, Schenectady, New York* (141)

R. BARON, *Hughes Research Laboratory, Malibu, California* (201)

MURRAY A. LAMPERT, *Princeton University and RCA Laboratories, Princeton, New Jersey* (1)

J. W. MAYER,[1] *Hughes Research Laboratory, Malibu, California* (201)

W. RUPPEL, *University of Karlsruhe, Karlsruhe, Germany* (315)

RONALD B. SCHILLING, *RCA Semiconductors Division, Princeton, New Jersey* (1)

RICHARD WILLIAMS, *RCA Laboratories, Princeton, New Jersey* (97)

[1] Present address: California Institute of Technology, Pasadena, California.

Preface

The extensive research that has been devoted to the physics of semiconductors and semimetals has been very effective in increasing our understanding of the physics of solids in general. This progress was made possible by significant advances in material preparation techniques. The availability of a large number of semiconductors with a wide variety of different and often unique properties enabled the investigators not only to discover new phenomena but to select optimum materials for definitive experimental and theoretical work.

In a field growing at such a rapid rate, a sequence of books which provide an integrated treatment of the experimental techniques and theoretical developments is a necessity. The books must contain not only the essence of the published literature, but also a considerable amount of new material. The highly specialized nature of each topic makes it imperative that each chapter be written by an authority. For this reason the editors have obtained contributions from a number of such specialists to provide each volume with the required detail and completeness. Much of the information presented relates to basic contributions in the solid state field which will be of permanent value. While this sequence of volumes is primarily a reference work covering related major topics, certain chapters will also be useful in graduate study. In addition, a number of the articles concerned with applications of specific phenomena will be of value to workers in various specialized areas of device development.

Because of the important contributions which have recently resulted from studies of the III–V compounds, the first few volumes of this series have been devoted to the physics of these materials: Volume I reviews key features of the III–V compounds, with special emphasis on band structure, magnetic field phenomena, and plasma effects. Volume 2 emphasizes physical properties, thermal phenomena, magnetic resonances, and photoelectric effects, as well as radiative recombination and stimulated emission. Volume 3 is concerned with optical properties, including lattice effects, intrinsic absorption, free carrier phenomena, and photoelectronic effects. Volume 4 includes thermodynamic properties, phase diagrams, diffusion, hardness, and phenomena in solid solutions as well as the effects of strong electric fields, hydrostatic pressure, nuclear irradiation, and nonuniformity of impurity distributions on the electrical and other properties of III–V compounds. Volume 5 is devoted to infrared detectors, and is the first of a number of volumes to deal specifically with applications of semiconductor properties. The present volume is concerned with injection phenomena in solids, including current injection and filament formation, double injection, internal photoemission, and photoconductor–metal contacts.

In addition to a second volume on infrared detection, subsequent volumes of *Semiconductors and Semimetals* will be devoted to other applications such as high-temperature diodes and power rectifiers, field-effect transistors, IMPATT diodes, tunnel diodes, and applications of bulk negative resistance. Volumes will also deal with such fundamental phenomena as lattice dynamics, galvanomagnetic effects, luminescence, and nonlinear optical phenomena, as well as electro-, thermo-, piezo-, and magnetooptical effects.

The editors are indebted to the many contributors and their employers who made this series possible. They wish to express their appreciation to the Bell and Howell Company and the Battelle Memorial Institute for providing the facilities and the environment necessary for such an endeavor. Thanks are also due to the U.S. Air Force Offices of Scientific Research and Aerospace Research and the U.S. Navy Office of Naval Research and the Corona Laboratories, whose support has enabled the editors to study many features of compound semiconductors. The assistance of Rosalind Drum, Martha Karl, and Inez Wheldon in handling the numerous details concerning the manuscripts and proofs is gratefully acknowledged. Finally, the editors wish to thank their wives for their patience and understanding.

R. K. WILLARDSON
ALBERT C. BEER

Semiconductors and Semimetals

Volume 1 **Physics of III–V Compounds**

C. Hilsum, Some Key Features of III–V Compounds
Franco Bassani, Methods of Band Calculations Applicable to III–V Compounds
E. O. Kane, The $k \cdot p$ Method
V. L. Bonch-Bruevich, Effect of Heavy Doping on the Semiconductor Band Structure
Donald Long, Energy Band Structures of Mixed Crystals of III–V Compounds
Laura M. Roth and Petros N. Argyres, Magnetic Quantum Effects
S. M. Puri and T. H. Geballe, Thermomagnetic Effects in the Quantum Region
W. M. Becker, Band Characteristics near Principal Minima from Magnetoresistance
E. H. Putley, Freeze-Out Effects, Hot Electron Effects, and Submillimeter Photoconductivity in InSb
H. Weiss, Magnetoresistance
Betsy Ancker-Johnson, Plasmas in Semiconductors and Semimetals

Volume 2 **Physics of III–V Compounds**

M. G. Holland, Thermal Conductivity
S. I. Novikova, Thermal Expansion
U. Piesbergen, Heat Capacity and Debye Temperatures
G. Giesecke, Lattice Constants
J. R. Drabble, Elastic Properties
A. U. Mac Rae and G. W. Gobeli, Low Energy Electron Diffraction Studies
Robert Lee Mieher, Nuclear Magnetic Resonance
Bernard Goldstein, Electron Paramagnetic Resonance
T. S. Moss, Photoconduction in III–V Compounds
E. Antončík and J. Tauc, Quantum Efficiency of the Internal Photoelectric Effect in InSb
G. W. Gobeli and F. G. Allen, Photoelectric Threshold and Work Function
P. S. Pershan, Nonlinear Optics in III–V Compounds
M. Gershenzon, Radiative Recombination in the III–V Compounds
Frank Stern, Stimulated Emission in Semiconductors

Volume 3 **Optical Properties of III–V Compounds**

Marvin Hass, Lattice Reflection
William G. Spitzer, Multiphonon Lattice Absorption
D. L. Stierwalt and R. F. Potter, Emittance Studies
H. R. Philipp and H. Ehrenreich, Ultraviolet Optical Properties
Manuel Cardona, Optical Absorption above the Fundamental Edge
Earnest J. Johnson, Absorption near the Fundamental Edge
John O. Dimmock, Introduction to the Theory of Exciton States in Semiconductors
B. Lax and J. G. Mavroides, Interband Magnetooptical Effects

Volume 4 **Physics of III–V Compounds**

Volume 5 **Infrared Detectors**

SEMICONDUCTORS AND SEMIMETALS

VOLUME 6

Injection Phenomena

CHAPTER 1

Current Injection in Solids: The Regional Approximation Method

Murray A. Lampert and Ronald B. Schilling

I. Introduction

A large number of interesting problems of current injection into solids cannot be solved analytically. Here, we are not referring to problems involving strange or complicated electrode shapes, that is, problems essentially in the realm of applied mathematics; we are talking about problems with one-dimensional current-flow geometry. This class of analytically unsolvable problems includes most planar-flow, double-injection problems, that is, planar-flow problems in which, simultaneously, electrons are injected at the cathode and holes at the anode, and essentially all problems, both single and double injection, of radial current flow, either cylindrical or spherical. What shall be done about such problems? One line of attack lies in the use of high-speed, high-capacity digital computers to obtain numerical solutions for specific choices of the parameters in each problem. If numerical accuracy of the solution is required, say, accuracy to within a few percent, this approach will likely be the only satisfactory one. However, this approach has a severe drawback : the near-total absence of physical insight accompanying a purely numerical solution. The science of current injection in solids is still in its early stages, and the unsolvable problems we shall be talking about are all fairly basic ones. It is very difficult to see how this science can be constructed solely on an edifice of numerical solutions. Even if this could be done, it surely would be the hard way to do it, and not likely a way that would please many practitioners. It is further relevant to note that the current state of the art in materials preparation of insulators is not such as to necessitate or justify a quest for extreme accuracy in the solution of injection problems. For what profit a man if his numerical solution to a homogeneous problem is good to a few percent, and the sample inhomogeneities are invariably on the order of 100% or greater?

In this chapter, we follow a different line of attack. We take the view that very substantial progress can be made by accepting a loss of accuracy in favor of an approximation scheme which not only brings the previously intractable problems back into the fold of analytic tractability, but emphasizes the underlying physics of each problem in so doing. This scheme, which we call the "regional approximation method," is based on the simple observation that where there are several terms in an equation "competing" with each other, each being a function of position, it is generally possible to divide up the volume of the solid into separate regions, in each of which either a single term or a couple of terms dominate. Within each such region, the basic approximation is made of neglecting all terms within the competing group except the one or two terms which dominate in that region. Since everything in this chapter is based on the use of this approximation, it is well to illustrate it by a few concrete examples.

a. First Example

The current equation characterizing one-carrier, planar current flow is

$$J = e\mu n(x)\mathscr{E}(x) = \text{const}, \qquad n(x) = n_i(x) + n_0, \tag{1}$$

with J the current density, e the electronic charge, μ the free-carrier drift mobility, $\mathscr{E}(x)$ the field intensity at position x, $n(x)$ the total free-carrier density at position x, $n_i(x)$ the injected free-carrier density at position x, and n_0 the thermally-generated density of free carriers, assumed independent of x. In writing Eq. (1), clearly, the diffusion contribution to the current flow has been neglected. Taking the injecting electrode at $x = 0$ and the collecting electrode at $x = L$, the spatial variation of $n_i(x)$, at "low" current levels, will resemble that shown in the schematic plot of Fig. 1; namely, $n_i(x)$ decreases monotonically from a high value near the injecting electrode to a low value near the collecting electrode, crossing the value n_0 at some plane $x_a(J)$ whose position will depend on the current J, as indicated. Between $x = 0$ and $x = x_a(J)$, denoted by region B, $n_i(x) > n_0$; between $x = x_a(J)$ and $x = L$, denoted by region A, $n_i(x) < n_0$. The regional approximation consists, in the above equation (1), of neglecting n_0 in region B and $n_i(x)$ in region A; thus,

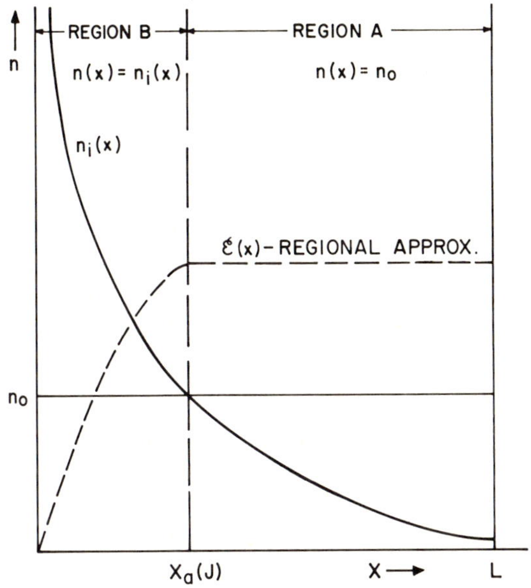

FIG. 1. Schematic spatial variations of n_i and $\mathscr{E}$ for a one-carrier, planar-flow problem.

Region A:

$$J = e\mu n_0 \mathscr{E}(x) = \text{const}, \qquad \mathscr{E} = J/e\mu n_0 = \text{const}; \tag{2a}$$

Region B:

$$J = e\mu n_i(x)\mathscr{E}(x) = \text{const}. \tag{2b}$$

The boundary $x_a(J)$ between the two regions is clearly given by $n_i(x_a) = n_0$. Region A may appropriately be called an ohmic region, region B a space–charge region, that is, a region dominated by injected space charge. These considerations apply to *any* planar flow, one-carrier injection problem irrespective of the other features of the problem, such as electron trapping. In the next example, we consider a specific problem involving a particular set of electron traps.

b. Second Example

Suppose that in the solid, in addition to the thermally generated free carriers n_0, there are also present a significant density N_t of electron traps lying at the energy level E_t above the Fermi level F_0, as illustrated in the schematic energy-band diagram of Fig. 2. For a complete characterization of the problem, in addition to the current equation (1), the following Poisson

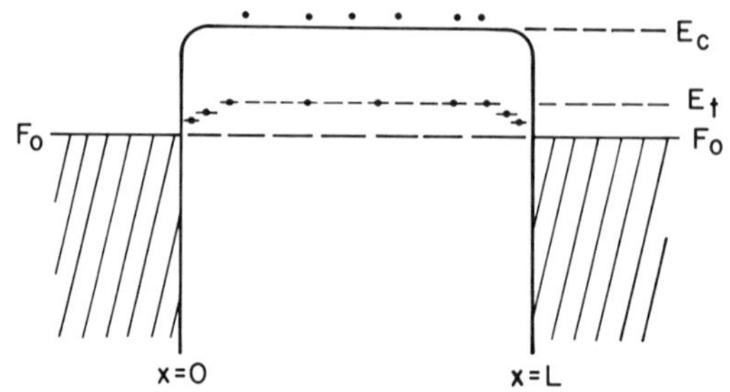

FIG. 2. Schematic energy-band diagram for the problem of an insulator with a single (significant) set of traps lying above the Fermi level. Note that contacts are ohmic (downward-bending) for electron injection.

equation must also be taken into account, again written for planar electron flow:

$$\frac{\varepsilon}{e}\frac{d\mathscr{E}(x)}{dx} = n_{\rm i}(x) + n_{\rm t,i}(x); \qquad n_{\rm i}(x) = n(x) - n_0, \qquad n_{\rm t,i}(x) = n_{\rm t}(x) - n_{\rm t,0}, \quad (3)$$

$$n_{\rm t}(x) = \frac{N_{\rm t}}{1 + [N/gn(x)]}, \qquad n_{\rm t,0} = \frac{N_{\rm t}}{1 + (N/gn_0)}, \qquad N = N_{\rm c} \exp\frac{E_{\rm t} - E_{\rm c}}{kT}, \quad (4)$$

with ε the static dielectric constant, $n_{\rm t,i}(x)$ and $n_{\rm t}(x)$ the injected and total trapped-electron densities, respectively, at position x, and $n_{\rm t,0}$ the thermal equilibrium value of $n_{\rm t}$, assumed independent of x. Equation (4) is a particular way of writing the familiar Fermi–Dirac statistical relationship between $n_{\rm t}(x)$ and $N_{\rm t}$, based on the assumption that the free and trapped electrons at each position x remain in quasithermal equilibrium in the presence of an applied field; g is the statistical weight of the trap, $N_{\rm c}$ the effective density of states in the conduction band, k is Boltzmann's constant, and T is the absolute lattice temperature. Since we are discussing a planar-flow, one-carrier problem, Eq. (1) is applicable, and therefore, at low currents, the discussion for the first example is applicable. There are two regions in the solid, region B extending from $x = 0$ to $x = x_a(J)$ over which n_0 can be neglected, and region A extending from $x = x_a(J)$ to $x = L$ over which $n_{\rm i}$ can be neglected. In region A, the problem is solved, within our approximation, by Eq. (2a). In region B, not only can n_0 be neglected compared to $n_{\rm i}$, but, concomitantly, for this particular problem, so can $n_{\rm t,0}$ be neglected compared to $n_{\rm t,i}$. Thus, the equations characterizing the problem in region B are

$$J = e\mu n_{\rm i}(x)\mathscr{E}(x) \qquad \text{or} \qquad J = e\mu n(x)\mathscr{E}(x), \quad (5a)$$

$$\frac{\varepsilon}{e}\frac{d\mathscr{E}}{dx} = n_{\rm i}(x) + n_{\rm t,i}(x) \qquad \text{or} \qquad \frac{\varepsilon}{e}\frac{d\mathscr{E}}{dx} = n(x) + n_{\rm t}(x), \quad (5b)$$

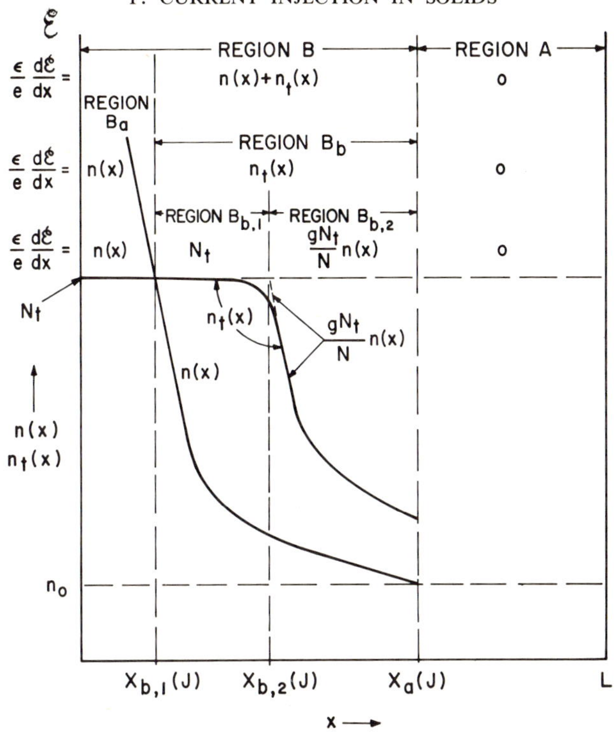

FIG. 3. Schematic spatial variations of n_i and $n_{t,i}$ for a one-carrier, planar-flow problem involving a single set of electron traps lying above the Fermi level. [In accordance with the procedure followed throughout the chapter, the subscript "i" is dropped in region B—see Eqs. (5a,b).] The final breakdown of the problem into its component elements is a four-region problem at low current levels.

with $n_t(x)$ given by Eq. (4). The equations can be written in either of the two alternative ways because $n_i(x) \approx n(x)$ and $n_{t,i}(x) \approx n_t(x)$ in region B. From this point on, throughout the text, we shall use the right-hand alternative. The two regions A and B are illustrated in Fig. 3. The boundary $x_a(J)$ between them is given by $n(x_a) = n_0$.

Since the right-hand side of the Poisson equation (5b) contains the sum of the two smoothly varying quantities $n(x)$ and $n_t(x)$, this equation is also a good candidate for application of the regional approximation method. Referring to the plots of these quantities in Fig. 3, we see that for $0 < x < x_{b,1}(J)$, referred to as region B_a, $n(x) > n_t(x)$, and for $x_{b,1}(J) < x < x_a(J)$, referred to as region B_b, $n_t(x) > n(x)$. Thus, the regional approximation method, applied to the Poisson equation, gives

Region B_a:

$$\frac{\varepsilon}{e}\frac{d\mathscr{E}}{dx} = n(x)\,; \tag{6a}$$

Region B_b:

$$\frac{\varepsilon}{e}\frac{d\mathscr{E}}{dx} = n_t(x), \qquad n_t(x) = \frac{N_t}{1 + [N/gn(x)]}. \tag{6b}$$

The functional dependence of $n_t(x)$ upon $n(x)$, as exhibited in Eq. (6b), suggests a further simplification: Why not apply the regional approximation to the denominator in the expression for $n_t(x)$? Region B_b is then further subdivided into two regions as illustrated in Fig. 3. For $x_{b,1}(J) < x < x_{b,2}(J)$, referred to as region $B_{b,1}$, $n_t(x) = N_t$, and for $x_{b,2}(J) < x < x_a(J)$, referred to as region $B_{b,2}$, $n_t(x) = gN_t n(x)/N$:

Region $B_{b,1}$:

$$\frac{\varepsilon}{e}\frac{d\mathscr{E}}{dx} = N_t, \tag{6c}$$

Region $B_{b,2}$:

$$\frac{\varepsilon}{e}\frac{d\mathscr{E}}{dx} = \frac{gN_t}{N}n(x). \tag{6d}$$

This last approximation is nothing more than a sharpening-up of the Fermi–Dirac distribution function in the neighborhood of the quasi-Fermi level when it crosses the trap level; it corresponds to extending the Boltzmann exponential tail right up to the Fermi sea instead of allowing for the more gradual transition dictated by nature.

We see that, by successive application of the regional approximation method, we have broken up the problem, at low current levels, into a four-region problem, as illustrated in Fig. 3. This problem is treated in detail in Section 1. There it is shown that as the current increases from low levels, a critical current $J_{cr,1}$ is reached at which region A leaves the solid, that is, at which $x_a(J_{cr,1}) = L$. For $J > J_{cr,1}$ the problem is then a three-region problem, until a new critical current $J_{cr,2}$ is reached at which region $B_{b,2}$ leaves the solid, that is, at which $x_{b,2}(J_{cr,2}) = L$. For $J > J_{cr,2}$ the problem is then a two-region problem, until a final critical current $J_{cr,3}$ is reached at which region $B_{b,1}$ leaves the solid, that is, at which $x_{b,1}(J_{cr,3}) = L$. Beyond $J_{cr,3}$, there is only the one region B_a in the solid.

A final simplification is still possible. At low currents, $J < J_{cr,1}$, where there are nominally four regions in the solid, it is found that negligible error is made in the current–voltage characteristic if the first two regions, namely, regions B_a and $B_{b,1}$ are ignored completely. That is, for $J < J_{cr,1}$ region $B_{b,2}$ is artificially extended, on the left, so as to reach the cathode: $x_{b,2} = 0$. Then, for $J_{cr,1} < J < J_{cr,2}$, again two regions suffice to determine the current–voltage characteristic; now the two regions are regions $B_{b,1}$ and $B_{b,2}$, region B_a being neglected: $x_{b,1} = 0$. It is to be noted that this final simplification procedure also works satisfactorily for one-carrier, spherical-flow problems, although here some interesting physics may be lost in its relentless application.

c. Third Example

The regional approximation method can also be used to advantage in dealing with equations that do not have the direct physical significance that Eqs. (1), (3), and (4) have. An example of this is furnished by the problem of the constant-lifetime plasma injected into a solid, illustrated schematically, for the semiconductor case, by the energy-band diagram of Fig. 4. The

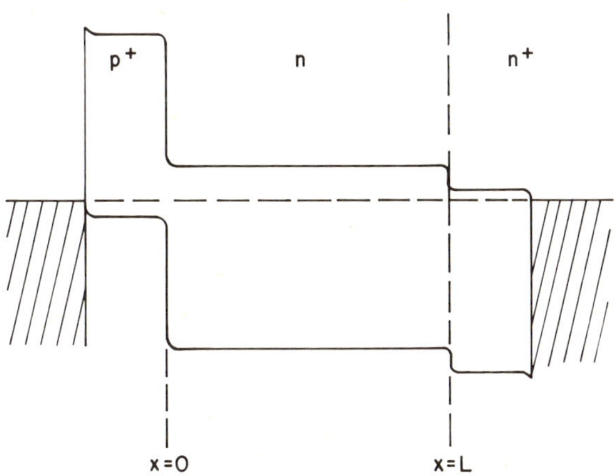

FIG. 4. Schematic energy-band diagram for the problem of a plasma injected into an N-type semiconductor. The hole-injecting contact is the p^+–n junction at $x = 0$, the electron-injecting contact is the n^+–n junction at $x = L$.

theoretical study of this problem is most conveniently transacted working with the master equation

$$-\frac{\varepsilon}{e}\frac{d}{dx}\left(\mathscr{E}\frac{d\mathscr{E}}{dx}\right) + (n_0 - p_0)\frac{d\mathscr{E}}{dx} + 2V_{\mathrm{T}}\frac{d^2 n}{dx^2} = \frac{(b+1)(n - n_0)}{\mu_n \tau}, \tag{7}$$

where p_0 is the thermal-equilibrium density of holes, V_{T} the thermal voltage kT/e, b the electron-to-hole mobility ratio μ_n/μ_p, and τ the plasma lifetime, assumed constant, independent of injection level. This equation is readily derived from the fundamental current, Poisson, and particle-conservation equations, as shown in Section 3.

If the solid is a good semiconductor, the middle term on the left-hand side (LHS) of Eq. (7) always dominates the first term, which is a pure space-charge term deriving from the Poisson equation; thus, the first term may be safely ignored:

Semiconductor:

$$(n_0 - p_0)\frac{d\mathscr{E}}{dx} + 2V_T\frac{d^2n}{dx^2} = \frac{(b+1)(n-n_0)}{\mu_n\tau}. \tag{8a}$$

On the other hand, if the solid is a good insulator, namely, with n_0 and p_0 negligible, then the middle term on the LHS can be safely ignored:

Insulator:

$$-\frac{\varepsilon}{e}\frac{d}{dx}\left(\mathscr{E}\frac{d\mathscr{E}}{dx}\right) + 2V_T\frac{d^2n}{dx^2} = \frac{(b+1)n}{\mu_n\tau}. \tag{8b}$$

For the sake of concreteness, we consider the semiconductor problem characterized by Eq. (8a). Because the p^+–n junction at $x = 0$, in Fig. 4, presents an energetic barrier to the egress of electrons, there is an accumulation of electrons and holes (plasma) in the neighborhood of this contact. For this region, the plasma dynamics are governed by the familiar diffusive flow equation:

Near the contacts:

$$2V_T\frac{d^2n}{dx^2} = \frac{(b+1)(n-n_0)}{\mu_n\tau}. \tag{9}$$

This equation follows from (8a) when the first term on the LHS is neglected in favor of the second term. Since the n^+–n junction at $x = L$ likewise presents an energetic barrier to the egress of holes, there is a comparable accumulation of plasma in the neighborhood of this contact. And so, in this region also, the governing equation is (9).

In the bulk of the semiconductor, some number of ambipolar diffusion lengths removed from the contacts, the diffusive current flow is no longer important and the first term on the LHS of (8a) dominates the second term, which is forthwith neglected:

Away from the contacts:

$$(n_0 - p_0)\,\frac{d\mathscr{E}}{dx} = \frac{(b+1)(n-n_0)}{\mu_n\tau}. \tag{10}$$

We therefore have a three-region problem, as schematically sketched in Fig. 5. In this problem, as current J increases, the diffusion-dominated regions adjacent to the contacts, regions I and III in Fig. 5, grow at the expense of the middle region II. Finally, a critical current J_{cr} is reached at which region II shrinks to a plane. For $J > J_{cr}$, there is only a single diffusion-dominated

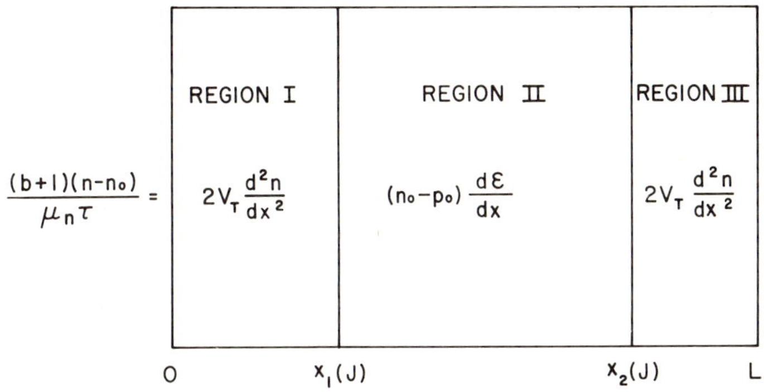

FIG. 5. Schematic regional approximation diagram for the problem of a constant-lifetime plasma injected into a semiconductor.

region in the semiconductor, consisting of the two previously separated regions I and III, now merged. A detailed discussion of this problem is given in Section 3.

The three examples cited above give an idea of what the basic ingredients of the regional approximation method look like in practice. In each case, the thrust of the method is simplification: the regions are chosen so that the governing equation in each region simplifies down to analytically manageable proportions. It then remains only to tie together the analytic solutions for the separate regions, and this is a straightforward process accomplished simply by the requirement of continuity of, say, the electric field in passing from one region to another. As was already brought out in citing the above examples, the regions are generally not fixed in extent, but vary with the magnitude of the current. A region may disappear from the solid at some critical current, either exiting the solid at an electrode (second example) or simply contracting to zero width in the interior of the solid (third example).

The regional approximation method leans totally on the underlying physics of a problem. It is the underlying physics which dictates the dominance of a particular term in a particular region, and the number of essential regions at any current level. This outstanding feature of the method should be clear from the three examples cited above, despite the cursory nature of the discussion. Therefore, the use of the method rests on physical insight, even perhaps to the extent that its success can only be assured if substantial insight is initially brought to bear on a problem.

The broad philosophy of the regional approximation method has been simple to state. However, it is the authors' experience that its application in practice is not merely an exercise in triviality. For one thing, the analytic intractability of the original equations is exchanged for a considerable, and sometimes quite formidable, amount of algebra, a great deal of it transcendental algebra. In the interest of minimizing this algebra, it is highly desirable to make all further simplifications in the problem that are consonant with the basic approximation defining the method. (The second example cited above is an illustration of the relentless pursuit of simplification.) It is a matter of experience that judgment must inevitably be brought into play in applying the method usefully. It is hard to imagine, at this stage of the game, that the method could presently be completely automated, that is, that a programming routine could be written such that *any* new, one-dimensional current-injection problem, one- or two-carrier, could be solved on a digital computer without further ado. (On the other hand, the computer can be an aid in determining approximations that are far from obvious.) For these various reasons, the regional approximation method is best conveyed by way of detailed examples, and that is the plan followed in this review chapter.

The remainder of this review is divided into three parts. Parts II and III are devoted to basic, prototype injection problems, Part IV to a device–design problem. In Part II, we treat one-carrier problems with planar-flow and spherical, radial-flow geometries, respectively. In Part III, we treat two-carrier, planar-flow problems, first injected plasmas and then varying-lifetime, negative-resistance problems. In Part IV, we treat a planar-flow transistor-design problem with base-widening as the key feature.

The emphasis throughout the article is on the methodology of problem solving. We are not attempting in this article to build up from scratch the theory of current injection into solids. It takes an entire volume[1] to do this properly. Rather, it is assumed that the reader already has some acquaintance with the physics of current injection.

A few historical words are in order. The basic concepts of charge injection into an insulator go back to Mott and Gurney.[2] The theory was first given realistic content in two classic papers by Rose.[3,4] In the early, subsequent theoretical investigations,[5–10] sufficiently simple models were studied that exact analysis proved feasible in handling the problems. The first instance of the use of the regional approximation method in this field, known to the

[1] M. A. Lampert and P. Mark, "Current Injection in Solids," Academic Press, New York, 1970.

[2] N. F. Mott and R. W. Gurney, "Electronic Processes in Ionic Crystals," Oxford Univ. Press (Clarendon), London and New York, 1940.

[3] A. Rose, *RCA Rev.* **12**, 362 (1951).

[4] A. Rose, *Phys. Rev.* **97**, 1538 (1955).

[5] M. A. Lampert, *Phys. Rev.* **103**, 1648 (1956).

[6] M. A. Lampert, *J. Appl. Phys.* **29**, 1082 (1958).

[7] R. H. Parmenter and W. Ruppel, *J. Appl. Phys.* **30**, 1548 (1959).

[8] M. A. Lampert, *RCA Rev.* **20**, 682 (1959).

[9] M. A. Lampert and A. Rose, *Phys. Rev.* **121**, 26 (1961).

[10] M. A. Lampert, *Phys. Rev.* **125**, 126 (1962).

authors, is by Patrick.[11] The extraordinary power of the method has been established more recently in the study of somewhat more difficult injection problems.[12–17] In any case, the method is such a logical one to use that, no doubt, it has a long history in various fields of applied mathematics.

II. One-Carrier Problems

Part II of this review considers problems of one-carrier injection. Since each injected carrier contributes one excess charge to the solid, space charge plays a vital role, via the Poisson equation, in the behavior of all one-carrier injection currents.

The first group of problems, the planar-flow problems, have the unusual distinction of being the one category that can be handled by purely analytic means. However, the analytic solutions tend to excessive unwieldiness, and so, even here, the regional approximation can be used to advantage. The second group of problems, the spherical, radial-flow problems, are analytically intractable, and, generally, what understanding we have of these problems has come from liberal use of the regional approximation method.

1. Planar-Flow Problems

For the sake of definiteness, we take the current carriers to be electrons. Assuming the possibility of electron trapping by a single set of electron traps, the equations characterizing the problem are

$$J = e\mu n(x)\mathscr{E}(x) = \text{const}, \tag{11}$$

$$\frac{\varepsilon}{e}\frac{d\mathscr{E}}{dx} = [n(x) - n_0] + [n_t(x) - n_{t,0}], \tag{12}$$

$$n_t(x) = \frac{N_t}{1 + [N/gn(x)]}, \qquad N = N_c \exp\frac{E_t - E_c}{kT}, \tag{13}$$

subject to the boundary condition

$$\mathscr{E}(0) = 0. \tag{14}$$

All of the quantities appearing in Eqs. (11)–(14) have previously been defined in Part I. If more than one set of traps are important, then the above equations

[11] L. Patrick, *J. Appl. Phys.* **28**, 765, Appendix A (1957).

[12] M. A. Lampert, A. Many, and P. Mark, *Phys. Rev.* **135**, A1444 (1964).

[13] A. Waxman and M. A. Lampert, *Phys. Rev.* (to be published).

[14] L. Rosenberg and M. A. Lampert, *J. Appl. Phys.* (to be published).

[15] R. B. Schilling and M. A. Lampert, *J. Appl. Phys.* (to be published).

[16] E. Rossiter, P. Mark, and M. A. Lampert, *Solid State Electron.* (to be published).

[17] R. B. Schilling, *IEEE Trans.-Education* **E12**, 152 (1969).

are trivially generalized by use of an additional subscript j on the quantities n_t, $n_{t,0}$, N_t, N, and g and by summation over j in Eq. (12). In writing Eqs. (11)–(14), it is assumed that the electrons are injected at the cathode at $x = 0$, move to the right, and are collected at the anode at $x = L$, as indicated on the appropriate energy-band diagram, Fig. 2. Since electron flow is involved, the relation between the vector $\mathbf{J}$ and J is $\mathbf{J} = -J\hat{\mathbf{x}}$, with $\hat{\mathbf{x}}$ being a unit vector along the x axis; that between $\boldsymbol{\mathscr{E}}$ and $\mathscr{E}$ is $\boldsymbol{\mathscr{E}} = -\mathscr{E}\hat{\mathbf{x}}$.

Note that the diffusion-current contribution to J has been neglected in Eq. (11), leading to what we call a "simplified theory" of current flow. Such a theory gives an unphysical description of the details of the current flow, namely, $n(0) = \infty$ from Eqs. (14) and (11), in the immediate vicinity of the cathode, as well as the anode. However, under almost all conditions of physical interest, the results of the simplified theory are generally useful.[18]

a. Problem 1: The Trap-Free Solid with Thermal Free Carriers

This problem is mathematically characterized by taking $n_t(x) = n_{t,0} = 0$ in Eq. (12):

$$\frac{\varepsilon}{e}\frac{d\mathscr{E}}{dx} = n(x) - n_0 . \tag{15}$$

Equations (11), (15), and (14) define the problem. How the regional approximation applies to this problem has already been discussed at length under the first example in Part I. At "low" currents, there are two regions in the solid, meeting at the plane $x_1 = x_1(J)$ [called $x_a(J)$ in Part I]. Equations (11) and (15) become, respectively, in these regions:

Region I $(0 \leqslant x \leqslant x_1)$: n_0 neglected:

$$J = e\mu n(x)\mathscr{E}(x) = \text{const}, \tag{16}$$

$$\frac{\varepsilon}{e}\frac{d\mathscr{E}}{dx} = n(x). \tag{17}$$

Since n_0 plays no role in this region, region I may appropriately be called a "perfect insulator" region.

Region II $(x_1 \leqslant x \leqslant L)$: $n(x) - n_0 = n_i(x)$ neglected:

$$J = e\mu n_0\mathscr{E} \rightarrow \mathscr{E} = \frac{J}{e\mu n_0} = \text{const}, \tag{18}$$

$$\frac{\varepsilon}{e}\frac{d\mathscr{E}}{dx} = 0. \tag{19}$$

[18] R. B. Schilling and H. Schachter, *J. Appl. Phys.* **38**, 841 (1967).

Since only n_0 plays a role in this region, region II may appropriately be called an "ohmic" region.

The plane $x = x_1$ connecting the two regions is characterized by

$$n(x_1) = n_0. \tag{20}$$

The solutions in the two regions are joined by requiring continuity of the electric field intensity at the connecting plane:

$$\mathscr{E}(x_1^-) = \mathscr{E}(x_1^+), \tag{21}$$

where $\mathscr{E}(x_1^-)$ denotes the value of $\mathscr{E}(x)$ as x approaches x_1 from below, i.e., from within region I, and $\mathscr{E}(x_1^+)$ denotes the value of $\mathscr{E}(x)$ as x approaches x_1 from above, i.e., from within region II. The regional breakdown of the problem is illustrated in Fig. 1, except that, in the interest of systematization, we have changed the notation: Region B is now region I, region A is region II, and $x_a = x_a(J)$ is $x_1 = x_1(J)$. At low currents, $x_1/L \ll 1$, so that most of the solid is in the ohmic region II, and Ohm's law will obtain. At some critical current J_{cr}, $x_1(J_{cr}) = L$; for $J \geqslant J_{cr}$, all of the solid is in the region I, so that the well-known perfect-insulator square-law[2] will obtain.

For further discussion of this problem, as well as for the remaining problems in Section 1, it is convenient to go over to dimensionless variables:

$$u = \frac{n_0}{n(x)} = \frac{en_0\mu\mathscr{E}(x)}{J}; \qquad w = \frac{e^2n_0^2\mu x}{\varepsilon J}; \qquad v = \frac{e^3n_0^3\mu^2V(x)}{\varepsilon J^2}. \tag{22}$$

Note that u, w, and v are all functions of x and depend parametrically on J. Letting subscript "a" on a quantity denote the value of that quantity evaluated at the anode, $x = L$,

$$\frac{1}{w_a} = \frac{\varepsilon J}{e^2n_0^2\mu L}, \qquad \frac{v_a}{w_a^2} = \frac{\varepsilon V}{en_0L^2}, \tag{23}$$

where $V = V_a = V(L)$. Thus, a plot of $1/w_a$ vs. v_a/w_a^2 is a dimensionless form of the current–voltage characteristic.

At low currents, $J < J_{cr}$, the two regions of the solid are now characterized by:

Region I $(0 \leqslant w \leqslant w_1)$:

$$du/dw = 1/u \rightarrow u\,du = dw. \tag{24}$$

Region II $(w_1 \leqslant w \leqslant L)$:

$$du/dw = 0, \tag{25}$$

being the dimensionless versions of (17) and (19), respectively. The current equation (16) is subsumed in the definition of u in (22).

In the dimensionless notation, the continuity condition (21) becomes

$$u(w_1^-) = u(w_1^+) = u_1 . \tag{26}$$

From Eqs. (20) and (26), it is clear that the transition plane $x = x_1$ is characterized by

$$u_1 = 1. \tag{27}$$

Noting from Eq. (22) that

$$V(x) = \int_0^x \mathscr{E}\, dx \rightarrow v = \int_0^w u\, dw = \int_0^u u \frac{dw}{du} du , \tag{28}$$

we readily obtain

Region I:

$$w = \tfrac{1}{2}u^2 , \qquad v = \tfrac{1}{3}u^3 , \tag{29}$$

satisfying the boundary condition $u = 0$ at $w = 0$, corresponding to Eq. (14).

At the connecting plane between regions I and II, we have, from Eqs. (26), (27) and (29),

$$x = x_1 : \qquad u_1 = 1 , \qquad w_1 = \tfrac{1}{2}, \qquad v_1 = \tfrac{1}{3}, \qquad x_1 = \varepsilon J/2e^2 n_0{}^2 \mu . \tag{30}$$

Note the linear dependence of x_1 on J.

From Eqs. (25) and (28), we obtain

Region II:

$$u = 1 , \qquad v = v_1 + \int_{w_1}^{w} u\, dw = v_1 + (w - w_1) = w - \tfrac{1}{6}, \tag{31}$$

using Eq. (30).

From Eq. (31), we obtain the dimensionless current–voltage characteristic at low currents:

$$J \leqslant J_{\text{cr}} : \qquad \frac{v_{\text{a}}}{w_{\text{a}}{}^2} = \frac{1}{w_{\text{a}}} - \frac{1}{6}\left(\frac{1}{w_{\text{a}}}\right)^2 . \tag{32}$$

This is plotted as the dotted curve in Fig. 6. In the limit of very low currents, Ohm's law obtains:

$$w_{\text{a}} \gg 1 : \qquad \frac{v_{\text{a}}}{w_{\text{a}}{}^2} = \frac{1}{w_{\text{a}}} \rightarrow J = e n_0 \mu \frac{V}{L} . \tag{33}$$

This is plotted as the lower branch of the dashed curve in Fig. 6.

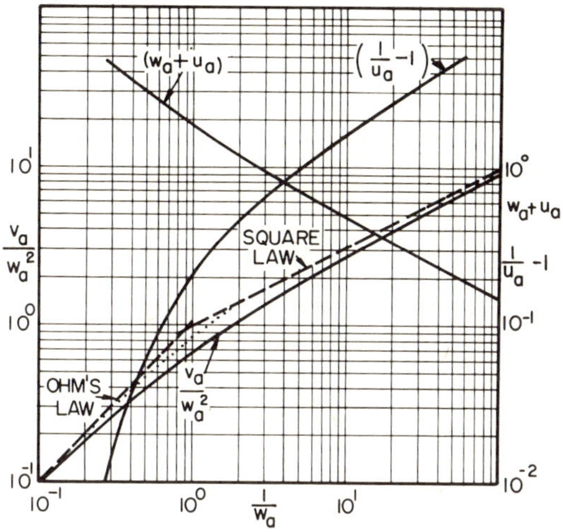

FIG. 6. Universal curves for space-charge-limited current injection into a trap-free insulator with thermal free carrier. Here, $J \propto 1/w_a$ and $V \propto v_a/w_a^2$; $(1/u_a) - 1 = (n_a - n_0)/n_0$ and $w_a + u_a = t/t_\Omega$, with t the free-carrier transit time and t_Ω the ohmic relaxation time: $t_\Omega = \varepsilon/en_0\mu$. The dotted curve represents the regional approximation solution discussed in the text, and the solid curve labeled v_a/w_a^2 is the exact solution. Note that the dotted curve coincides with the dashed square-law curve for $1/w_a > 2$.

Since $x_1 = L$ at the critical current and critical voltage, it follows from Eq. (30) that

$$J_{cr} = \frac{2e^2 n_0{}^2 \mu L}{\varepsilon}, \qquad V_{cr} = \frac{4en_0 L^2}{3\varepsilon}, \tag{34}$$

where the latter relation follows from Eqs. (22) and (30) and the former relation. In the dimensionless variables, from Eq. (30),

$$(1/w_a)_{cr} = 2, \qquad (v_a/w_a{}^2)_{cr} = \tfrac{4}{3}. \tag{35}$$

For $J > J_{cr}$, there is only region I in the solid, and Eq. (29) holds throughout the solid. It follows immediately from (29), with all quantities evaluated at the anode, that

$$J \geqslant J_{cr}: \qquad \frac{1}{w_a} = \frac{9}{8}\left(\frac{v_a}{w_a{}^2}\right)^2 \rightarrow J = \frac{9}{8}\varepsilon\mu\frac{V^2}{L^3}. \tag{36}$$

This is the famous perfect-insulator square law.[2] It is plotted as the upper branch of the dashed line in Fig. 6. The dotted curve in Fig. 6 is the complete regional solution. It merges with the dashed, square-law line at $1/w_a = 2$.

This problem need not be done by the regional approximation method. Equation (15) becomes, in dimensionless variables,

$$\frac{u\,du}{1-u} = dw, \tag{37}$$

with solution

$$w = -u - \ln(1 - u). \tag{38}$$

From Eq. (28), the potential is given by

$$v = -\tfrac{1}{2}u^2 - u - \ln(1 - u). \tag{39}$$

Evaluating Eqs. (38) and (39) at the anode, we obtain an implicit relation between $1/w_a$ and v_a/w_a^2, a relation involving the additional variable u_a. The variable u_a cannot be eliminated analytically, but only through numerical computation. The final result for the dimensionless current–voltage characteristic is plotted as the solid curve in Fig. 6. It is seen that agreement with the results of the regional approximation method (the dotted curve) is quite good.

b. Problem 2: A Single Set of Traps Lying below the Fermi Level

This problem is schematically illustrated by the energy-band diagram of Fig. 7. The Poisson equation (12) is here more conveniently written in the form

$$\frac{\varepsilon}{e}\frac{d\mathscr{E}}{dx} = [n(x) - n_0] + [p_{t,0} - p_t(x)], \tag{40}$$

$$p_t(x) = \frac{N_t N}{g n(x)}, \qquad p_{t,0} = \frac{N_t N}{g n_0}, \qquad N = N_c \exp\frac{E_t - E_c}{kT}, \tag{41}$$

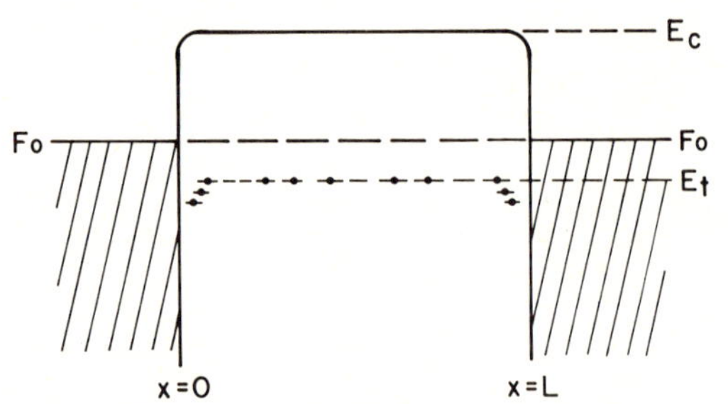

FIG. 7. Schematic energy-band diagram for the problem of an insulator with a single (significant) set of traps lying below the Fermi level. The contacts are ohmic (downward-bending) for electrons.

where $p_t(x) = N_t - n_t(x)$ denotes the hole occupancy of the traps. Note that the relations (41) for $p_t(x)$ and $p_{t,0}$ are valid only if the traps are deep, that is, if $(F_0 - E_t)/kT > 1$.

At low currents, there will be the usual ohmic region on the anode side of the solid and space-charge region on the cathode side. In the space-charge region, not only is n_0 negligible compared to the injected density of free carriers, $n(x) - n_0$, but $p_t(x)$ is negligible compared to $p_{t,0}$. The reason for this is that the injected carriers "fill" the initially empty traps as soon as they are sufficient in number to compete with the thermal carriers. Thus, in the space-charge region, the Poisson equation (40) can be written as

Space-charge region:

$$\frac{\varepsilon}{e}\frac{d\mathscr{E}}{dx} = n(x) + p_{t,0}. \tag{42}$$

The form of Eq. (42) immediately suggests a further breakup of the space-charge region into two subregions, in the first of which, near the cathode, $n(x)$ dominates the right-hand side (RHS) of Eq. (42), and in the second of which, $p_{t,0}$ dominates the RHS. This leads, at low currents, to a three-region problem, as illustrated schematically in Fig. 8. The equations characterizing

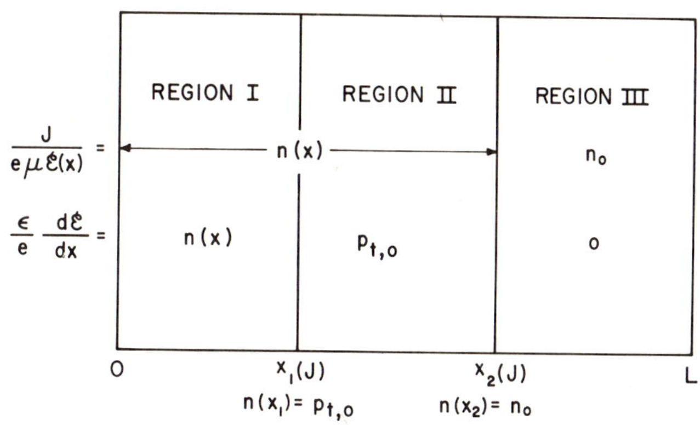

FIG. 8. Schematic regional approximation diagram for the problem of an insulator with a single (significant) set of traps lying below the Fermi level.

the three regions are, in both physical variables and dimensionless variables (22):

Region I $\quad 0 \leqslant x \leqslant x_1 \quad (0 \leqslant w \leqslant w_1)$: $\qquad n_0 \ll n(x), \quad p_{t,0} \ll n(x)$:

$$J = e\mu n(x)\mathscr{E}(x), \tag{43}$$

$$\frac{\varepsilon}{e}\frac{d\mathscr{E}}{dx} = n(x) \rightarrow \frac{du}{dw} = \frac{1}{u}. \tag{44}$$

Region II $x_1 \leqslant x \leqslant x_2$ $(w_1 \leqslant w \leqslant w_2)$: $n_0 \ll n(x) \ll p_{t,0}$:

$$\frac{\varepsilon}{e}\frac{d\mathscr{E}}{dx} = p_{t,0} \rightarrow \frac{du}{dw} = A, \qquad A = \frac{p_{t,0}}{n_0} \gg 1, \tag{45}$$

and J given by Eq. (43).

Region III $x_2 \leqslant x \leqslant L$ $(w_2 \leqslant w \leqslant w_a)$: $n(x) - n_0 \ll n_0$:

$$J = e\mu n_0 \mathscr{E} \rightarrow \mathscr{E} = \frac{J}{e\mu n_0}, \tag{46}$$

$$\frac{\varepsilon}{e}\frac{d\mathscr{E}}{dx} = 0 \rightarrow \frac{du}{dw} = 0. \tag{47}$$

The transition planes connecting the separate regions are defined by

$$n(x_1) = p_{t,0} \rightarrow u_1 = u(w_1^{\,-}) = u(w_1^{\,+}) = \frac{1}{A} \ll 1, \tag{48}$$

$$n(x_2) = n_0 \rightarrow u_2 = u(w_2^{\,-}) = u(w_2^{\,+}) = 1, \tag{49}$$

where the continuity of the $\mathscr{E}$-field (u-field) crossing these planes has been noted.

Integrating the corresponding equations for the separate regions, we obtain:

Region I:

$$w = \tfrac{1}{2}u^2, \qquad v = \tfrac{1}{3}u^3. \tag{50}$$

satisfying the boundary condition $u = 0$ at $w = 0$, corresponding to Eq. (14).

At the connecting plane between regions I and II, we have, from Eqs. (48) and (50),

$$x = x_1: \quad u_1 = \frac{1}{A}, \quad w_1 = \frac{1}{2A^2}, \quad v_1 = \frac{1}{3A^3}, \quad x_1 = \frac{\varepsilon J}{2A^2 e^2 n_0^{\,2} \mu}. \tag{51}$$

Region II:

$$u - u_1 = A(w - w_1) \rightarrow w = \frac{1}{A}u - \frac{1}{2A^2}, \tag{52}$$

$$v = v_1 + \int_{u_1}^{u} u\frac{dw}{du}du \rightarrow v = \frac{1}{2A}u^2 - \frac{1}{6A^3}. \tag{53}$$

At the connecting plane between regions II and III we have, from Eqs. (49), (52) and (53),

$$x = x_2: \qquad u_2 = 1, \qquad w_2 = \frac{1}{A} - \frac{1}{2A^2}, \qquad v_2 = \frac{1}{2A} - \frac{1}{6A^3},$$

$$x_2 = \frac{\varepsilon J}{Ae^2 {n_0}^2 \mu}. \tag{54}$$

Note the linear dependence of x_1 and x_2 on J.

Region III:

$$u = 1, \qquad v = v_2 + (w - w_2) = w - \frac{1}{2A} + \frac{1}{2A^2} - \frac{1}{6A^3}. \tag{55}$$

From Eq. (55), we obtain the dimensionless current–voltage characteristic at low currents:

$$J < J_{\mathrm{cr},1}: \qquad \frac{v_a}{{w_a}^2} = \frac{1}{w_a} - \frac{1}{2A}\left(1 - \frac{1}{A} + \frac{1}{3A^2}\right)\left(\frac{1}{w_a}\right)^2 \approx \frac{1}{w_a} - \frac{1}{2A}\left(\frac{1}{w_a}\right)^2. \tag{56}$$

At very low currents, this gives Ohm's law, Eq. (33). The $J_{\mathrm{cr},1}$ is the critical current at which region III exits at the anode: $x_2(J_{\mathrm{cr},1}) = L$. From Eqs. (54) and (22), we find that $J_{\mathrm{cr},1}$ and the corresponding voltage $V_{\mathrm{cr},1}$ are given by:

$$J_{\mathrm{cr},1} \approx \frac{Ae^2 {n_0}^2 \mu L}{\varepsilon}, \qquad V_{\mathrm{cr},1} \approx \frac{Aen_0 L^2}{2\varepsilon} = \frac{ep_{t,0}L^2}{2\varepsilon}. \tag{57}$$

It is very useful to note that for $J < J_{\mathrm{cr},1}$, region I makes a negligible contribution to the current–voltage characteristic. Thus, if region I is neglected altogether and region II extended leftward up to the cathode, so that, in effect, Eq. (51) is replaced by $u_1 = 0$, $w_1 = 0$, $v_1 = 0$, then, in place of Eq. (56), there is obtained

$$J < J_{\mathrm{cr},1} \quad \text{(Region I omitted)}: \qquad \frac{v_a}{{w_a}^2} = \frac{1}{w_a} - \frac{1}{2A}\left(\frac{1}{w_a}\right)^2, \tag{58}$$

a result which is practically indistinguishable from Eq. (56).

For $J > J_{\mathrm{cr},1}$, there are only the regions I and II in the solid. Noting, from Eq. (52), that $v_a = ({u_a}^2/2A) - (1/6A^3)$, and, from Eq. (52), that $u_a = Aw_a + (1/2A)$, we obtain for the current–voltage characteristic

$$J_{\mathrm{cr},1} \leqslant J \leqslant J_{\mathrm{cr},2}: \qquad \frac{v_a}{{w_a}^2} = \frac{A}{2} + \frac{1}{2A}\frac{1}{w_a} - \frac{1}{24A^3}\left(\frac{1}{w_a}\right)^2 \approx \frac{A}{2} + \frac{1}{2A}\frac{1}{w_a}, \tag{59}$$

this being the trap-filled-limit (TFL) law.[5]

$J_{cr,2}$ is the critical current at which region II exists at the anode: $x_1(J_{cr,2}) = L$. From Eqs. (51) and (22), we find that $J_{cr,2}$ and the corresponding voltage $V_{cr,2}$ are given by:

$$J_{cr,2} = \frac{2A^2e^2n_0{}^2\mu L}{\varepsilon}, \qquad V_{cr,2} = \frac{4Aen_0L^2}{3\varepsilon} = \frac{4ep_{t,0}L^2}{3\varepsilon}. \tag{60}$$

Finally, for $J > J_{cr,2}$, there is only the perfect-insulator region, region I, in the solid, and the well-known square law, Eq. (36), obtains.

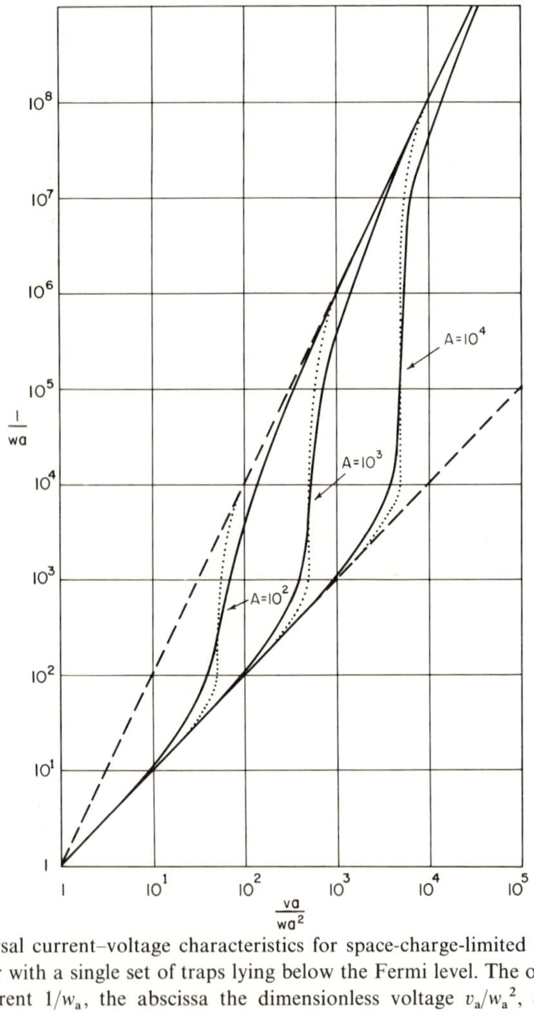

FIG. 9. Universal current–voltage characteristics for space-charge-limited current injection into an insulator with a single set of traps lying below the Fermi level. The ordinate is the dimensionless current $1/w_a$, the abscissa the dimensionless voltage $v_a/w_a{}^2$, and $A = p_{t,0}/n_0$. The dotted curves are the solutions obtained by the regional approximation, and the solid curves are the exact solutions. The upper dashed line is the trap-free square law, the lower dashed line is Ohm's law. (Figure supplied by P. Mark.)

The results of the regional approximation calculation for this problem, namely, Eqs. (56), (59), and (36), are plotted as the dotted curves in Fig. 9 for the parameter values $A = 10^2$, 10^3, and 10^4.

As with the previous problem, this problem also need not be done by the regional approximation method. Equation (40) becomes, in dimensionless variables,

$$\frac{1}{1 + A}\left[\frac{1}{1 - u} - \frac{1}{1 + Au}\right] du = dw, \tag{61}$$

with solution

$$w = \frac{1}{1 + A}\left[-\ln(1 - u) - \frac{1}{A}\ln(1 + Au)\right]. \tag{62}$$

From Eq. (28), the potential is given by

$$v = \frac{1}{1 + A}\left[-\left(1 + \frac{1}{A}\right)u - \ln(1 - u) + \frac{1}{A^2}\ln(1 + Au)\right]. \tag{63}$$

Evaluating Eqs. (62) and (63) at the anode, we obtain an implicit relation between $1/w_a$ and $v_a/w_a{}^2$, involving the additional variable u_a. As in the previous problem, u_a cannot be eliminated analytically, and so the dimensionless current–voltage characteristic must be constructed numerically. The final results obtained from this exact solution are plotted as the solid lines in Fig. 9 for the parameter values $A = 10^2$, 10^3, and 10^4.

c. Problem 3: A Single Set of Traps Lying above the Fermi Level

This problem is schematically illustrated by the energy-band diagram of Fig. 2. The appropriate equations for the discussion of this problem are Eqs. (11)–(13). At low currents, outside the ohmic region, that is, in the space-charge region on the cathode side of the solid, both n_0 and $n_{t,0}$ can be neglected in the Poisson equation:

Space-charge region:

$$\frac{\varepsilon}{e}\frac{d\mathscr{E}}{dx} = n(x) + \frac{N_t}{1 + [N/gn(x)]}. \tag{64}$$

The form of Eq. (64) leads to a further division of the space-charge region into three subregions, as discussed at length under the second example in Part I and illustrated in Fig. 3. In accordance with the more systematic notation here, we reproduce several of the main features of Fig. 3 in Fig. 10. It is seen that at low currents, we have a four-region problem. The equations characterizing the four regions are, in both physical variables and the dimensionless

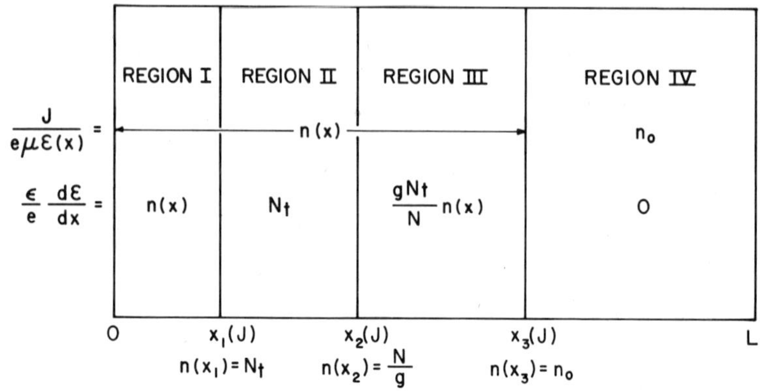

FIG. 10. Schematic regional approximation diagram for the problem of an insulator with a single (significant) set of traps lying above the Fermi level.

variables, Eq. (22):

Region I $0 \leqslant x \leqslant x_1$ $(0 \leqslant w \leqslant w_1)$: $n_0 \ll n(x)$, $N_t \ll n(x)$:

$$J = e\mu n(x)\mathscr{E}(x), \tag{65}$$

$$\frac{\varepsilon}{e}\frac{d\mathscr{E}}{dx} = n(x) \rightarrow \frac{du}{dw} = \frac{1}{u}. \tag{66}$$

Region II $x_1 \leqslant x \leqslant x_2$ $(w_1 \leqslant w \leqslant w_2)$: $n_0, N/g \ll n(x) \ll N_t$:

$$\frac{\varepsilon}{e}\frac{d\mathscr{E}}{dx} = N_t \rightarrow \frac{du}{dw} = B, \qquad B = \frac{N_t}{n_0}, \tag{67}$$

and J given by Eq. (65).

Region III $x_2 \leqslant x \leqslant x_3$ $(w_2 \leqslant w \leqslant w_3)$: $n_0 \ll n(x) \ll N/g \ll N_t$:

$$\frac{\varepsilon}{e}\frac{d\mathscr{E}}{dx} = \frac{n}{\theta} \rightarrow \frac{du}{dw} = \frac{1}{\theta u}, \qquad \theta = \frac{N}{gN_t} \ll 1, \tag{68}$$

and J given by Eq. (65).

Region IV $x_3 \leqslant x \leqslant L$ $(w_3 \leqslant w \leqslant w_a)$: $n(x) - n_0 \ll n_0$:

$$J = e\mu n_0\mathscr{E} \rightarrow \mathscr{E} = \frac{J}{e\mu n_0}, \tag{69}$$

$$\frac{\varepsilon}{e}\frac{d\mathscr{E}}{dx} = 0 \rightarrow \frac{du}{dw} = 0. \tag{70}$$

The transition planes connecting the separate regions are defined by

$$n(x_1) = N_t \to u_1 = u(w_1^-) = u(w_1^+) = \frac{1}{B} \ll 1. \tag{71}$$

$$n(x_2) = \frac{N}{g} \to u_2 = u(w_2^-) = u(w_2^+) = \frac{1}{\theta B} \ll 1. \tag{72}$$

$$n(x_3) = n_0 \to u_3 = u(w_3^-) = u(w_3^+) = 1. \tag{73}$$

Integrating the corresponding equations for the separate regions, we obtain:

Region I:

$$w = \tfrac{1}{2}u^2, \qquad v = \tfrac{1}{3}u^3. \tag{74}$$

At the connecting plane between regions I and II, we have, from Eqs. (71) and (74),

$$x = x_1: \quad u_1 = \frac{1}{B}, \quad w_1 = \frac{1}{2B^2}, \quad v_1 = \frac{1}{3B^3}, \quad x_1 = \frac{\varepsilon J}{2B^2 e^2 n_0{}^2 \mu}. \tag{75}$$

Region II:

$$u - u_1 = B(w - w_1) \to w = \frac{1}{B}u - \frac{1}{2B^2}, \tag{76}$$

$$v = v_1 + \frac{1}{2B}(u^2 - u_1{}^2) \to v = \frac{1}{2B}u^2 - \frac{1}{6B^3}. \tag{77}$$

At the connecting plane between regions II and III, we have, from Eqs. (72), (76), and (77),

$$x = x_2: \quad u_2 = \frac{1}{\theta B}, \quad w_2 = \frac{1}{B^2}\left(\frac{1}{\theta} - \frac{1}{2}\right), \quad v_2 = \frac{1}{2B^3}\left(\frac{1}{\theta^2} - \frac{1}{3}\right),$$
$$x_2 = \frac{\varepsilon J}{\theta e^2 n_0{}^2 \mu B^2}. \tag{78}$$

Region III:

$$w - w_2 = \frac{\theta}{2}(u^2 - u_2{}^2) \to w = \frac{\theta}{2}u^2 + \frac{1}{2B^2}\left(\frac{1}{\theta} - 1\right), \tag{79}$$

$$v = v_2 + \frac{\theta}{3}(u^3 - u_2{}^3) \to v = \frac{\theta}{3}u^3 + \frac{1}{6B^3}\left(\frac{1}{\theta^2} - 1\right). \tag{80}$$

At the connecting plane between regions III and IV, we have, from Eqs. (73), (79), and (80),

$$x = x_3: \qquad u_3 = 1, \qquad w_3 = \frac{\theta}{2} + \frac{1}{2B^2}\left(\frac{1}{\theta} - 1\right),$$

$$v_3 = \frac{\theta}{3} + \frac{1}{6B^3}\left(\frac{1}{\theta^2} - 1\right), \qquad x_3 = \frac{\theta \varepsilon J}{2e^2 n_0{}^2 \mu}. \tag{81}$$

Region IV:

$$u = 1, \qquad v = v_3 + (w - w_3) = w - \frac{\theta}{6} - \frac{1}{2B^2\theta} + \frac{1}{2B^2} + \frac{1}{6B^3\theta^2} - \frac{1}{6B^3}. \tag{82}$$

From (82), we obtain the dimensionless current–voltage characteristic at low currents,

$$J < J_{\mathrm{cr},1}: \qquad \frac{v_\mathrm{a}}{w_\mathrm{a}{}^2} = \frac{1}{w_\mathrm{a}} - \frac{\theta}{6}\left(\frac{1}{w_\mathrm{a}}\right)^2, \tag{83}$$

which is Ohm's law with a small correction, and where we have ignored even smaller corrections. Here, $J_{\mathrm{cr},1}$ is the critical current at which region IV exits at the anode: $x_3(J_{\mathrm{cr},1}) = L$. From Eqs. (81) and (22), we find that $J_{\mathrm{cr},1}$ and the corresponding voltage $V_{\mathrm{cr},1}$ are given by

$$J_{\mathrm{cr},1} = \frac{2e^2 n_0{}^2 \mu L}{\theta \varepsilon}, \qquad V_{\mathrm{cr},1} = \frac{4en_0 L^2}{3\varepsilon\theta}. \tag{84}$$

Note that for $J < J_{\mathrm{cr},1}$, regions I and II both make a negligible contribution to the current–voltage characteristic. Thus, if they are both ignored and region III is extended right up to the cathode, then the result Eq. (83) is still obtained. [The presence of regions I and II contributes to Eq. (83) even smaller corrections, which were neglected in passing from Eq. (82) to (83).]

For $J > J_{\mathrm{cr},1}$, there are only the regions I, II, and III in the solid. Evaluating Eqs. (79) and (80) at the anode and substituting $u_\mathrm{a} = \{(2/\theta)[w_\mathrm{a} - (1/2B^2\theta)]\}^{1/2}$ into (80), we readily obtain

$$J_{\mathrm{cr},1} < J < J_{\mathrm{cr},2}: \qquad \left[\frac{v_\mathrm{a}}{w_\mathrm{a}{}^2} - \frac{1}{6B^3\theta^2}\left(\frac{1}{w_\mathrm{a}}\right)^2\right]^2 = \frac{8}{9\theta}\frac{1}{w_\mathrm{a}}\left(1 - \frac{1}{2B^2\theta}\frac{1}{w_\mathrm{a}}\right)^3, \tag{85}$$

which is just the shallow-trap square law[5]

$$\frac{1}{w_\mathrm{a}} = \frac{9}{8}\theta\left(\frac{v_\mathrm{a}}{w_\mathrm{a}{}^2}\right)^2 \rightarrow J = \frac{9}{8}\theta\varepsilon\mu\frac{V^2}{L^3}, \tag{86}$$

with small corrections. Here, $J_{cr,2}$ is the critical current at which region III exits at the anode: $x_2(J_{cr,2}) = L$. From Eqs. (78) and (22), $J_{cr,2}$ and the corresponding voltage $V_{cr,2}$ are given by

$$J_{cr,2} = \frac{\theta B^2 e^2 n_0{}^2 \mu L}{\varepsilon}, \qquad V_{cr,2} = \frac{Ben_0 L^2}{2\varepsilon} = \frac{eN_t L^2}{2\varepsilon}. \tag{87}$$

As in the lower-current regime, $J < J_{cr,1}$, here too only the two regions closest to the anode, namely, regions II and III, contribute significantly to the current–voltage characteristic; if region I is ignored and region II extended right up to the cathode, essentially the same result, namely Eq. (85) with negligible correction, is obtained.

For $J > J_{cr,2}$, there are only the regions I and II in the solid. Since Eqs. (76) and (77) are identical to (52) and (53), respectively, except for B replacing A, it follows that the current–voltage characteristic is now given by Eq. (59), with B replacing A:

$$J_{cr,2} < J < J_{cr,3}: \qquad \frac{v_a}{w_a{}^2} = \frac{B}{2} + \frac{1}{2B}\frac{1}{w_a}. \tag{88}$$

Here, $J_{cr,3}$ is the critical current at which region II exits at the anode: $x_1(J_{cr,3}) = L$. From Eqs. (75) and (22), $J_{cr,3}$ and the corresponding voltage $V_{cr,3}$ are given by

$$J_{cr,3} = \frac{2B^2 e^2 n_0{}^2 \mu L}{\varepsilon} = \frac{2e^2 N_t{}^2 \mu L}{\varepsilon}, \qquad V_{cr,3} = \frac{4eN_t L^2}{3\varepsilon}. \tag{89}$$

Finally, for $J > J_{cr,3}$ there is only the region I in the solid, and the square law, Eq. (36), obtains.

The results of the regional-approximation calculation for a specific problem are plotted as the short-dashed curve in Fig. 11. The parameter values chosen for this problem are $n_0 = 10^6\ \text{cm}^{-3}$, $N_t = 10^{14}\ \text{cm}^{-3}$, $E_c - E_t = 0.6$ eV, $g = 2$, $N_c = 10^{19}\ \text{cm}^{-3}$, $\varepsilon/\varepsilon_0 = 11$, and $\mu = 200\ \text{cm}^2/\text{V-sec}$. These parameters correspond to $B = 10^8$, $\theta = 5 \times 10^{-6}$, and $\rho_0 = 3 \times 10^{10}$ ohm-cm.

As with the previous problems, this problem also has an exact analytic solution within the framework of the simplified theory. Equation (12) becomes, in dimensionless variables,

$$\frac{u(1 + Cu)}{(1 - u)(1 + Gu)}\,du = dw, \tag{90}$$

with $C = B\theta$, $G = C + D$, $D = BC/(1 + C)$, and B, and θ given by Eqs. (67) and (68), respectively. Using the well-known method of expanding in partial

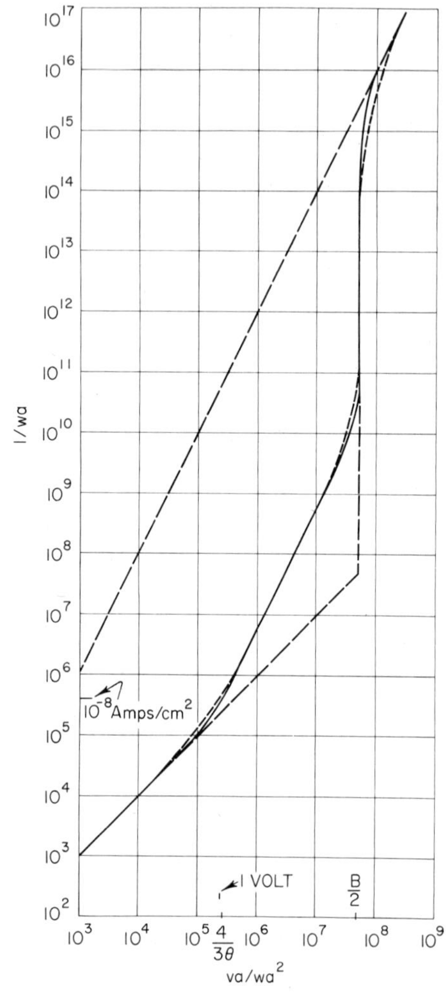

FIG. 11. Current–voltage characteristic for a particular case of space-charge-limited current injection into an insulator with a single set of traps lying above the Fermi level. The ordinate is the dimensionless current $1/w_a$, the abscissa the dimensionless voltage v_a/w_a^2. $\theta = \{N_c \exp(E_t - E_c)/kT\}/gN_t = 5 \times 10^{-6}$, and $B = N_t/n_0 = 10^8$. The lower, short-dashed curve is the solution obtained by the regional approximation, and the solid curve is the exact solution. The upper long-dashed line is the trap-free square law, and the lower long-dashed line is Ohm's law. The vertical long-dashed line marks the trap-filled-limit voltage. (Figure supplied by P. Mark.)

fractions, Eq. (90) yields, at the anode,

$$w_a = -\frac{C}{G}u_a - R\ln(1 - u_a) - \frac{S}{G}\ln(1 + Gu_a). \tag{91}$$

Equations (28) and (90) yield, for the potential at the anode,

$$v_a = -\frac{C}{2G}u_a{}^2 - \left(R + \frac{S}{G}\right)u_a - R\ln(1 - u_a) + \frac{S}{G^2}\ln(1 + Gu_a), \tag{92}$$

with $R = (C + 1)/(G + 1)$ and $S = D/G(G + 1)$. From Eqs. (92) and (91), the direct relation of $1/w_a$ to v_a/w_a^2 must be calculated numerically. The resultant plot of $1/w_a$ versus v_a/w_a^2 for $B = 10^8$ and $\theta = 5 \times 10^{-6}$ is shown as the solid curve in Fig. 11.

2. Spherical, Radial-Flow Problems

As in Section 1, we take the current carriers to be electrons. The equations characterizing the problem are

$$I = 4\pi e\mu n(r) r^2 \mathscr{E}(r) = \text{const}, \tag{93}$$

$$\begin{aligned} \frac{\varepsilon}{e}\frac{1}{r^2}\frac{d}{dr}(r^2\mathscr{E}) &= n(r) - n_0 + \sum_j [n_{tj}(r) - n_{tj,0}] \\ &= n(r) - n_0 + \sum_j [p_{tj,0} - p_{tj}(r)], \end{aligned} \tag{94}$$

$$\begin{gathered} n_{tj}(r) = \frac{N_{tj}}{1 + [N_j/gn(r)]}, \qquad p_{tj}(r) = N_{tj} - n_{tj}(r) = \frac{N_{tj}N_j/gn(r)}{1 + [N_j/gn(r)]}, \\ N_j = N_c \exp\frac{E_{tj} - E_c}{kT}, \end{gathered} \tag{95}$$

subject to the boundary condition

$$\mathscr{E}(r_c) = 0 \tag{96}$$

where r_c is the radius of the cathode. Equations (93) and (94) are the spherical, radial analogs of Eqs. (11) and (12), respectively, except that here we have explicitly included the possibility of more than one set of traps by use of the additional subscript j and summation over j. In Eq. (93), I is the total current, as compared to the current density J which appears in Eq. (11). In writing Eqs. (93) and (94), it is assumed that the electrons are injected at the cathode, at radius $r = r_c$, move outward radially, and are collected at the anode, at radius $r = r_a \gg r_c$. Thus, the relations between corresponding vector and scalar quantities are $\mathbf{I} = -I\hat{\mathbf{r}}$ and $\boldsymbol{\mathscr{E}} = -\mathscr{E}\hat{\mathbf{r}}$, where $\hat{\mathbf{r}}$ is a unit radial vector.

The spherical, radial-flow geometry is of practical significance, in that it describes current injection at a point contact.

a. Problem 1: The 3/2-Power Law for the Trap-Free Solid with Thermal Free Carriers

This problem is the spherical analog of problem 1 of Section 1. [See Eq. (15).] In Eq. (94), we take $n_{tj}(r) = n_{tj,0} = 0$:

$$\frac{\varepsilon}{e}\frac{1}{r^2}\frac{d}{dr}(r^2\mathscr{E}) = n(r) - n_0. \tag{97}$$

Equations (93), (97), and (96) define the problem.

The discussion under the first example in Part I is just as applicable to this problem as to the corresponding planar-flow problem. At "low" currents, there will be the usual two regions in the solid, a space-charge region I extending from the cathode out to the radius $r_x = r_x(I)$, and the ohmic region II extending from r_x out to the anode r_a.

Region I $(r_c \leqslant r \leqslant r_x)$: $\quad n_0$ neglected:

$$I = 4\pi e\mu n(r) r^2 \mathscr{E}(r) = \text{const}, \tag{98}$$

$$\frac{\varepsilon}{e}\frac{1}{r^2}\frac{d}{dr}(r^2\mathscr{E}) = n(r). \tag{99}$$

Region I ends at radius r_x defined by

$$n(r_x) = n_0, \tag{100}$$

this being the spherical analog of Eq. (20).

Elimination of $n(r)$ from Eqs. (98) and (99) yields the differential equation:

$$\frac{d}{d(r^3)}(r^2\mathscr{E})^2 = \frac{I}{6\pi\varepsilon\mu}, \tag{101}$$

with solution

$$\mathscr{E}(r) = \left(\frac{I}{6\pi\varepsilon\mu}\right)^{1/2}\frac{(r^3 - r_c^{\,3})^{1/2}}{r^2}, \qquad n(r) = \left(\frac{3\varepsilon I}{8\pi e^2\mu}\right)^{1/2}\frac{1}{(r^3 - r_c^{\,3})^{1/2}}, \tag{102}$$

satisfying the boundary condition (96).

Using Eq. (100) in (102), we now obtain

$$r_x = \left[\frac{3\varepsilon I}{8\pi e^2\mu n_0^{\,2}} + r_c^{\,3}\right]^{1/3}. \tag{103}$$

It is clear that if I is large enough, we can neglect r_c in Eq. (103):

$$I \gg \frac{8\pi e^2\mu n_0^{\,2} r_c^{\,3}}{3\varepsilon}; \qquad r_x \approx \left[\frac{3\varepsilon I}{8\pi e^2\mu n_0^{\,2}}\right]^{1/3} \tag{104}$$

From Eq. (102), it follows that the field intensity increases very rapidly from 0 at $r = r_c$, reaches a maximum at $r = r_m = 4^{1/3} r_c$, with $\mathscr{E}_m = \mathscr{E}(r_m) = (I/2^{11/3}\pi\varepsilon\mu r_c)^{1/2}$, and thereafter decreases as $1/\sqrt{r}$ in this region. A schematic plot of $\mathscr{E}(r)$ versus r is given in Fig. 12.

For the voltage across region I, namely, $V_{c,x} = \int_{r_c}^{r_x} \mathscr{E}(r)\,dr$, we have

$$r_x \gg r_c: \qquad V_{c,x} \approx \left(\frac{2Ir_x}{3\pi\varepsilon\mu}\right)^{1/2} \approx \left[\frac{I^2}{3\pi^2\varepsilon e\mu^2 n_0}\right]^{1/3}, \tag{105}$$

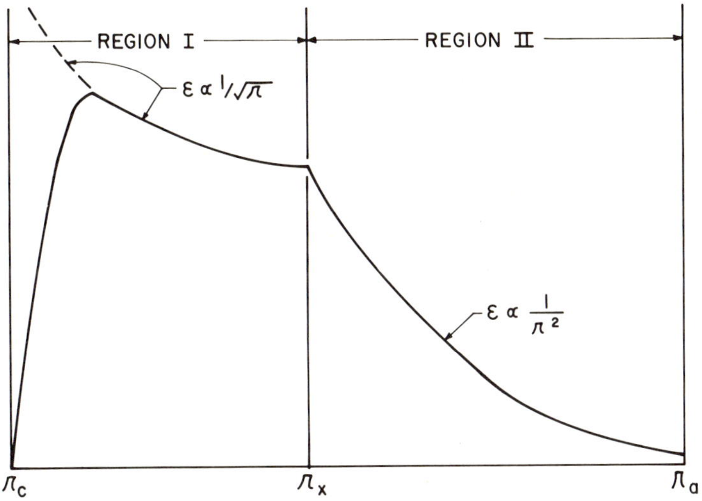

FIG. 12. Schematic radial variation of $\mathscr{E}$ for the one-carrier, spherical-flow problem for a trap-free solid with thermal free carriers.

where we have taken $\mathscr{E} \approx (I/6\pi\varepsilon\mu r)^{1/2}$ throughout *all* of region I, and also have used Eq. (104).

In region II, we have:

Region II $(r_x \leqslant r \leqslant r_a)$: $\quad n_i(r) = n(r) - n_0$ neglected:

$$I \approx 4\pi e\mu n_0 r^2 \mathscr{E}(r) \rightarrow \mathscr{E}(r) \approx \frac{I}{4\pi e\mu n_0 r^2}, \tag{106}$$

$$\frac{\varepsilon}{e}\frac{1}{r^2}\frac{d}{dr}(r^2\mathscr{E}) = 0, \tag{107}$$

where Eq. (107) is, of course, redundant with Eq. (106).

For the voltage across region II, namely, $V_{x,a} = \int_{r_x}^{r_a} \mathscr{E}(r)\, dr$, we have, using Eqs. (106), (104), and (105),

$$r_x \ll r_a: \qquad V_{x,a} \approx \frac{I}{4\pi e\mu n_0 r_x} \approx \left[\frac{I^2}{24\pi^2 \varepsilon e\mu^2 n_0}\right]^{1/3} = \frac{1}{2} V_{c,x}. \tag{108}$$

For the full voltage across the crystal, we have, from Eqs. (105) and (108),

$$r_c \ll r_x \ll r_a: \qquad V = V_{c,x} + V_{x,a} \approx \frac{3}{2} V_{c,x} \approx \frac{3}{2}\left[\frac{I^2}{3\pi^2 \varepsilon e\mu^2 n_0}\right]^{1/3}, \tag{109}$$

giving

$$r_c \ll r_x \ll r_a, \quad I_{cr,1} < I < I_{cr,2}: \qquad I \approx \frac{2\pi\sqrt{2}}{3}\mu(\varepsilon e n_0)^{1/2} V^{3/2}, \tag{110}$$

where we have indicated that the 3/2-power law is valid only over a certain range of currents, namely, between $I_{\mathrm{cr},1}$ and $I_{\mathrm{cr},2}$. The exact, computer-determined solution[12] differs from Eq. (110) only in that the numeric 1.06 replaces $2\sqrt{2}/3 = 0.94$.

For $I < I_{\mathrm{cr},1}$, Ohm's law for the spherical geometry holds:

$$I < I_{\mathrm{cr},1}: \qquad I = 4\pi e \mu n_0 r_c V. \tag{111}$$

For $I > I_{\mathrm{cr},2}$, the perfect-insulator square law for the spherical geometry holds:

$$I > I_{\mathrm{cr},2}: \qquad I = \frac{3\pi}{2} \varepsilon \mu \frac{V^2}{r_a}, \tag{112}$$

as follows directly from Eq. (105), taking $r_x = r_a$. Note that Eq. (112), although based on the approximation $\mathscr{E} \propto r^{-1/2}$ to Eq. (102), agrees precisely with the exact result of Meltzer[19] in the limit $r_a \gg r_c$.

Clearly, $I_{\mathrm{cr},1}$ is given by the intersection of the Ohm's law (111) and 3/2-power law (110); in like manner, $I_{\mathrm{cr},2}$ is given by the intersection of (110) and the square law (112),

$$I_{\mathrm{cr},1} = 72\pi \frac{e^2 \mu n_0{}^2 r_c{}^3}{\varepsilon}, \qquad I_{\mathrm{cr},2} = \frac{2^9 \pi}{3^7} \frac{e^2 \mu n_0{}^2 r_a{}^3}{\varepsilon}. \tag{113}$$

From (113), it follows that the total range of validity of the 3/2-power law is $I_{\mathrm{cr},2}/I_{\mathrm{cr},1} \approx (2^6/3^9)(r_a/r_c)^3$, which can be a large ratio for $r_a \gg r_c$.

The 3/2-power law is a transition regime between the Ohm's law regime (111) and the perfect-insulator, square-law regime (112). In the comparable planar-flow situation, problem 1 of Section 1, the transition between the two end regimes (33) and (36) occurs over such a relatively narrow range of voltages and currents that it does not constitute a separate regime in the current–voltage characteristic. That the transition in the spherical case takes place over a sufficiently extended range to constitute a separate regime is due to two circumstances which are unique to the spherical geometry: First, the transition surface separating the space-charge region I from the ohmic region II is a very slowly varying function of current: $r_x \propto I^{1/3}$, from Eq. (104), as compared to the analogous result $x_1 \propto J$, from Eq. (30), for the planar case. [The "motion" of transition planes is linear with J for all planar-flow, one-carrier problems, as seen in Eqs. (51), (54), (75), (78), and (81).] Second, the division of the applied voltage between regions I and II is independent of the current for $I_{\mathrm{cr},1} < I < I_{\mathrm{cr},2}$, namely, $V_{c,x}/V_{x,a} = 2$, from Eq. (108). These differences between the spherical, radial-flow, and planar-flow geometries for the same problem of the trap-free solid with thermal free carriers

[19] B. Meltzer, *J. Electron. Control* **8**, 171 (1960).

stem from the radically different spatial variations of the electric field intensity, as seen by comparing Fig. 1 with Fig. 12.

b. Problem 2: The $\frac{3}{2}$-Power Law as a Universal Law

The $\frac{3}{2}$-power law following the Ohm's law has been derived above for the particular case of the trap-free solid with thermal free carriers. Actually, it comes out that $I \propto V^{3/2}$ is a much more general result for the initial space-charge regime following Ohm's law; it is universally valid, independent of the absence or presence of electron traps and their detailed properties. In order to establish this law, we will once again resort to the regional approximation method, only we shall now use it in a somewhat different way than it has been used up to this point; for the application must now be independent of trapping details, in marked contrast to our previous uses of the method.

If we write the universal $\frac{3}{2}$-power law in the form: $I = KV^{3/2}$, then specific effects of the electron traps appear in the constant K, as is shown below. However, a remarkable, and potentially quite useful, fact is that K depends neither on the cathode radius r_c nor on the anode radius r_a.

Our basic strategy in handling the general case is to stick with the simple, two-region approximation as first spelled out in the first example of Part I and then used in problem 1 of Section 1 and again in problem 1 of Section 2.

Region I

$$(r_c \leqslant r \leqslant r_x,\ r_c \ll r_x): \qquad \text{space-charged dominated.} \tag{114}$$

Region II

$$(r_x \leqslant r \leqslant r_a,\ r_x \ll r_a): \qquad \text{ohmic.} \tag{115}$$

$$n_x = n(r_x) = n_0\,. \tag{116}$$

Obviously, we are assuming that $r_c \ll r_a$, a situation of practical interest (point-contact geometry) and a necessary condition for the existence of the $\frac{3}{2}$-power-law regime in the current–voltage characteristic.

Region II, being ohmic, offers no difficulties. In region I, the full Poisson equation (94), in all of its generality, applies. Since this equation cannot be handled head-on, we circumvent it in favor of a dimensional analysis of its main consequences; namely, we write for the voltage $V_{c,x}$ across region I and the total injected charge Q_x contained in region I, respectively:

Region I $(r_c \leqslant r \leqslant r_x, r_c \ll r_x)$:

$$V_{c,x} = \int_{r_c}^{r_x} \mathscr{E}\, dr = c_1 r_x \mathscr{E}_x, \qquad \mathscr{E}_x \equiv \mathscr{E}(r_x), \tag{117}$$

and

$$Q_x = 4\pi e \int_{r_c}^{r_x} [n_i(r) + n_{t,i}(r)]r^2\,dr = c_2\left(\frac{4\pi}{3}\right)r_x{}^3 e(n_0 + n_{t,0}), \tag{118}$$

with

$$n_i(r) = n(r) - n_0, \qquad n_{t,i}(r) = \sum_j [n_{tj}(r) - n_{tj,0}], \tag{119}$$

and where $n_{t,0}$ is the injected, excess, trapped-electron density in quasi-thermal equilibrium with an injected, excess free-electron density $n_i = n_0$ (that is, $n_{t,0}$ is the excess, trapped electron density corresponding to a motion of the quasi-Fermi level upward in the forbidden gap, from its thermal-equilibrium position, by an energy of 0.7 kT).

Significant content is given to Eqs. (117) and (118) by the assertion that

$$c_1 \quad \textit{and} \quad c_2 \quad \textit{are constants of order unity}. \tag{120}$$

Arguments supporting this assertion are presented below. Here, we note the particular physical significance of Eqs. (117) and (118). Equation (117) states that the main contribution to the voltage in region I comes from the neighborhood of the effective anode for this region, namely, $r = r_x$, rather than from the neighborhood of the cathode, in direct contrast to the situation in the ohmic region II. Similarly, since $n_i(r_x) = n_0$ and $n_{t,i}(r_x) = n_{t,0}$, Eq. (118) states that the total injected charge Q_x in region I is adequately estimated by assuming that the excess charge density at the effective anode, namely, $e(n_0 + n_{t,0})$, is uniformly distributed through region I. Since $n_i(r)$ decreases monotonically with increasing r, $c_2 \geqslant 1$ necessarily.

Region II is a strictly ohmic region with its effective cathode at $r = r_x$. Therefore, in this region, we can write

Region II $(r_x \leqslant r \leqslant r_a,\ r_x \ll r_a)$:

$$V_{x,a} = \int_{r_x}^{r_a} \mathscr{E}\,dr = r_x \mathscr{E}_x \tag{121}$$

$$I = 4\pi e \mu n_0 r_x V_{x,a} \tag{122}$$

Adding Eqs. (117) and (121), we obtain for the full voltage V across the solid

$$V = r_x \mathscr{E}_x (1 + c_1). \tag{123}$$

Substituting for $V_{x,a}$ from Eqs. (121) and (123) into (122), we obtain

$$I = 4\pi e \mu n_0 r_x \frac{V}{1 + c_1}. \tag{124}$$

It remains to determine r_x. The exact integral of the Poisson equation (94) in region I, subject to boundary condition (96) is

$$r_x{}^2\mathscr{E}_x = \frac{Q_x}{4\pi\varepsilon} = r_x\frac{V}{1 + c_1}, \tag{125}$$

where we have also used Eq. (123). Combining Eqs. (125) and (118), we get

$$r_x = \left[\frac{3\varepsilon V}{(1 + c_1)c_2 e(n_0 + n_{t,0})}\right]^{1/2}. \tag{126}$$

The universal $\frac{3}{2}$-power law now follows from substitution of Eq. (126) into Eq. (124):

$$I = 4\pi e\mu n_0\left[\frac{3\varepsilon}{c_2 e(n_0 + n_{t,0})}\right]^{1/2}\left[\frac{V}{1 + c_1}\right]^{3/2}. \tag{127}$$

In practical units,

$$I = 2.6 \times 10^{-15}\left[\frac{\mu^2 n_0{}^2 K}{c_2(n_0 + n_{t,0})}\right]^{1/2}\left[\frac{V}{1 + c_1}\right]^{3/2} \text{A}, \tag{128}$$

with μ in $\text{cm}^2/\text{V-sec}$, n_0 and $n_{t,0}$ in cm^{-3}, and V in volts, and where K is the relative dielectric constant. Note that Eqs. (127) and (128) refer to the full spherical geometry. For a hemispherical geometry, ignoring surface effects, divide the RHS of each equation by 2.

The critical current $I_{cr,1}$ and voltage $V_{cr,1}$ for the onset of the $\frac{3}{2}$-power law are given by the intersection of the $\frac{3}{2}$-power law (127) with the Ohm's law (111):

$$I_{cr,1} = \frac{4\pi e^2\mu n_0(n_0 + n_{t,0})r_c{}^3 c_2(1 + c_1)^3}{3\varepsilon},$$
$$V_{cr,1} = \frac{e(n_0 + n_{t,0})r_c{}^2 c_2(1 + c_1)^3}{3\varepsilon} \tag{129}$$

The capacitance C relating the total injected charge Q_x to the applied voltage is readily obtained from Eq. (125):

$$C = \frac{Q_x}{V} = \frac{4\pi\varepsilon r_x}{1 + c_1} = \left[\frac{48\pi^2\varepsilon^3 V}{(1 + c_1)^3 c_2 e(n_0 + n_{t,0})}\right]^{1/2} \tag{130}$$

where Eq. (126) has also been used.

As with the special case studied in problem 1, the transition radius r_x varies slowly with current: $r_x \propto V^{1/2} \propto I^{1/3}$, and the voltage division between regions I and II is independent of current, $V_{c,x}/V_{x,a} = c_1$. These are the basic features underlying the $\frac{3}{2}$-power law, and they derive from the crucial fact that the physical properties of region I are determined in the neighborhood of the effective anode of region I, whereas the physical properties of region II are determined in the neighborhood of the effective cathode of region II. It is for this reason that neither the cathode nor anode radius appears in the universal $\frac{3}{2}$-power law, Eq. (127).

It will now be argued that the expected ranges of c_1 and c_2 are

$$\tfrac{1}{2} \leqslant c_1 \leqslant 2 \qquad \text{and} \qquad 1 \leqslant c_2 \leqslant 2, \tag{131}$$

which may be regarded as a sharpening up of Eq. (120). The lower limits in the inequalities hold for the problem of deep trapping, and the upper limits hold for shallow trapping and no trapping. [Note that taking $c_1 = c_2 = 2$ in Eq. (127) reduces it to (110).] The corresponding variation in the factor $1/c_2^{1/2}(1 + c_1)^{3/2}$ appearing in Eqs. (127) and (128) covers the range from $(2/3)^{3/2} = 0.545$ at the lower limits in Eq. (131) to $1/3\sqrt{6} = 0.136$ at the upper limits in Eq. (131), a factor of four between the two extremes.

The important result Eq. (131) is a consequence of the following properties of the constants c_1 and c_2:

$$\mathscr{E} \propto r^n \qquad \text{and} \qquad n > -1 \rightarrow c_1 = 1/(n+1) \tag{132}$$

which follows directly from the definition of $V_{c,x}$ in Eq. (117), and

$$(n_i + n_{t,i}) \propto r^m \qquad \text{and} \qquad 0 \geqslant m \geqslant -3 \rightarrow c_2 = 3/(m+3) \tag{133}$$

which follows directly from the definition of Q_x in Eq. (118), with n_i and $n_{t,i}$ defined in Eq. (119). Note that the exponent m cannot, under any conditions, be positive, since $(n_i + n_{t,i})$ must be nonincreasing with increasing r. The conditions $n > -1$ and $m > -3$ enable us, in doing the specified integrations for $V_{c,x}$ in Eq. (117) and Q_x in Eq. (118), to neglect r_c^{n+1} relative to r_x^{n+1}, and r_c^{m+3} relative to r_x^{m+3}, respectively: these latter are approximations crucial to the validity of the basic relations (117) and (118). In practice, c_1 and c_2 are determined by first determining the exponents n and m in (132) and (133) respectively.

The key step in utilizing Eqs. (132) and (133) to establish Eq. (131) is the representation of the functional relationship between $n_{t,i}$ and n_i by the simple form

$$0 \leqslant p \leqslant 1: \qquad n_{t,i} \propto n_i{}^p \rightarrow \frac{dn_{t,i}}{dn_i} = p\frac{n_{t,i}}{n_i}. \tag{134}$$

The maximum change of $n_{t,i}$ with n_i occurs if the traps are shallow, namely, $n_{t,i} = n_i/\theta$, with θ a constant given by Eq. (68), so that $p = 1$: the minimum change of $n_{t,i}$ with n_i occurs if the traps are deep, namely, no change at all, $n_{t,i} = \text{const}$, so that $p = 0$. In general, for a single trap level,

$$n_t = \frac{N_t}{1 + (N/gn)} \rightarrow \frac{dn_t}{dn} = q\frac{n_t}{n}, \qquad \text{with} \qquad q = \frac{1}{1 + (gn/N)}\cdot \tag{135}$$

Obviously, $0 \leqslant q \leqslant 1$, the left-hand limit holding for $n = \infty$ (the extreme deep-trap limit) and the right-hand limit for $n = 0$ (the extreme shallow-trap limit). A comparison of Eqs. (134) and (135) establishes the inequality limits on p.

Writing Eq. (94) as

$$\frac{\varepsilon}{e}\frac{1}{r^2}\frac{d}{dr}(r^2\mathscr{E}) = n_{\mathrm{i}}(r) + n_{\mathrm{t,i}}(r) \approx n_{\mathrm{t,i}}(r) \tag{136}$$

with $n_{\mathrm{i}}(r)$ and $n_{\mathrm{t,i}}(r)$ defined by Eq. (119), and noting, from Eqs. (93) and (134), that $n_{\mathrm{t,i}} \propto (r^2\mathscr{E})^{-p}$, Eq. (136) then gives $(r^2\mathscr{E})^p\, d(r^2\mathscr{E}) \propto r^2\, dr$. Thus, $(r^2\mathscr{E})^{p+1} \propto r^3$, so that

$$\mathscr{E} \propto r^{(1-2p)/(1+p)} \qquad \text{and} \qquad n_{\mathrm{t,i}} \propto r^{-3p/(1+p)}. \tag{136a}$$

As p ranges from 0 to 1: $(1 - 2p)/(1 + p) = n$ ranges from 1 to $-\frac{1}{2}$, and, from Eq. (132), c_1 ranges from $\frac{1}{2}$ to 2, as asserted in (131); and $(-3p)/(1 + p) = m$ ranges from 0 to $-\frac{3}{2}$, and, from Eq. (133), c_2 ranges from 1 to 2, as asserted in Eq. (131).

The above argument is rigorous only to the extent that the representation Eq. (134) is valid. In the case of a single discrete trapping level, we showed above that this representation is, indeed, strictly valid only in the limit of deep trapping or shallow trapping. For intermediate positions of the Fermi level, the "equivalent p" is a function of position, and hence of n_{i}. However, since the "equivalent p" lies in the very restricted range $0 < p < 1$ (as is likewise true for an exponential distribution of traps), and since the corresponding ranges of c_1 and c_2 are also relatively narrow, namely, as given in (131), it is obviously plausible that the $\frac{3}{2}$-power law (127) is indeed generally valid, only with the coefficients c_1 and c_2 having a very weak, inconsequential, voltage dependence, both still being confined within the relatively narrow limits of (131).

d. Some Specific Trap Configurations

We have seen above that the $\frac{3}{2}$-power law Eq. (127) is given in terms of a trap-density parameter $n_{\mathrm{t,0}}$ and two dimensionless constants, c_1 and c_2 that depend "weakly" on the trap configuration, being of order unity. The three quantities $n_{\mathrm{t,0}}$, c_1, and c_2 are defined in Eqs. (117) and (118). In the following sections, we consider four specific trap configurations and compute the relevant quantities for each of them. The four configurations are: trap-free case, shallow traps, deep traps, and an exponential distribution of traps. In establishing the dependence of current on voltage, in each case it is only the region I, $r_{\mathrm{c}} \leqslant r \leqslant r_x$, with which we need be concerned. Here again, we shall assume that $r_x \gg r_{\mathrm{c}}$. Where $r_x < 3r_{\mathrm{c}}$, we are dealing with the transition from Ohm's law to the $\frac{3}{2}$-power law, and we cannot expect a simplified treatment to describe accurately such a transition.

Since we shall use Eqs. (132) and (133) to obtain c_1 and c_2, respectively, we shall need to know the radial variation of $\mathscr{E}$ and $n_i + n_{t,i}$ in region I. In obtaining these we shall simplify, by approximation, to the greatest possible extent, thereby sacrificing details which are not crucial in arriving at the current–voltage characteristic.

(1) *Trap-Free Case* ($n_{t,0} = 0$). Since the $\frac{3}{2}$-power law has already been determined for this case in problem 1, Eq. (110), it remains only to verify that Eq. (127) gives the same answer. Throughout region I, $\mathscr{E} \propto r^{-1/2}$ is a sufficiently good approximation to Eq. (102), as already pointed out in obtaining Eq. (105). From Eq. (93), it then follows that $n_i \propto r^{-3/2}$. Thus, with reference to Eqs. (132) and (133),

$$n = -\tfrac{1}{2} \rightarrow c_1 = 2; \qquad m = -\tfrac{3}{2} \rightarrow c_2 = 2. \tag{137}$$

It is readily checked that insertion of Eq. (137) into (127) yields Eq. (110).

(2) *The Shallow-Trap Case* ($E_t - F_0 \gg kT$). Here it is assumed that the only effective traps are a single set of shallow traps, that is, traps located in the forbidden gap at an energy E_t well above the Fermi level F_0: $E_t - F_0 \gg kT$. As injection proceeds, the quasi-Fermi level F rises in the forbidden gap. So long as $E_t - F > kT$, the ratio of free-to-trapped carrier concentrations is a constant: $n_i/n_{t,i} = \theta = N/gN_t$, with N, g, and N_t the customary trap parameters. Assume that this condition is realized throughout most of region I, except close to the cathode, where $F > E_t$ necessarily. Following the usual line of simplification, we take the shallow-trap condition to hold everywhere in region I. Then the RHS of the Poisson equation (94), namely, $n_i + n_{t,i}$, becomes n_i/θ, taking $\theta \ll 1$, which is the interesting case. Thus, mathematically, the shallow-trap problem has been made formally identical to the trap-free problem if ε is everywhere replaced by $\theta\varepsilon$, e.g., in (110):

$$I \approx \frac{2\pi\sqrt{2}}{3}\mu(\theta\varepsilon e n_0)^{1/2} V^{3/2}. \tag{138}$$

In a meticulous treatment of this problem, we would closely parallel the route taken in the discussion of problem 3 in Section 1, since this is the identical problem, only with spherical, radial flow. Thus, in place of region I above there would be three regions, namely, the exact analogs of regions I, II, and III of the planar-flow problem. However, for both problems, so long as we confine our discussion to the regime in the current–voltage characteristic immediately following the Ohm's-law regime, the first two regions, namely, those closest to the cathode, absorb negligible voltage, and so make no substantive contribution to the current–voltage characteristic. The important region, which is called region I above, is the analog of region III for the planar-flow problem, characterized by Eq. (68).

(3) *The Deep-Trap Case* ($F_0 - E_t > kT$). In this problem, it is assumed that the only effective traps are a single set of deep traps, that is, traps lying below the Fermi level F_0: $F_0 - E_t > kT$. This problem is the spherical analog of the planar-flow problem 2 of Section 1. When the quasi-Fermi level has been raised, by injection, by more than kT, the previously unoccupied traps, of density $p_{t,0}$ are filled with electrons; thus, in region I, $n_{t,i} = p_{t,0}$ independent of radius. Taking $p_{t,0} \gg n_0$, which is the interesting case, it follows that $n_{t,i} > n_i$ everywhere in region I except very close to the cathode. Again, we simplify the analysis considerably by taking $n_{t,i} > n_i$ throughout all of region I. Then the RHS of the Poisson equation (94) is now $n_t + n_{t,i} \approx p_{t,0}$. Straightforward integration now gives

Region I:

$$\mathscr{E} \approx \frac{ep_{t,0}}{3\varepsilon}\left(\frac{r^3 - r_c{}^3}{r^2}\right) \approx \frac{ep_{t,0}}{3\varepsilon} r \qquad \text{for} \qquad r \gg r_c. \tag{139}$$

Since $n_{t,i} \approx p_{t,0} \propto r^0$, it follows that, referring to Eqs. (132) and (133),

$$n = 1 \rightarrow c_1 = \tfrac{1}{2}; \qquad m = 0 \rightarrow c_2 = 1. \tag{140}$$

Insertion of Eq. (140) into (127) yields

$$I = \frac{8\pi\sqrt{2}}{3} e\mu n_0 \left(\frac{\varepsilon}{ep_{t,0}}\right)^{1/2} V^{3/2}. \tag{141}$$

This problem is the precise spherical analog of problem 2 of Section 1. We see, from the discussion there, that, strictly speaking, we are dealing with a three-region problem. However, consistent with our philosophy of simplification, we have neglected the region adjacent to the cathode, which, as always, is a region dominated by excess free charge. Since it can be shown that the voltage absorbed by this region is down by a factor of the order of $(n_0/p_{t,0})^{2/3}$ from the voltages absorbed by the two main regions, our simplification procedure is justified. Nonetheless, as our study of problem 3 below shows, some interesting physical behavior has indeed been lost by the extreme simplification.

(4) *The Exponential Distribution of Traps.* A set of traps need not be so precisely localized in energy in the forbidden gap as we have been assuming up to this point. Because of random structural disorder or random chemical imperfections, the environment of a given kind of defect, and, correspondingly,

its energy level, might vary somewhat throughout the solid. A convenient representation for traps distributed in energy is the exponential trap distribution[3,4]:

$$N_t(E) = N_0 \exp \frac{E - E_c}{kT_t} \tag{142}$$

where $N_t(E)$ is the density of traps per unit energy range, k is Boltzmann's constant, and T_t is simply a temperature parameter characterizing the distribution.

In order to obtain the parameters in the $\frac{3}{2}$-power law, it is sufficient to establish the functional relationship between $n_{t,i}$ and n_i, as written in the form (134). From Eqs. (142) and (95), it readily follows[3,4,12] that

$$n_{t,i} = kT_tN_0(n_i/N_c)^{1/l}, \qquad l = T_t/T > 1\,. \tag{143}$$

Only the case $l > 1$ is of interest. For $l < 1$, the present analysis breaks down: in effect, the energy states nearest the band edge dominate the trapping and the entire scheme reduces to the shallow-trap case already discussed. Taking $p = 1/l$ in Eqs. (134) and (136a), it follows that, referring to Eqs. (132) and (133),

$$n = \frac{l-2}{l+1} \to c_1 = \frac{l+1}{2l-1}; \qquad m = -\frac{3}{l+1} \to c_2 = \frac{l+1}{l}. \tag{144}$$

Insertion of (144) into (127) gives

$$I = 4\pi e\mu n_0 \left(\frac{3\varepsilon l}{ekT_tN_0(l+1)} \right)^{1/2} \left(\frac{2l-1}{3l} \right)^{3/2} V^{3/2}. \tag{145}$$

A final comment is in order. Since $\mathscr{E} \propto r^n$, it follows from (144) that for l large ($l \to \infty$) $n \to 1$ which is precisely the radial dependence of $\mathscr{E}$ in the deep-trapping case. Indeed, (145) reduces to (141) for $l \to \infty$, taking $kT_tN_0 = p_{t,0}$. Thus, in the limit of large l, the behavior is that of deep trapping: in the limit of small l ($l < 1$), the behavior is that of shallow trapping. It is therefore appropriate to regard the usual case ($1 < l < 10$) as a case intermediate between deep and shallow trapping.

d. Problem 3: The Unusual Field Distribution Associated with Deep Trapping

For deep trapping, as for all other cases, the regime in the current–voltage characteristic immediately following the initial Ohm's-law regime is the $\frac{3}{2}$-power law; indeed, we have obtained above the specific parameters for that law, namely (141). In carrying out the analysis yielding (141) we have, as usual, resorted to the maximum simplification possible, namely, we have ignored any region not absorbing a significant fraction of the applied voltage.

This turns out to be the case in which some interesting physics was lost in the simplification.

We shall now show, still using a regional approximation analysis, only a more detailed one, that in the $\frac{3}{2}$-power-law regime, the radial distribution of electric field is unusual in that it has three extrema, two maxima and one minimum. It is actually easy to see this using the basic properties of spherical, radial flow that we have already established. We refer to the previous two-region approximation, in region I, dominated by injected *trapped* charge, $\mathscr{E} \propto r$, as given by Eq. (139). Let us now call this region I_b. Previously, we arbitrarily extended this region right up to the cathode. In actuality, we know that near the cathode, the space charge is dominated by injected *free* charge. Let us call this region I_a. This region, extending out to $r_{x'}$ is precisely the same as that labeled region I in problem 1, and has the field distribution described there following Eq. (104). Summarizing the field distribution in the separate regions:

Region I_a

$$(r_\mathrm{c} \leqslant r \leqslant r_{x'}): \quad \mathscr{E} \text{ rises from } 0 \text{ at } r = r_\mathrm{c} \text{ to a maximum at } r = r_\mathrm{m} = 4^{1/3} r_\mathrm{c}; \text{ thereafter, } \mathscr{E} \propto 1/\sqrt{r}. \tag{146a}$$

Region I_b

$$(r_{x'} \leqslant r \leqslant r_x): \quad \mathscr{E} \propto r. \tag{146b}$$

Region II

$$(r_x \leqslant r \leqslant r_\mathrm{a}): \quad \mathscr{E} \propto 1/r^2. \tag{146c}$$

Clearly, three extrema are required to encompass the radial variations in $\mathscr{E}$ specified by Eqs. (146a)–(146c), as is seen in the schematic sketch of Fig. 13. Note that Fig. 13 is obtained from Fig. 12 by insertion, between regions I and II, of the new region I_b.

We now present a more quantitative analysis based on the regional approximation method. The three regions and their connecting surfaces are characterized by:

Region I_a: perfect insulator region, $r_\mathrm{c} \leqslant r \leqslant r_{x'}$:

$$\frac{\varepsilon}{e} \frac{1}{r^2} \frac{d}{dr}(r^2 \mathscr{E}) = n \tag{147}$$

$$n_{x'} = n(r_{x'}) = p_{\mathrm{t},0} \tag{148}$$

Region I_b: TFL (trap-filled limit) region, $r_{x'} \leqslant r \leqslant r_x$:

$$\frac{\varepsilon}{e} \frac{1}{r^2} \frac{d}{dr}(r^2 \mathscr{E}) = p_{\mathrm{t},0} \tag{149}$$

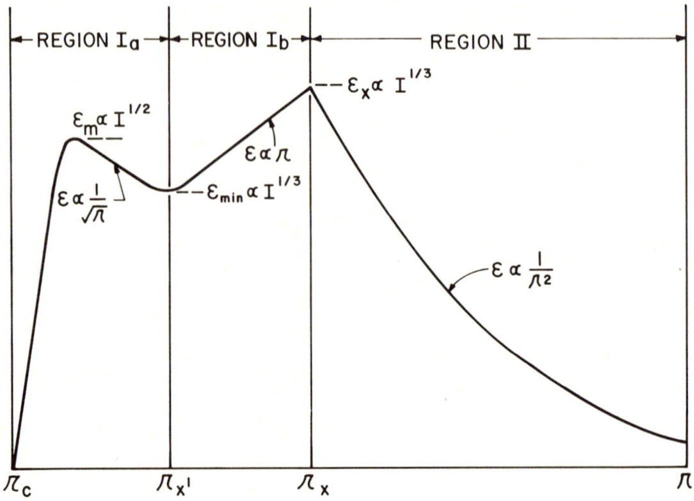

FIG. 13. Schematic radial variation of $\mathscr{E}$ for the one-carrier, spherical-flow, deep-trap problem in the $\frac{3}{2}$-power-law regime.

$$n_x = n(r_x) = n_0 \tag{150}$$

Region II: ohmic region, $r_x \leqslant r \leqslant r_a$:

$$I = 4\pi e\mu n_0 r^2 \mathscr{E} \tag{151}$$

The solution for the electric field is now as follows:

(1) *Region I_a*. Integration of Eq. (147) gives

$$\mathscr{E}(r) = \left(\frac{I}{6\pi\varepsilon\mu}\right)^{1/2} \frac{(r^3 - r_c{}^3)^{1/2}}{r^2}, \qquad n(r) = \left(\frac{3\varepsilon I}{8\pi e^2\mu}\right)^{1/2} \frac{1}{(r^3 - r_c{}^3)^{1/2}}, \tag{152}$$

which is just Eq. (102). Using (148) in (152), we obtain for $r_{x'}$

$$r_{x'} = r_c(1 + \mathscr{I})^{1/3}, \qquad \mathscr{I} = \frac{3\varepsilon I}{8\pi e^2 \mu p_{t,0}^2 r_c{}^3} \tag{153}$$

which is just Eq. (103) with $p_{t,0}$ replacing n_0.

The maximum electric field, occurring at $r = r_m$, is given by

$$r_m = 4^{1/3} r_c: \qquad \mathscr{E}_m = \mathscr{E}(r_m) = \left(\frac{I}{2^{11/3}\pi\varepsilon\mu r_c}\right)^{1/2} = \frac{e p_{t,0} r_c}{2\varepsilon}\left(\frac{2^{4/3}}{3}\mathscr{I}\right)^{1/2} \tag{154}$$

In the sense of the "unusual" field distribution of Fig. 13, region I_a is fully developed for current levels such that $r_{x'} > r_m$. Letting $r_{x'} = r_m = 4^{1/3} r_c$ in (153), this gives a minimum current

$$\mathscr{I} = 3: \qquad I_{\min} = \frac{8\pi e^2 \mu p_{t,0}^2 r_c^{\ 3}}{\varepsilon} \tag{155}$$

(2) *Region I_b*. Integration of Eq. (149) gives

$$\mathscr{E}(r) = Cr + \frac{D}{r^2}; \qquad C = \frac{ep_{t,0}}{3\varepsilon}, \qquad D = Cr_c^{\ 3}(\mathscr{I} - 1) \tag{156}$$

with D determined from the boundary condition (148), using (153).

From (156), $d\mathscr{E}/dr = C - (2D/r^3)$, so that $\mathscr{E}$ has an extremum (minimum) $\mathscr{E} = \mathscr{E}_{\min}$ at $r = r_{\min}$ characterized by

$$\begin{aligned} D > 0, \qquad r_{\min} &= \left(\frac{2D}{C}\right)^{1/3} = r_c[2(\mathscr{I} - 1)]^{1/3}, \\ \mathscr{E}_{\min} &= \frac{3}{2} C r_{\min} = \frac{ep_{t,0} r_c}{2\varepsilon}[2(\mathscr{I} - 1)]^{1/3} \end{aligned} \tag{157}$$

From Eqs. (154) and (157),

$$\mathscr{I} \geqslant 3: \qquad \frac{\mathscr{E}_{\max}(\text{region I}_a)}{\mathscr{E}_{\min}(\text{region I}_b)} = \frac{2^{1/3}}{3^{1/2}} \frac{\mathscr{I}^{1/2}}{(\mathscr{I} - 1)^{1/3}}, \tag{158}$$

the ratio being unity at $\mathscr{I} = 3$, corresponding to $r_{\min} = 4^{1/3} r_c = r_m$. The ratio increases with current, varying with $\mathscr{I}^{1/6} \propto I^{1/6}$ for $\mathscr{I} \gg 1$.

Region I_b ends at $r = r_x$ given by

$$r_x = r_c\left[\mathscr{I}\left(\frac{2p_{t,0}}{n_0} - 1\right) + 1\right]^{1/3} \approx r_c\left[\frac{2p_{t,0}}{n_0}\mathscr{I}\right]^{1/3} \tag{159}$$

where we have used boundary condition (150), and (93) and (156). From Eqs. (156) and (159),

$$\mathscr{E}_x = \mathscr{E}(r_x) \approx Cr_x = \frac{ep_{t,0} r_c}{2\varepsilon}\left[\frac{16}{27}\frac{p_{t,0}}{n_0}\mathscr{I}\right]^{1/3} \tag{160}$$

Although not an extremum, since $(d\mathscr{E}/dr)_x \neq 0$, $\mathscr{E}_x$ is the maximum value of $\mathscr{E}(r)$ in region I_b, and is indeed the second maximum of $\mathscr{E}$ in the solid, since $\mathscr{E} \propto 1/r^2$ in region II. Although $\mathscr{E}$ is made continuous in crossing from region I_b to II, $d\mathscr{E}/dr$ is discontinuous. The discontinuity in the slope of $\mathscr{E}$ at the connecting surfaces is an inevitable concomitant of the regional approximation method (for if the slope of $\mathscr{E}$ were also continuous, we would have, essentially, the exact solution!).

The ratio of the two field maxima is, using Eqs. (154) and (160),

$$\frac{\mathscr{E}_x}{\mathscr{E}_m} = \left(\frac{16}{27\mathscr{I}}\right)^{1/6}\left(\frac{p_{t,0}}{n_0}\right)^{1/3}, \tag{161}$$

which varies as $I^{-1/6}$. Thus, although initially $\mathscr{E}_x > \mathscr{E}_{\mathrm{m}}$ because of the factor $(p_{t,0}/n_0)^{1/3}$, at large enough current, $\mathscr{E}_{\mathrm{m}}$ overtakes and exceeds $\mathscr{E}_x$.

III. Two-Carrier Problems

The next part of this review takes up problems of two-carrier injection. Because carriers of both signs of charge are injected, neutralization of space charge is possible. This neutralization may be only partial, or it may be essentially complete. Further, a new phenomenon enters the picture, recombination: injected electrons and holes may recombine in the bulk before reaching their respective collector. Finally, the phenomenon which lends richness to the one-carrier injection problem, namely, trapping of injected carriers, also plays a role in double injection. Further, there are now two kinds of traps to worry about, electron traps and hole traps. In effect, a considerably richer set of ingredients constitutes the raw materials of the double-injection problem and, correspondingly, the double-injection problem is far more complex than the single-injection problem. The double-injection problem, in fact, is so complex that a single, nicely packaged approach, handling all facets, is simply not available. We must be satisfied with the separate packaging of restricted parts of the problem.

Two classes of double-injection problems are studied in our review. Most completely understood is the class of injected-plasma problems, which we take up first. Here, the injected carriers remain free and, in most cases of interest, largely neutralize each other. The second class, namely, varying-lifetime, negative-resistance problems, is incompletely understood, but has yielded somewhat to investigation. In all cases, we restrict ourselves to planar-flow problems. Further, we always take the electrode configuration such that the hole-injecting contact (anode) is at $x = 0$ and the electron-injecting contact (cathode) is at $x = L$. Thus, holes flow from left to right, and conversely for electrons.

3. Injected-Plasma Problems

The injected-plasma problems are defined as those in which the injected electrons and holes remain free. The *only* role permitted the defect states, in these problems, is that of determining the recombination lifetime. Population changes in the defect-state occupancies, as a result of injection, are considered negligible.

The simplest injected-plasma problem conceptually (although not mathematically) is that of double injection into a perfect insulator.

a. Problem 1: The Perfect-Insulator Injected Plasma

The perfect insulator, by definition, is free of defect states. Hence, the injected electrons and holes, in the bulk, can only recombine directly. The equations characterizing this problem are the current-flow equations

$$J_n = e\mu_n n\mathscr{E}, \tag{162}$$

$$J_p = e\mu_p p\mathscr{E}, \tag{163}$$

the Poisson equation

$$\frac{\varepsilon}{e}\frac{d\mathscr{E}}{dx} = p - n, \tag{164}$$

and the particle-conservation equations

$$\frac{1}{e}\frac{dJ_n}{dx} = \mu_n\frac{d}{dx}(n\mathscr{E}) = r = \langle v\sigma_{\mathrm{R}}\rangle pn, \tag{165}$$

$$-\frac{1}{e}\frac{dJ_p}{dx} = -\mu_p\frac{d}{dx}(p\mathscr{E}) = r = \langle v\sigma_{\mathrm{R}}\rangle pn, \tag{166}$$

with Eqs. (165) and (166) combining to give

$$J = J_n + J_p = \text{const}. \tag{167}$$

In the above equations, J_n and J_p are the electron and hole current densities respectively, μ_n and μ_p the free-electron and free-hole drift mobilities respectively, n and p the free-electron and free-hole densities respectively, r the recombination rate density, v the microscopic, relative velocity of electron and hole, σ_{R} the mutual recombination cross section (which depends on v), and the angular brackets denote an average over the velocity distribution of both electrons and holes.

From Eqs. (162) and (163), it is clear that we are ignoring diffusion-current flow, and so we take the usual boundary conditions for such a simplified theory at the injecting contacts:

$$\mathscr{E} = 0 \quad \text{at} \quad x = 0 \quad \text{and at} \quad x = L. \tag{168}$$

The set of equations (162)–(168) has been studied analytically by Parmenter and Ruppel,[7] who succeeded in obtaining the current–voltage characteristic, Eq. (189). However, the mathematics of the problem are so complicated that the spatial dependence of $\mathscr{E}$, p, and n cannot be obtained from their work. We are therefore led to apply the regional approximation method to a study of the problem.

Several of the regions can be identified immediately. Near the anode, the flow of holes completely dominates the picture. This, then, is a region adequately described as a one-carrier, hole, space-charge current region. By symmetry, there will be a corresponding electron-dominated space-charge region near the cathode. Since $p \gg n$ near the anode (at $x = 0$), and since $n \gg p$ near the cathode (at $x = L$), there must be a crossing plane x_m at which $n_m = n(x_m) = p_m = p(x_m)$. Encompassing x_m is a plasma region throughout which n and p differ, at any plane, by less than a factor of two. Whether or not there is an additional region depends on the magnitude of the mobility ratio $b = \mu_n/\mu_p$. For $\frac{1}{2} \leqslant b \leqslant 2$, there is no additional region. For $b > 2$, there is an additional region lying between the hole-dominated space-charge region and the plasma region. This is an unusual region in which the current is dominated by the electrons, whereas the space charge is dominated by the holes. By symmetry, for $b < \frac{1}{2}$, the additional region lies between the electron-dominated space-charge region and the plasma region. In this case, the additional region is such that the current is dominated by the holes and the space charge by the electrons.

For the sake of definiteness, we take the case $b > 2$. We are then dealing with the following four-region problem, illustrated schematically in Fig. 14.

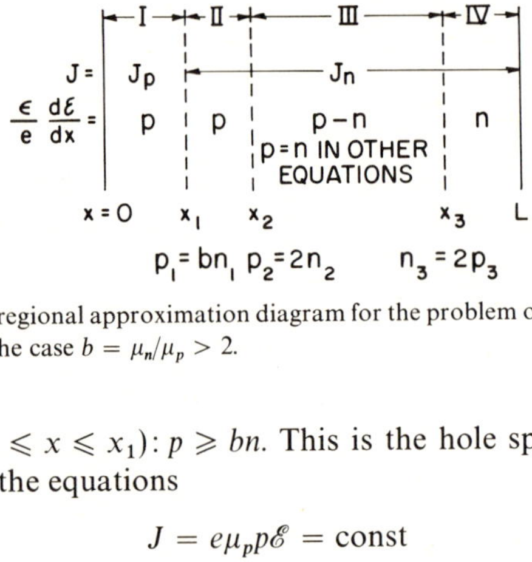

FIG. 14. Schematic regional approximation diagram for the problem of double injection into a perfect insulator for the case $b = \mu_n/\mu_p > 2$.

(1) *Region I* ($0 \leqslant x \leqslant x_1$): $p \geqslant bn$. This is the hole space-charge region, characterized by the equations

$$J = e\mu_p p\mathscr{E} = \text{const} \tag{169}$$

replacing (167), and

$$\frac{\varepsilon}{e}\frac{d\mathscr{E}}{dx} = p \tag{170}$$

replacing (164). Equations (169) and (170) are combined to give

$$\frac{d\mathscr{E}^2}{dx} = \frac{2J}{\varepsilon\mu_p}, \tag{171}$$

with solution $\mathscr{E} = (2Jx/\varepsilon\mu_p)^{1/2}$ satisfying boundary condition (168). The variation of p with x is gotten from (170), and the variation of n with x is obtained by solving (165) with the now-known variations of p and $\mathscr{E}$ with x.

Region I ends at x_1 given by

$$p_1 = p(x_1) = bn(x_1) = bn_1 \,. \tag{172}$$

For $x > x_1$, the main contributor to J is the electron current.

(2) *Region II* $(x_1 \leqslant x \leqslant x_2)$: $bn \geqslant p \geqslant 2n$. This is the "hybrid" region, characterized by the equations

$$J = e\mu_n n\mathscr{E} = \text{const} \tag{173}$$

replacing (167), and

$$\frac{\varepsilon}{e}\frac{d\mathscr{E}}{dx} = p \tag{174}$$

replacing (164).

Equation (166) is used unmodified. Substituting from (174) and (173) into (166), we get

$$-\mathscr{E}^2\frac{d^2\mathscr{E}^2}{dx^2} = A\frac{d\mathscr{E}^2}{dx}, \qquad A = \frac{\langle v\sigma_{\mathrm{R}}\rangle J}{e\mu_n\mu_p}. \tag{175}$$

This differential equation is readily solved[14] in terms of the well-known exponential integral function $E_1(y) = \int_y^\infty dt \exp(-t)/t$. The variation of p with x is then gotten from (174), and n with x from (173).

Region II ends at x_2 given by

$$p_2 = p(x_2) = 2n(x_2) = 2n_2 \,.$$

(3) *Region III* $(x_2 \leqslant x \leqslant x_3)$: $2n \geqslant p \geqslant n/2$. This is the plasma region, characterized by Eqs. (173) and (164), and with

$$\mu_n\frac{d}{dx}(n\mathscr{E}) = \langle v\sigma_{\mathrm{R}}\rangle n^2 \tag{176}$$

$$-\mu_p\frac{d}{dx}(p\mathscr{E}) = \langle v\sigma_{\mathrm{R}}\rangle n^2 \tag{177}$$

replacing Eqs. (165) and (166), respectively. Multiplying Eq. (177) by b and adding to Eq. (176) gives, after using Eq. (164) to eliminate $(p - n)$ and Eq. (173) to eliminate n,

$$-\mathscr{E}^2\frac{d^2\mathscr{E}^2}{dx^2} = B'; \qquad B' = \frac{4(b+1)\mu_{\mathrm{R}}J^2}{\varepsilon^2 b^3 \mu_p{}^3}, \qquad \mu_{\mathrm{R}} = \frac{\varepsilon\langle v\sigma_{\mathrm{R}}\rangle}{2e}. \tag{178}$$

The solution to this differential equation is the Gaussian integral (186). The variation of n with x is gotten from (173), and p with x from (164).

Region III ends at x_3 given by

$$p_3 = p(x_3) = \tfrac{1}{2}n(x_3) = \tfrac{1}{2}n_3 .$$

(4) *Region IV* ($x_3 \leqslant x \leqslant L$): $n/2 > p$. This is the electron space-charge region, characterized by Eq. (173), and

$$\frac{\varepsilon}{e}\frac{d\mathscr{E}}{dx} = -n \tag{179}$$

replacing (164). The combination of Eqs. (173) and (179) gives

$$\frac{d\mathscr{E}^2}{dx} = -\frac{2J}{\varepsilon\mu_n} \tag{180}$$

with solution $\mathscr{E} = [2J(L - x)/\varepsilon\mu_n]^{1/2}$ satisfying boundary condition (168). The variation of n with x is gotten from (179), and the variation of p with x is obtained by solving (166) with the now-known variations of n and $\mathscr{E}$ with x.

The solutions in the four regions are tied together in the usual way, requiring the $\mathscr{E}$-field to be continuous in crossing a transition plane:

$$\mathscr{E}(x_1^-) = \mathscr{E}(x_1^+), \qquad \mathscr{E}(x_2^-) = \mathscr{E}(x_2^+), \qquad \mathscr{E}(x_3^-) = \mathscr{E}(x_3^+). \tag{181}$$

A most unusual feature of this problem is the "static" quality of the regions: x_1, x_2, and x_3 are constant in position, independent of current. The reason

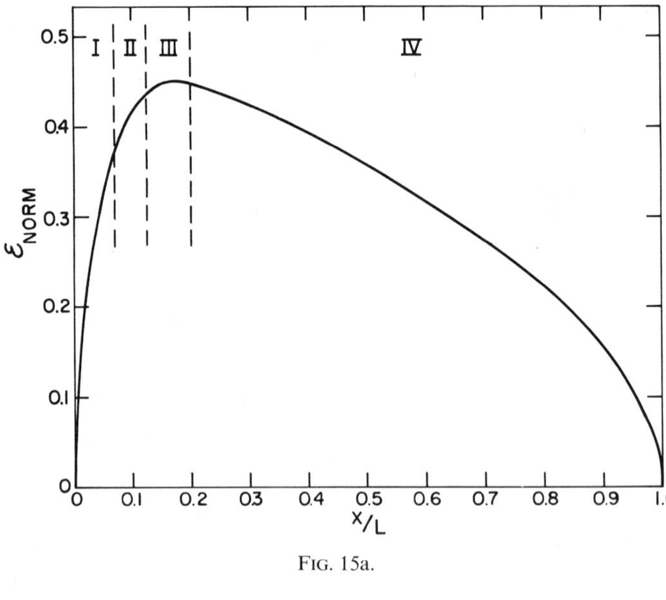

FIG. 15a.

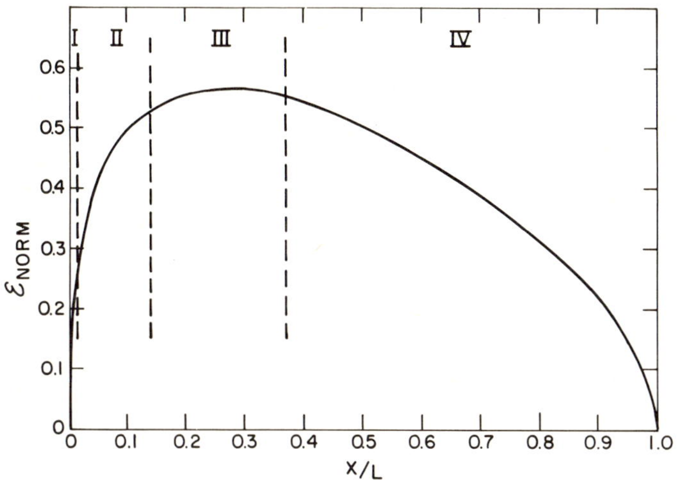

FIG. 15b.

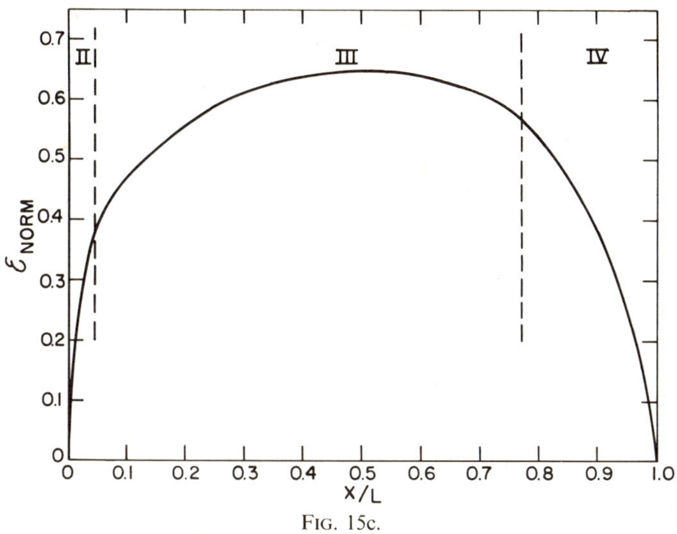

FIG. 15c.

FIG. 15. Variation of the (normalized) field with position for the problem of double injection into a perfect insulator. $\nu_p = \mu_p/\mu_R$, $\nu_n = \mu_n/\mu_R$, $\mu_R = \varepsilon\langle v\sigma_R\rangle/2e$, and $b = 8$ for all cases: (a) $\nu_p = \frac{1}{16}$, $\nu_n = \frac{1}{2}$; (b) $\nu_p = \frac{1}{4}$, $\nu_n = 2$; (c) $\nu_p = 2$, $\nu_n = 16$.

for this is that the basic equations (162)–(168) are such that the functional form of their solution, $[p(x), n(x), \mathscr{E}(x), V(x), J_p(x), J_n(x)]$ is independent of the applied voltage. That is, if the applied voltage is doubled, $p(x)$, $n(x)$, $\mathscr{E}(x)$, and $V(x)$ retain their functional form, but are each doubled in magnitude: $J_p(x)$ and $J_n(x)$ retain their functional form, but are each quadrupled. Thus $J \propto V^2$, a result obtained essentially by mere inspection of the equations!

We do not present further details of the regional approximation solution in this review. The interested reader who wishes a fuller picture is referred to the original article.[14] Some typical results for the spatial distribution $\mathscr{E}(x)$ are presented in Fig. 15. Figure 15a corresponds to a situation dominated by the one-carrier, electron space-charge-limited current, Fig. 15c to an injected-plasma situation, and Fig. 15b to a transitional situation. The positions x_1, x_2, and x_3 of the connecting planes between regions are marked on each curve. The computed current–voltage characteristic, that is, the coefficient of the square law, obtained by the regional approximation method is compared to the exact coefficient,[7] and to some approximate coefficients obtained analytically, in Fig. 16.

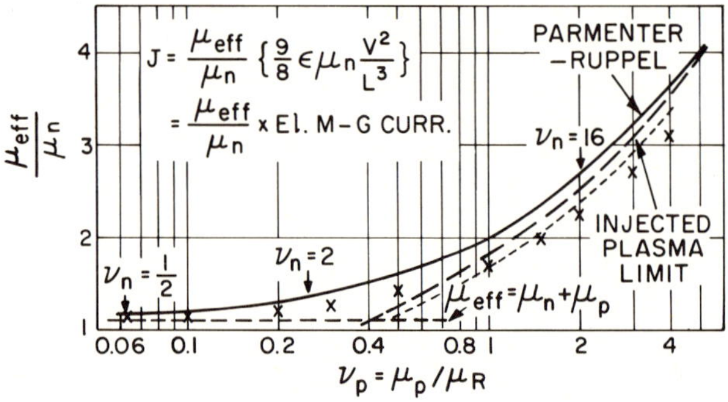

FIG. 16. The current "amplification" factor μ_{eff}/μ_n versus ν_p [μ_{eff} given by (189), ν_p and ν_n given in Fig. 15] for the problem of double injection into a perfect insulator ($b = 8$). The solid curve is the exact Parmenter–Ruppel result (189a). The long-dashed curve is the injected-plasma-limit result (188). The short-dashed curve is the long-dashed curve scaled by the multiplicative factor $b/(b + 1)$. The horizontal dashed line is the infinite-σ_R-limit result (182). The crosses are results obtained by the regional approximation. Because the latter neglects the hole contribution to the injected plasma current in region III, the regional approximation solution is asymptotic, at large ν_p, to the short-dashed curve. "El M–G CURR" denotes the electron Mott–Gurney (i.e., space-charge-limited, trap-free) current given in brackets immediately above.

b. The Large-Recombination Cross Section Limit

As the electron–hole recombination cross section becomes very large, namely, as $\sigma_R \to \infty$, it is obvious that regions II and III shrink to a plane, which we label x_m: $x_1 = x_2 = x_3 = x_m$. There can be no overlap of the

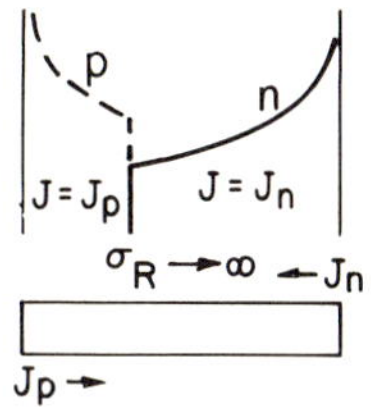

FIG. 17. Schematic variation of injected free-carrier densities with position for the problem of double injection into a perfect insulator for the cases of infinite electron–hole recombination cross section.

electron and hole spatial distributions, since the recombination current would then be infinite. In this limiting circumstance, we are left with a two-region problem, as illustrated schematically in Fig. 17, region I extending from the anode up to x_m and region IV extending from x_m up to the cathode. There are only injected holes in region I, and only injected electrons in region IV. The electrons and holes meet, and mutually annihilate, at plane x_m. We have here a unique case in which the regional approximation method gives the exact solution!

It is an elementary exercise to obtain the solution. Define $L_p = x_m$, $L_n = L - x_m$, $V_p = V(x_m)$, and $V_n = V - V(x_m)$. In region I, $J = J_p = 9\varepsilon\mu_p V_p^2/9L_p^3$; in region IV, $J = J_n = 9\varepsilon\mu_n V_n^2/8L_n^3$. Continuity of the electric field across x_m gives $V_p/L_p = V_n/L_n$. It is now easy to show that $V_n/V_p = L_n/L_p = (L - x_m)/x_m = \mu_n/\mu_p$ and that $(\mu_p + \mu_n)(V_p + V_n)^2/(L_p + L_n)^3 = \mu_p V_p^2/L_p^3 = \mu_n V_n^2/L_n^3$. These results then yield the desired current–voltage characteristic:

$$\sigma_R \to \infty: \qquad J = \frac{9}{8}\varepsilon(\mu_n + \mu_p)\frac{V^2}{L_3}. \tag{182}$$

c. The Small-Recombination Cross Section Limit

If the electron–hole recombination cross section is small, then electrons and holes both traverse the solid with only small attrition, and, in effect, an injected plasma fills the solid. Regions I and II occupy a negligible fraction of the bulk near the anode, and region IV a negligible fraction near the cathode. Since only the injected-plasma region, namely, region III, is important, we follow the usual line of simplification and extend it right up to the electrodes. The problem is reduced to a relatively simple, one-region problem, for which we now give the solution.

The equations characterizing the problem are (162)–(164), (167), (168), (176), and (177). Note that for the current itself we are using the less restrictive Eq. (167) rather than (173) which was previously used in the discussion of region III. The use of the latter was dictated by the assumption that $b > 2$ and our treatment of region II. With our neglect of regions I, II, and IV, this restriction is no longer necessary and our treatment is valid for arbitrary b, if σ_R is sufficiently small. The basic approximation made in treating the problem is that of taking $p = n$ everywhere except in the Poisson equation (164), that is, in (167) and in the recombination rate expression, $r = \langle v\sigma_R \rangle np$. Thus, Eqs. (165) and (166) are replaced by (176) and (177), respectively. This approximation, obviously valid under injected-plasma conditions, is crucial to the achievement of analytic tractability.

Using the same algebraic manipulations that yielded Eq. (178) for region III, we obtain here

$$-\mathscr{E}^2 \frac{d^2\mathscr{E}^2}{dx^2} = B, \qquad B = \frac{4\mu_R J^2}{\varepsilon^2 b(b+1)\mu_p{}^3}, \tag{183}$$

with μ_R defined in Eq. (178). We obtain B from B' in Eq. (178) by replacing b^3 by $(b+1)^2 b$.

For further discussion, it is convenient to use dimensionless variables:

$$w = \frac{8e^2\mu_R^{5/2}}{JL^6\varepsilon[b(b+1)\mu_p{}^3]^{1/2}} x, \qquad u = \frac{2\mu_R}{(b+1)\mu_p L^3} \frac{1}{n} = \frac{2e\mu_R}{JL^3}\mathscr{E},$$
$$v = \frac{(2\pi)^{3/4}\varepsilon^{1/2}[b(b+1)\mu_p{}^3]^{1/4}}{2J^{1/2}L^{3/2}\mu_R^{1/4}} V. \tag{184}$$

Substitution of Eq. (184) into (183) yields the equation

$$u^2 \frac{d^2u^2}{dw^2} = -1. \tag{185}$$

The solution to Eq. (185) is conveniently expressed in terms of an additional, dummy variable s:

$$u = \left(\frac{w_c{}^2}{2\pi}\right)^{1/4} \exp\left(-\frac{s^2}{2}\right), \qquad \frac{w}{w_c} = \frac{x}{L} = \frac{1}{\sqrt{\pi}} \int_{-\infty}^{s} ds \exp(-s^2) \tag{186}$$

where $s = -\infty$ corresponds to $w = 0$ and $s = +\infty$ to $w = w_c$. Note that u vanishes at these limits, thereby satisfying the boundary conditions (168). The relations $du^2/ds = -2su^2$, $dw/ds = u^2\sqrt{2}$, $du^2/dw = (du^2/ds)/(dw/ds) = -s\sqrt{2}$, and $(d/dw)(du^2/dw) = -\sqrt{2}\, ds/dw = -1/u^2$ are all useful in establishing that (186) is a solution to (185). Since

$$V(x) = \int_0^x \mathscr{E}\, dx \propto \int_0^w u\, dw \propto \int_{-\infty}^{s} u(dw/ds)\, ds \propto \int_{-\infty}^{s} u^3\, ds,$$

we have

$$v = v(s) = \int_{-\infty}^{s} ds \exp(-\tfrac{3}{2}s^2), \qquad v_c = v(\infty) = (2\pi/3)^{1/2}. \tag{187}$$

Substitution for v_c from (187) into the defining relation (184) gives the desired current–voltage characteristic:

$$J = \left(\frac{9\pi}{8}\right)^{1/2} \varepsilon \left[\frac{\mu_p\mu_n(\mu_p + \mu_n)}{\mu_R}\right]^{1/2} \frac{V^2}{L^3}. \tag{188}$$

The exact result of Parmenter and Ruppel[7] is

$$J = \frac{9}{8}\varepsilon\mu_{\text{eff}}\frac{V^2}{L^3}, \tag{189}$$

with

$$\mu_{\text{eff}} = \frac{4}{9}\mu_R\nu_n\nu_p\left\{\frac{[\frac{3}{2}(\nu_n + \nu_p) - 1]!}{(\frac{3}{2}\nu_n - 1)!\,(\frac{3}{2}\nu_p - 1)!}\right\}^2\left\{\frac{(\nu_n - 1)!\,(\nu_p - 1)!}{(\nu_n + \nu_p - 1)!}\right\}^2 \tag{189a}$$

and $\nu_n = \mu_n/\mu_R$ and $\nu_p = \mu_p/\mu_R$, where μ_R is given in (178).

It is straightforward to show that in the small-σ_R limit, that is, with $\nu_n \gg 1$ and $\nu_p \gg 1$, Eq. (189) reduces to (188) [using Stirling's approximation: $n! \approx (n/e)^n(2\pi n)^{1/2}$], and that in the large-σ_R limit, that is, $\nu_n \ll 1$ and $\nu_p \ll 1$, Eq. (189) reduces to (182) [using $(\delta - 1)! \approx 1/\delta$ for $0 < \delta \ll 1$].

The relative electric field and potential distributions for the injected-plasma limit, obtained from Eqs. (186) and (187), plotted vs. x/L, are exhibited in Fig. 18. Corresponding exact distributions are not available for comparison.

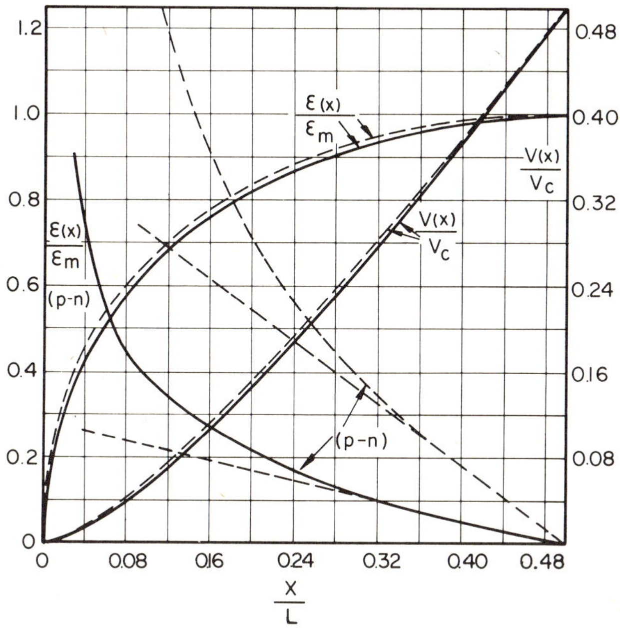

FIG. 18. Theoretical normalized electric field intensity, potential, and space-charge distributions for a plasma injected into an insulator. The solid curves pertain to the constant-lifetime case, the dashed curves to the bimolecular recombination case. The quantity labeled "$(p - n)$" is actually, for the constant-lifetime case, $(2/3)[\tau/(\bar{t}_n + \bar{t}_p)][(p - n)/n_m]$, and, for the bimolecular recombination case, $[(2\mu_R/\mu_p) + (2\mu_R/\mu_n)]^{-1}[(p - n)/n_m]$. Here, τ is the (constant) lifetime, $\bar{t}_n$ and $\bar{t}_p$ the average free-electron and free-hole transit times, respectively, and n_m the free-electron density at the midplane, $x = L/2$. The dashed straight lines are linear approximations to the "$(p - n)$" curves.

Also plotted is a quantity proportional to the local space charge, namely, the quantity $2\mu_R[(1/\mu_p) + (1/\mu_n)]^{-1}(p - n)/n_m$, with n_m the electron density at the mid-plane $x = L/2$. It is seen that the linear approximation (the dashed line) is a good one over most of the insulator. This linear approximation has proven very useful[1] in developing a simple phenomenological theory for the injected plasma.

d. The Constant-Lifetime Injected Plasma

Problem 1, studied above, is a highly idealized problem, in that it ignores entirely the defect states which are inevitably present in a real solid. A simple, yet realistic way (familiar, for example, from transistor physics) in which defect states can affect the current flow is by enforcing a constant lifetime for the injected carriers, this being the *only* significant effect of the defect states. We shall now study the injected plasma under this constant-lifetime condition. The equations characterizing this problem are the current-flow equations

$$J_n = e\mu_n n\mathscr{E} + eV_T\mu_n\frac{dn}{dx}, \tag{190}$$

$$J_p = e\mu_p p\mathscr{E} - eV_T\mu_p\frac{dp}{dx}, \tag{191}$$

with V_T the "thermal voltage," $V_T = kT/e$, and where we have now included the diffusion contributions neglected in the previous problem; the Poisson equation

$$\frac{\varepsilon}{e}\frac{d\mathscr{E}}{dx} = (p - p_0) - (n - n_0), \tag{192}$$

where p_0 and n_0 are the thermal densities of holes and electrons, respectively; and the particle-conservation equations

$$\frac{1}{e}\frac{dJ_n}{dx} = \mu_n\frac{d}{dx}(n\mathscr{E}) + V_T\mu_n\frac{d^2n}{dx^2} = r = \frac{n - n_0}{\tau} = \frac{p - p_0}{\tau}, \tag{193}$$

$$-\frac{1}{e}\frac{dJ_p}{dx} = -\mu_p\frac{d}{dx}(p\mathscr{E}) + V_T\mu_p\frac{d^2p}{dx^2} = r = \frac{n - n_0}{\tau} = \frac{p - p_0}{\tau}, \tag{194}$$

where τ is the common lifetime of the injected electrons and holes.

Addition of Eqs. (190) and (191) gives

$$J = J_n + J_p = e\mu_p(bn + p)\mathscr{E} + eV_T\mu_p\frac{d}{dx}(bn - p) = \text{const.} \tag{195}$$

Multiplication of Eq. (194) by b and addition to (193) gives

$$-\frac{\varepsilon}{e}\frac{d}{dx}\left(\mathscr{E}\frac{d\mathscr{E}}{dx}\right) + (n_0 - p_0)\frac{d\mathscr{E}}{dx} + 2V_T\frac{d^2n}{dx^2} = \frac{(b + 1)(n - n_0)}{\mu_n\tau}, \tag{196}$$

where Eq. (192) has been used and $d^2(p + n)/dx^2$ has been replaced by $2d^2n/dx^2$.

In the problems of interest here, the diffusive contribution to the total, conserved current J in Eq. (195) can be neglected compared to the drift contribution. Further, we shall be interested in high injection levels: $n \approx p \gg n_0, p_0$. Under these conditions, Eq. (195) simplifies to

$$J \approx e\mu_p(b + 1)n\mathscr{E} = \text{const}. \tag{197}$$

The use of Eq. (197) in place of (195) is not strictly justified within the first ambipolar diffusion length of each contact. However, the resulting errors are not significant in the problems discussed below.

We shall apply the above equations to the study of two limiting cases of the injected-plasma: injection into a semiconductor and injection into an insulator.

e. Problem 2: The Semiconductor Injected-Plasma with Diffusion Corrections

A plasma injected into a semiconductor is characterized by a high degree of local neutrality, so that the first term on the LHS of Eq. (196) is negligible. (If we take exact local neutrality, $n - n_0 = p - p_0$, then the first term vanishes altogether.) Thus, in place of Eq. (196), we have the simpler

$$(n_0 - p_0)\frac{d\mathscr{E}}{dx} + 2V_T\frac{d^2n}{dx^2} = \frac{(b + 1)n}{\mu_n\tau}, \tag{198}$$

where we have also incorporated the high-injection-level condition: $n - n_0 \approx n$.

The theory is characterized by Eqs. (197) and (198) and the boundary conditions

$$n_a = n(0) = \frac{JL_a}{2e\mu_pV_T}, \qquad n_c = n(L) = \frac{JL_a}{2e\mu_nV_T}, \qquad L_a = \left(\frac{2V_T\mu_n\tau}{b + 1}\right)^{1/2}. \tag{199}$$

These boundary conditions correspond to highly efficient, injecting contacts —more precisely, to the conditions $J_n(0) = 0$ and $J_p(L) = 0$, with J_n and J_p given by Eqs. (190) and (191), respectively, and with the drift term in (198) neglected in the vicinity of the contacts. The quantity L_a is the famous ambipolar diffusion length, and it is the scale length for pure diffusion processes. If there are very many such lengths contained between the anode and cathode, that is, if $L/L_a \gg 1$, then diffusion processes will play a minor or insignificant role in the current flow. However, if L/L_a is not large, then the diffusion corrections can be important.

Because the egress of carriers is blocked at a good injecting contact (that is, the egress of electrons at the hole-injecting contact, and the egress of holes

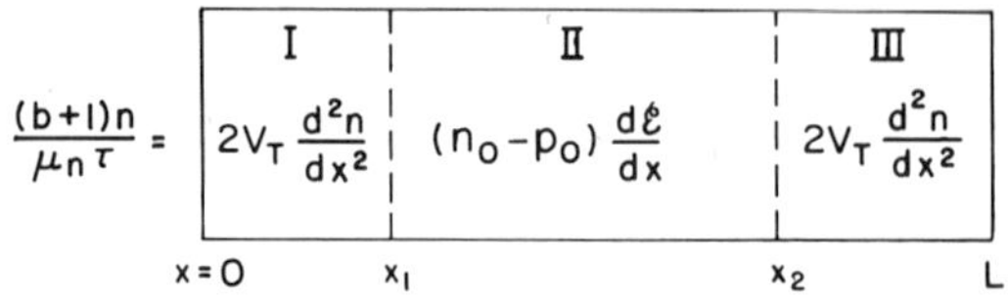

FIG. 19. Schematic regional approximation diagram for the problem of a constant-lifetime plasma injected into a semiconductor. [In writing the "recombination term" as $(b + 1)n/\mu_n\tau$, it is assumed that $n > 2n_0$ everywhere.]

at the electron-injecting contact), there is a large buildup of plasma density near the contact, and the plasma properties are governed by the diffusion equation. On the other hand, in the middle of the solid, away from the contacts, the plasma behavior is governed by drift processes. We thus have a three-region problem as illustrated schematically in Fig. 19.

(1) *Region I* ($0 \leqslant x \leqslant x_1$): *Diffusion Term Dominant.* Here, the diffusion term, $2V_T\, d^2n/dx^2$, is dominant on the LHS of Eq. (198), and so the drift term, $(n_0 - p_0)\, d\mathscr{E}/dx$, is neglected. The plasma density decays exponentially with distance from the anode, and so, likewise, does the diffusion term. But, from Eq. (197), if n is falling exponentially with distance, then $\mathscr{E}$ must be growing exponentially with distance, and so, therefore, must the neglected drift term, which is positive and proportional to $d\mathscr{E}/dx$. Where the neglected drift term catches up with the retained diffusion term, namely, at plane $x = x_1$, it can obviously no longer be neglected. This marks the end of region I.

(2) *Region II* ($x_1 \leqslant x \leqslant x_2$): *Drift Term Dominant.* Here, the drift term is dominant on the LHS of (198), and so the diffusion term is neglected. This leads to $\mathscr{E}(x)$ increasing monotonically going from x_1 to x_2. The characterization of x_2 requires discussion of region III.

(3) *Region III* ($x_2 \leqslant x \leqslant L$): *Diffusion Term Dominant.* This region, just as region I, is adjacent to a contact; therefore, the diffusion term is dominant in Eq. (198) and the drift term is neglected. Since $\mathscr{E}$ is large at x_2, and must become relatively small near $x = L$ (see the discussion of region I), $\mathscr{E}$ must go through a maximum in region III, namely, at plane $x = x_m$ near $x = x_2$: $(d\mathscr{E}/dx)_{x=xm} = 0$. To the right of x_m, the neglected drift term in Eq. (198) is negative; to the left of x_m, it is positive. The plane $x = x_2 < x_m$ marks the location where the neglected drift term, now positive, has caught up to the retained diffusion term in region III. In order that $\mathscr{E}$ have a maximum in region III (compare with region I), it is necessary that both increasing and decreasing exponentials be retained in the solution to the diffusion equation.

The solutions in the different regions are matched up by requiring that the drift term be continuous in crossing x_1 and x_2. This procedure is equivalent to requiring continuity of $\mathscr{E}$, and likewise n, in crossing the connecting planes.

It is convenient to carry out further mathematical discussion using dimensionless variables:

$$u = \frac{\mathscr{E}}{\mathscr{E}^*}, \qquad s = \frac{n}{n^*}, \qquad w = \frac{x}{x^*}, \qquad v = \frac{V}{V^*},$$

$$\mathscr{E}^* = \left[\frac{2V_T J^2}{e^2 \mu_p{}^3 b(b+1)\tau(n_0 - p_0)^2}\right]^{1/4}, \qquad x^* = L_a = \left[\frac{2V_T \mu_n \tau}{b+1}\right]^{1/2}, \tag{200}$$

$$n^* = \frac{J}{e\mu_p(b+1)\mathscr{E}^*} = \left[\frac{b\tau(n_0 - p_0)^2 J^2}{2e^2\mu_p(b+1)^3 V_T}\right]^{1/4},$$

$$V^* = \mathscr{E}^* x^* = \left[\frac{8V_T{}^3 J^2 b\tau}{e^2\mu_p(b+1)^3(n_0 - p_0)^2}\right]^{1/4}.$$

The defining equations (197) and (198) become, respectively,

$$1 = su, \tag{201}$$

$$\frac{du}{dw} + \frac{d^2 s}{dw^2} = s = \frac{1}{u}. \tag{202}$$

The current–voltage characteristic is given by a plot of $s_c{}^2$ versus $v_c s_c$, since

$$s_c{}^2 = J\frac{[(b+1)\tau]^{1/2}}{e(n_0 - p_0)[2\mu_p V_T b^3]^{1/2}}; \qquad v_c s_c = V\frac{b+1}{2bV_T}. \tag{203}$$

Here, V is now the potential of the cathode, i.e., the total voltage across the specimen. In arriving at Eq. (203), use was made of Eq. (199). The separate regions of the problem now are as follows:

Region I ($0 \leqslant w \leqslant w_1$). The drift term is dropped from Eq. (202), giving

$$d^2 s/dw^2 = s. \tag{204}$$

The other defining equation is (201).

Only the exponentially decaying solution to (204) is needed:

$$s = s_a e^{-w}, \qquad u = u_a e^{w}, \qquad u_a = 1/s_a. \tag{205}$$

The neglected drift term is $du/dw = u_a e^w = e^w/s_a$. Region I ends, at $w = w_1$, where this term overtakes the diffusion term $d^2 s/dw^2 = s$:

$$\exp w_1 = s_a: \qquad s_1 = s(w_1) = u_1 = u(w_1) = 1. \tag{206}$$

The voltage across region I is

$$v_I = \int_0^{w_1} u\,dw = 1 - (1/s_a). \tag{207}$$

Region II ($w_1 \leqslant w \leqslant w_2$). The diffusion term is dropped from Eq. (202), giving

$$\frac{du}{dw} = \frac{1}{u} \qquad \text{or} \qquad \frac{1}{2}\frac{du^2}{dw} = 1. \tag{208}$$

The other defining equation is (201). The solution to Eq. (208) is

$$u^2 - 1 = 2(w - w_1) \qquad \text{or} \qquad u = [2(w - w_1) + 1]^{1/2}, \tag{209}$$

where we have used (206), matching u across w_1.

The voltage across region II is

$$v_{\mathrm{II}} = \int_{w_1}^{w_2} u\, dw = \tfrac{1}{3}\{[2(w_2 - w_1) + 1]^{3/2} - 1\}. \tag{210}$$

Note that if we take region II to fill the entire semiconductor (this is equivalent to taking $w_1 \ll w_2$ and $w_2 \approx w_c \gg 1$), then Eq. (210) reduces to

$$v_c^{\,2} = \frac{8}{9} w_c^{\,3} \qquad \text{or} \qquad J = \frac{9}{8} e(n_0 - p_0)\mu_n \mu_p \tau \frac{V^2}{L^3}, \tag{211}$$

a characteristic which has been called the "semiconductor, injected-plasma square law."[9]

Region III ($w_2 \leqslant w \leqslant w_c$). As in region I, the drift term is dropped from (202), so that the defining equations are (204) and (201). However, for the reason given above (and in contrast to region I), we must now use both the increasing and decreasing exponential solutions to (204):

$$s = A \exp(w - w_c) + B \exp(w_c - w). \tag{212}$$

The maximum in the electric field intensity, at $w = w_m$, is determined from $d\mathscr{E}/dx = 0$ or $du/dw = d(1/s)/dw = 0$. From Eq. (202), it follows that

$$\exp 2(w_c - w_m) = A/B. \tag{213}$$

For $w < w_m$, $du/dw > 0$, and for $w > w_m$, $du/dw < 0$.

The left-hand end of region III, at $w = w_2$, occurs where the neglected drift term, du/dw, equals the diffusion term $d^2s/dw^2 = s$ (the same criterion which determined the right-hand end of region I). Thus,

$$\left(\frac{du}{dw}\right)_2 = \left(\frac{d^2 s}{dw^2}\right)_2 = s_2 \qquad \text{or} \qquad -\left(\frac{ds}{dw}\right)_2 = s_2^{\,3} \tag{214}$$

This gives the relation

$$\beta - \alpha = (\beta + \alpha)^3 \tag{215}$$

with

$$\beta = B\exp(w_c - w_2), \qquad \alpha = A\exp-(w_c - w_2).$$

Continuity of s (the density) across w_2 gives, using Eqs. (209) and (212),

$$\beta + \alpha = \frac{1}{[2(w_2 - w_1) + 1]^{1/2}}. \tag{216}$$

At the cathode, $w = w_c$, Eq. (212) reduces to

$$s_c = A + B. \tag{217}$$

The voltage across region III is, using Eq. (212),

$$v_{\rm III} = \int_{w_2}^{w_c} u\,dw = \frac{1}{(AB)^{1/2}}\left\{\tan^{-1}\left[\left(\frac{B}{A}\right)^{1/2}\exp(w_c - w_2)\right] - \tan^{-1}\left(\frac{B}{A}\right)^{1/2}\right\}. \tag{218}$$

We now discuss how the above results are put together to obtain the current–voltage characteristic. All relevant physical parameters of the semiconductor are assumed specified, as well as the cathode–anode spacing L. Thus, $w_c = L/L_a$ is known. A particular value of J is now chosen and the corresponding V is to be found. From Eq. (203), s_c is now known, whence, from Eq. (199), so likewise is $s_a = bs_c$. From Eq. (203), it remains to find v_c, where, obviously,

$$v_c = v_{\rm I} + v_{\rm II} + v_{\rm III}, \tag{219}$$

with the separate regional voltages given by Eqs. (207), (210), and (218), respectively. We obtain w_1 from Eq. (206). Next, using Eqs. (215) and (216), we obtain for A and B:

$$\begin{aligned} A &= \frac{w_2 - w_1}{[2(w_2 - w_1) + 1]^{3/2}}\exp(w_c - w_2), \\ B &= \frac{w_2 - w_1 + 1}{[2(w_2 - w_1) + 1)^{3/2}}\exp[-(w_c - w_2)]. \end{aligned} \tag{220}$$

Equations (217) and (220) now give

$$s_c = \frac{(w_2 - w_1)\{\exp(w_c - w_2) + \exp[-(w_c - w_2)]\} + \exp[-(w_c - w_2)]}{[2(w_2 - w_1) + 1]^{3/2}}. \tag{221}$$

With s_c and w_1 known, Eq. (221) is a transcendental equation determining w_2. The solution is substantially simplified if $\exp[-(w_c - w_2)]$ is neglected, this being a valid approximation over a large range of currents. With w_2 now known, we go back to Eq. (220) to determine A and B. All the necessary quantities, s_a, w_1, w_2, w_c, A, and B, are now known for the determination of v_c, using Eq. (219).

The above procedure has been carried out, taking $b = 2$, for the two cases $w_c = L/L_a = 12$ and $w_c = 25$, with results plotted in Fig. 20 as the open circles. Corresponding computer calculations of Baron[20] are also exhibited in Fig. 20 as the solid lines.

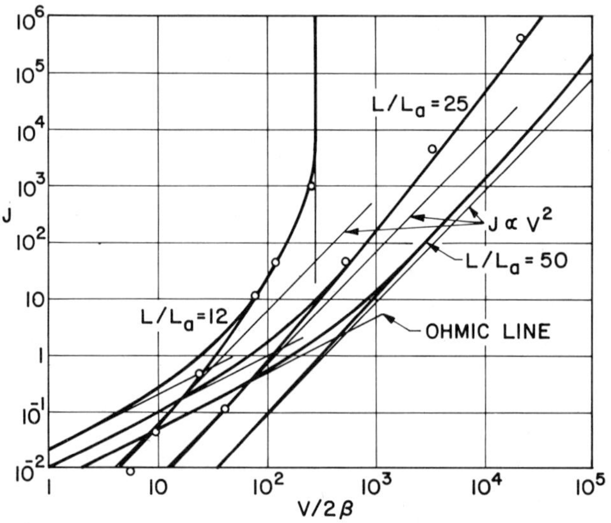

FIG. 20. Calculated current–voltage curves for the constant-lifetime plasma injected into a semiconductor. The heavy solid lines are computer characteristics obtained by Baron.[20] The open circles are points obtained by the regional approximation method for the two cases $L/L_a = 12$ and $L/L_a = 25$. The parameter β is given by $\beta \equiv V_T \equiv kT/e$.

Note that with increasing voltage, a critical voltage V_{cr} is reached at which region II shrinks to zero width: $w_1 = w_2$. From the above analysis, it follows that

$$V_{cr} \approx V_T \frac{\pi\sqrt{b}}{b+1} \exp \frac{L}{2L_a}. \tag{222}$$

For $V > V_{cr}$, J increases without further increase in voltage, as in the upper part of the curve labeled "$L/L_a = 12$" in Fig. 20.

f. Problem 3: The Insulator Injected-Plasma with Diffusion Corrections

Space charge plays a major role in determining the behavior of a plasma injected into an insulator. Consequently, it is the first term on the LHS of Eq. (196) that is important and the middle term that is negligible. (In the

[20] R. Baron, *J. Appl. Phys.* **39**, 1435, Fig. 2 (1968).

limit of infinitive resistivity, $n_0 = p_0 = 0$, the middle term obviously vanishes altogether.) Now, in place of Eq. (196), we have the simpler

$$-\frac{\varepsilon}{e}\frac{d}{dx}\left(\mathscr{E}\frac{d\mathscr{E}}{dx}\right) + 2V_{\mathrm{T}}\frac{d^2n}{dx^2} = \frac{(b+1)n}{\mu_n\tau}, \tag{223}$$

where we have dropped the n_0 from the RHS of Eq. (196).

Even though the presence of space charge implies local inequality between the injected electron and hole densities, nevertheless, in the plasma limit, $(n - p) \ll n, p$ throughout the solid, and therefore Eq. (197) is still useful. Likewise, the same boundary conditions (199) as used in the semiconductor problem are appropriate for the insulator problem. Thus, the equations defining the problem are (197), (223) and (199).

Just as in the semiconductor problem, and for the same reasons, we again have a three-region problem, illustrated schematically in Fig. 21.

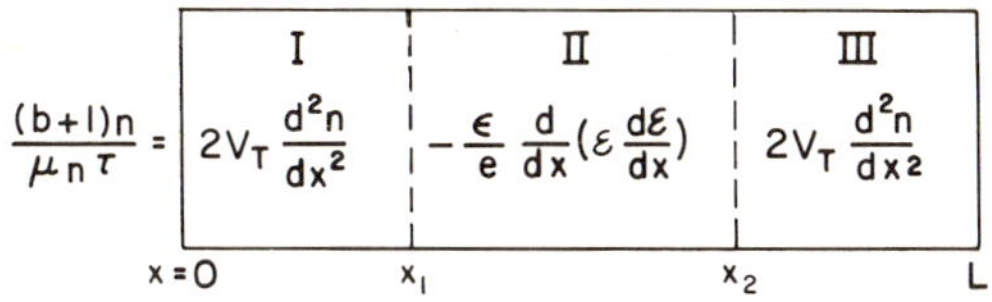

FIG. 21. Schematic regional approximation diagram for the problem of a constant-lifetime plasma injected into an insulator.

(1) *Region I* ($0 \leqslant x \leqslant x_1$): *Diffusion Term Dominant.* The diffusion term, $2V_{\mathrm{T}}\, d^2n/dx^2$, is dominant on the LHS of (223), and so the drift term, $-(\varepsilon/e)\, d[\mathscr{E}\, d\mathscr{E}/dx]/dx$, is neglected. Here, in contrast to region I of the semiconductor problem, we must use both the decaying and growing exponential solutions to the diffusion equation. The reason is that, with just the decaying solution, the neglected drift term is always negative; the admixture of solutions is needed to "turn the drift term around," that is, to bring it positive. Using both solutions, the drift term is negative out to some plane x_l where it goes through 0. For $x > x_l$, the drift term is positive, overtaking the diffusion term at $x = x_1$, which marks the end of region I.

(2) *Region II* ($x_1 \leqslant x \leqslant x_2$): *Drift Term Dominant.* Here, the drift term is dominant on the LHS of (223), and so the diffusion term is neglected. In contrast to region II of the semiconductor problem, this leads to $\mathscr{E}(x)$ rising to a maximum in the middle of the solid and thereafter decreasing (an essentially symmetrical distribution of field). Region II begins at plane $x = x_1$ and ends at plane $x = x_2$.

(3) *Region III* $(x_2 \leqslant x \leqslant L)$: *Diffusion Term Dominant.* Here, as in region I, the diffusion term is dominant, and so the drift term is neglected. This region is then similar to region I in every way (a reflection of the high degree of symmetry in the insulator problem as compared to the semiconductor problem). Again, both the decaying and growing exponential solutions must be used in order that the neglected drift term not be everywhere negative. Proceeding from the cathode into the solid, the drift term is initially negative, going through 0 at $x = x_r$. For $x < x_r$, the drift term is positive, overtaking the diffusion term at $x = x_2$, which marks the end of region III.

As with the semiconductor problem, the solutions in the different regions are matched up by requiring that the drift term, hence $\mathscr{E}$ and n, be continuous in crossing x_1 and x_2. However, these are now an insufficient number of conditions to specify a unique solution, because we now have a second-order differential equation in region II instead of a first-order equation, and we are using both solutions to the diffusion equation in region I instead of only the decaying exponential solution. We need two more conditions, and these are the requirement that the neglected diffusion term in region II equals the drift term at both boundaries, x_1 and x_2.

It is convenient to carry out the further mathematical discussion with the following dimensionless variables:

$$u = \frac{\mathscr{E}}{\mathscr{E}^\dagger}, \qquad s = \frac{n}{n^\dagger}, \qquad w = \frac{x}{x^\dagger}, \qquad v = \frac{V}{V^\dagger};$$

$$\mathscr{E}^\dagger = \left[\frac{2V_T J}{\varepsilon(b+1)\mu_p}\right]^{1/3}, \qquad x^\dagger = L_a = \left[\frac{2V_T\mu_n\tau}{b+1}\right]^{1/2},$$

$$n^\dagger = \frac{J}{e\mu_p(b+1)\mathscr{E}^\dagger} = \left[\frac{\varepsilon J^2}{2V_T e^3(b+1)^2\mu_p{}^2}\right]^{1/3}, \tag{224}$$

$$V^\dagger = \mathscr{E}^\dagger x^\dagger = \left[\frac{32b^3\mu_p V_T{}^5 J^2\tau^3}{\varepsilon^2(b+1)^5}\right]^{1/6}.$$

The defining equations (197) and (223) become, respectively,

$$1 = su \tag{225}$$

$$-\frac{d}{dw}\left(u\frac{du}{dw}\right) + \frac{d^2s}{dw^2} = s = \frac{1}{u}. \tag{226}$$

The current–voltage characteristic is given by a plot of $s_c{}^3$ versus $v_c s_c$:

$$s_c{}^3 = J\left[\frac{(b+1)\mu_p\tau^3}{2b^3\varepsilon^2 V_T}\right]^{1/2}, \qquad v_c s_c = V\frac{b+1}{2bV_T}. \tag{227}$$

Here again, V is the total voltage across the specimen.

The separate regions of the problem are as follows.

Region I ($0 \leqslant w \leqslant w_1$). The drift term is dropped from Eq. (226), giving

$$d^2s/dw^2 = s. \tag{228}$$

The other defining equation is (225). As discussed above, the solution to (228) is written as

$$s = Ce^{-w} + De^{w}, \tag{229}$$

which gives, at the anode ($w = 0$),

$$s_a = C + D \approx C. \tag{230}$$

The neglected drift term is

$$-\frac{d(u\,du/dw)}{dw} = -\frac{d^2(1/s^2)}{2\,dw^2} = \frac{1 - (3/s^2)(ds/dw)^2}{s^2}.$$

This term is negative near $w = 0$, goes through 0 inside region I and thereafter is positive, overtaking the diffusion term at $w = w_1$:

$$-\left[\frac{d}{dw}\left(u\frac{du}{dw}\right)\right]_1 = \left(\frac{d^2s}{dw^2}\right)_1 = s_1 = \frac{1}{s_1{}^2}\left[1 - \frac{3}{s_1{}^2}\left(\frac{ds}{dw}\right)_1^2\right]. \tag{231}$$

From Eq. (229),

$$s_1 = \gamma + \delta; \qquad \gamma = C\exp(-w_1), \qquad \delta = D\exp w_1. \tag{232}$$

Using Eq. (232) in (231) and noting that $(ds/dw)_1 = \delta - \gamma$, we obtain

$$(\delta + \gamma)^3 = 1 - 3\frac{(\delta - \gamma)^2}{(\delta + \gamma)^2}. \tag{233}$$

The voltage across region I is

$$v_{\mathrm{I}} = \int_0^{w_1} u\,dw = \frac{1}{(CD)^{1/2}}\left\{\tan^{-1}\left[\left(\frac{D}{C}\right)^{1/2}\exp w_1\right] - \tan^{-1}\left(\frac{D}{C}\right)^{1/2}\right\}, \tag{234}$$

in complete analogy with Eq. (218).

Region II ($w_1 \leqslant w \leqslant w_2$). The diffusion term is dropped from Eq. (226), giving

$$u\frac{d}{dw}\left(u\frac{du}{dw}\right) = -1. \tag{235}$$

The other defining equation is (201). In order to solve Eq. (235), it is convenient to make the substitution $u = dw/dy$, whence Eq. (235) becomes

$$\frac{d^2u}{dy^2} = -1, \tag{236}$$

with solution

$$u = -\tfrac{1}{2}y^2 + Ey + u_1, \qquad E = (1/y_2)[(u_2 - u_1) + \tfrac{1}{2}y_2{}^2], \tag{237}$$

satisfying the boundary conditions $y = y_2$ at $w = w_2$ and $u = u_2$, and $y = 0$ at $w = w_1$ and $u = u_1$, this last being a matter of convenience (since only differences in y are significant).

From $dw = u\,dy$, it follows that $w - w_1 = \int_0^y u\,dy$. Doing the integration, using Eq. (237), we obtain

$$w_2 - w_1 = \tfrac{1}{12}y_2{}^3 + \tfrac{1}{2}(u_2 + u_1)y_2. \tag{238}$$

The requirement that the neglected diffusion term equal the drift term at w_1 and w_2 leads, after some algebraic manipulation, to the equations

$$u_1{}^4 = u_1 + 3E^2, \qquad u_2{}^4 = u_2 + 3(E - y_2)^2. \tag{239}$$

The solution to this pair of equations compatible with (237) is

$$u_1 = u_2 (s_1 = s_2), \qquad y_2 = 2E. \tag{240}$$

Using (240) in (238), we get

$$w_2 - w_1 = \tfrac{1}{12}y_2{}^3 + u_1 y_2. \tag{241}$$

The voltage across region II is

$$v_{\mathrm{II}} = \int_{w_1}^{w_2} u\,dw = \int_0^{y_2} u\frac{dw}{dy}\,dy = \int_0^{y_2} u^2\,dy = \frac{1}{120}y_2{}^5 + \frac{1}{6}u_1 y_2{}^3 + u_1{}^2 y_2, \tag{242}$$

where Eqs. (237) and (240) have been used.

Note that if we take region II to fill the entire insulator (equivalent to taking $w_1 = 0$ and $w_2 = w_c, u_1 = u_2 = 0$), then Eqs. (242) and (241) together yield the current–voltage characteristic,

$$v_c{}^3 = \frac{18}{125}w_c{}^5 \qquad \text{or} \qquad J = \frac{125}{18}\varepsilon\mu_n\mu_p\tau\frac{V^3}{L^5}, \tag{243}$$

which has been called the "insulator, injected-plasma cube law."[9]

Region III ($w_2 \leqslant w \leqslant w_c$). This region is completely symmetrical with region I. The drift term is dropped from Eq. (226), so that the defining equations are (225) and (228). Analogous to Eq. (229), we have, for the solution to Eq. (228),

$$s = M\exp[-(w_c - w)] + N\exp(w_c - w), \tag{244}$$

which gives, at the cathode ($w = w_c$),

$$s_c = M + N \approx M. \tag{245}$$

The requirement that the (neglected) drift term equal the diffusion term at the end of region III, namely, at $w = w_2$, gives, analogously to Eq. (232),

$$s_a = \mu + \nu; \qquad \mu = M \exp[-(w_c - w_2)], \qquad \nu = N \exp(w_c - w_2). \tag{246}$$

In analogy with Eq. (233), we have

$$(\nu + \mu)^3 = 1 - 3\frac{(\nu - \mu)^2}{(\nu + \mu)^2}. \tag{247}$$

From Eqs. (232), (240), and (246), we have $\delta + \gamma = \nu + \mu$. In view of this, comparing Eq. (233) with (247), it is obvious that $\delta - \gamma = \nu - \mu$, and hence

$$\gamma = \mu, \qquad \delta = \nu. \tag{248}$$

The voltage across region III is, in analogy with Eq. (234),

$$v_{\text{III}} = \int_{w_2}^{w_c} u\, dw = \frac{1}{(MN)^{1/2}}\left\{\tan^{-1}\left[\left(\frac{N}{M}\right)^{1/2} \exp(w_c - w_2)\right] - \tan^{-1}\left(\frac{N}{M}\right)^{1/2}\right\}. \tag{249}$$

These results are now put together as follows: From Eqs. (199), (230), and (245),

$$s_c = s_a/b, \qquad M \approx C/b. \tag{250}$$

With L given, w_c is known. Choose a value for $s_1 = s_2 = \gamma + \delta = \mu + \nu$. Next, solve Eq. (233) for $\delta - \gamma = \nu - \mu$. Then γ, δ, μ, and ν are known. The value of the parameter $E = y_2/2$ is determined using Eq. (239). Next, $w_2 - w_1$ is found from Eq. (241). It remains to determine C, D, M, and N. From Eqs. (248), (232), (246), and (250), $\gamma\mu = \gamma^2 = bM^2 \exp(w_2 - w_1)\exp(-w_c)$, whence $M = (\gamma/b)^{1/2} \exp(w_c/2) \exp[-(w_2 - w_1)/2]$ and $C = bM$. Now note that Eqs. (234) and (249) can be rewritten, respectively, as

$$v_{\text{I}} = \frac{1}{(\gamma\delta)^{1/2}}\left[\tan^{-1}\left(\frac{\delta}{\gamma}\right)^{1/2} - \tan^{-1}\frac{(\gamma\delta)^{1/2}}{C}\right],$$

$$v_{\text{III}} = \frac{1}{(\gamma\delta)^{1/2}}\left[\tan^{-1}\frac{M}{(\gamma\delta)^{1/2}} - \tan^{-1}\left(\frac{\gamma}{\delta}\right)^{1/2}\right].$$

Using these results and Eq. (242), we finally have $v_c = v_{\text{I}} + v_{\text{II}} + v_{\text{III}}$. From Eq. (245), we have s_c and therefore, from Eq. (227), we have the current–voltage characteristic.

It may appear unusual that we start the calculation with s_1 rather than with the current variable s_a (compare with the procedure followed in the semiconductor problem). However, experience shows that s_1 is the most sensitive variable in the problem. From (239), it follows that $u_1 \geqslant 1$, whence $s_1 = 1/u_1 \leqslant 1$. At $s_1 = 1$, $w_1 = w_2$ and region II disappears. At $s_1 \approx 0.5$, $w_2 \approx w_c$.

Specific calculations taking $b = 2$ and for the case $L/L_a = 2y_2 = 16$ have been carried out following the above procedure, with results exhibited as the crosses in Fig. 22. Corresponding computer calculations of Baron[21] are also exhibited in Fig. 22, as the solid lines. As with the semiconductor injected-plasma problem, and for the identical reason, beyond the critical voltage V_{cr} at which region II disappears, the current is independent of voltage. As in that problem, V_{cr} is of the order of $V_T \exp(L/2L_a)$.

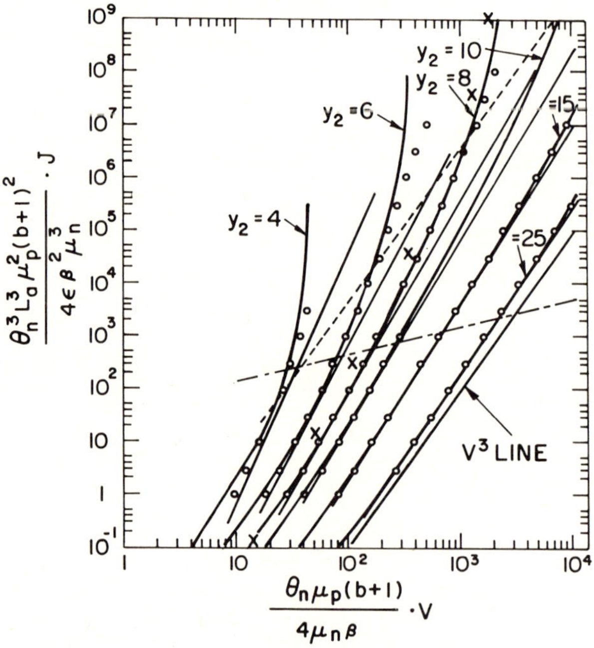

FIG. 22. Calculated current–voltage curves for the constant-lifetime plasma injected into an insulator. The heavy solid lines are computer characteristics (and the open circles are computed points) obtained by Baron.[21] The crosses are points obtained by the regional approximation method for the case $L/L_a = 2y_2 = 16$.

4. VARYING-LIFETIME, NEGATIVE-RESISTANCE PROBLEMS

The injected-plasma problems discussed in Section 3 are characterized, in part, by the minimal role assigned to the localized defect states: they could function as recombination centers, but changes in their occupancy were

[21] R. Baron, *Phys. Rev.* **137**, A272, Fig. 9 (1965).

of no account. The condition under which such a model is a realistic one is a rather stringent one: the injected–plasma density must exceed the density of deep-lying defect states. The realization of this condition is limited to a handful of materials, such as Ge, Si, and InSb, with which ultrapurification has been achieved. In the more usual cases of double injection, the injected free-carrier densities will be substantially smaller than the density of deep-lying states. In such cases, changes of occupancy of the deep-lying centers can be a controlling factor in the behavior of the double-injection currents. Since the lifetime of a free carrier depends on the density of unoccupied centers available to capture it, significant changes in occupancy of recombination centers can produce corresponding changes in a free-carrier lifetime. Thus, we are now faced with a new element in the double-injection problem—namely, a free-carrier lifetime varying with injection level.

In the two problems studied here, we deal with situations where a controlling free-carrier lifetime increases drastically with the injection level. These situations are of particular technological interest in that they lead to substantial current-controlled negative resistances. In both cases, we work with a particularly simple model, namely, one in which the defect states are just a single set of states which act as recombination centers. In problem 1, the centers are initially filled, whereas, in problem 2, the centers are initially only partially filled. Although the differences in these two problems appear, superficially, to be minor, they are actually quite striking. The resulting current–voltage characteristics are, in part, very different, and the detailed mathematics are correspondingly different. Hence, the problems are treated separately. In both problems, we neglect the thermal free carriers.

a. Problem 1: Recombination Centers Completely Filled: $\sigma_p \gg \sigma_n$

This problem is illustrated schematically by the energy-band diagram in Fig. 23. The centers lie sufficiently below the Fermi level that they are completely filled with electrons in thermal equilibrium. Further, it is assumed that the average cross section σ_p for capture of a free hole by a filled center greatly exceeds the average cross section σ_n for capture of a free electron by an empty center. This would be the case if the centers were acceptorlike, that is,

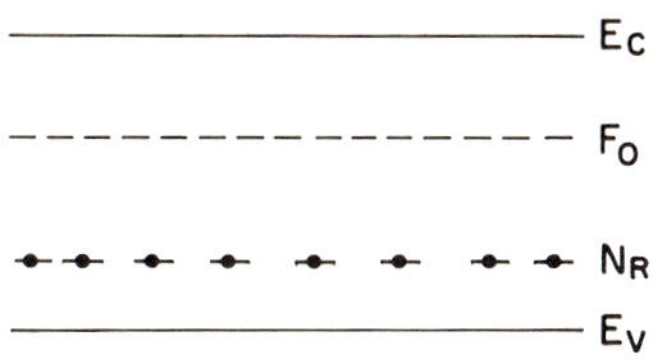

FIG. 23. Schematic energy-band diagram for the problem of double injection into an insulator with a single set of recombination centers completely filled in thermal equilibrium.

negatively charged when occupied by electrons. (Neutrality of the insulator would be achieved, for example, by the presence of an equal density of shallow donors which play no further role in the electrical behavior of the insulator.)

The equations characterizing this problem are: the current-flow equation

$$J = e\mu_n n\mathscr{E} + e\mu_p p\mathscr{E} = \text{const}, \tag{251}$$

where the diffusion currents have been neglected; the Poisson equation

$$\frac{\varepsilon}{e}\frac{d\mathscr{E}}{dx} = p + p_{\mathrm{R}} - n, \tag{252}$$

where p_{R} is the density of empty recombination centers; the particle-conservation equations

$$\mu_n \frac{d}{dx}(n\mathscr{E}) = r = -\mu_p \frac{d}{dx}(p\mathscr{E}); \tag{253}$$

and the recombination-kinetic expressions

$$r = \frac{n}{\tau_n} = \frac{p}{\tau_p}, \qquad \frac{1}{\tau_n} = \langle v\sigma_n \rangle p_{\mathrm{R}}, \qquad \frac{1}{\tau_p} = \langle v\sigma_p \rangle n_{\mathrm{R}}, \qquad p_{\mathrm{R}} + n_{\mathrm{R}} = N_{\mathrm{R}}, \tag{254}$$

where N_{R} is the density of recombination centers.

The two equations in (253) are readily combined to give

$$\frac{d}{dx}[(n - p)\mathscr{E}] = \frac{(a + 1)n}{\mu_p \tau_n} = \frac{(a + 1)p}{\mu_p \tau_p}; \qquad a = \frac{\mu_p}{\mu_n} = \frac{1}{b}. \tag{255}$$

The boundary conditions are the usual ones appropriate to a simplified theory neglecting diffusion:

$$\mathscr{E} = 0 \qquad \text{at} \quad x = 0 \quad \text{and at} \quad x = L, \tag{256}$$

where the hole-injecting contact is at $x = 0$ and the electron-injecting contact at $x = L$.

As might be expected, this problem is beyond the reach of exact analytic solution, and is therefore a prime candidate for the regional approximation method. In order to map out a strategy for the designation of the separate regions, we first consider the gross physical behavior expected of the model. At low injection levels, corresponding to low applied fields, the hole lifetime is very short, because of the large σ_p, and the electron lifetime is infinite, because there are no empty centers to capture the injected electrons. Penetration of holes into the insulator will be negligible, being confined to the immediate neighborhood of the anode. On the other hand, the electrons can transverse the solid as a one-carrier, space-charge-limited current, recombining with the holes just in front of the anode. The total current will be essentially just the electron current. As the current increases, the penetration of the holes into the insulator increases, at the expense of the region of pure electron-space-charge current. From this gross picture, we see that the two basic regions in the insulator are region A, a region dominated by hole penetration, and region B, a region dominated purely by free-electron space charge.

In region A, we can assume local charge neutrality since, for the most part, the injected holes are captured by the initially filled recombination centers, the electrons in these centers simply being transferred to the conduction band. It is further convenient to subdivide region A into two further regions, region I, over which the recombination centers have been largely depopulated by hole capture, so that $p_R \approx N_R$, $n > N_R$, and region II, over which the recombination centers are still largely filled, so that $n < N_R$, but local neutrality is still a good approximation. Between region A, dominated by hole injection and local neutrality, and region B, dominated by free-electron space charge, there must be a transition region over which electrons, holes, and net space charge all play an important role. Call this region III. Finally, for consistency of notation, relabel region B as region IV. In all, we have to deal with a four-region problem as illustrated in Fig. 24.

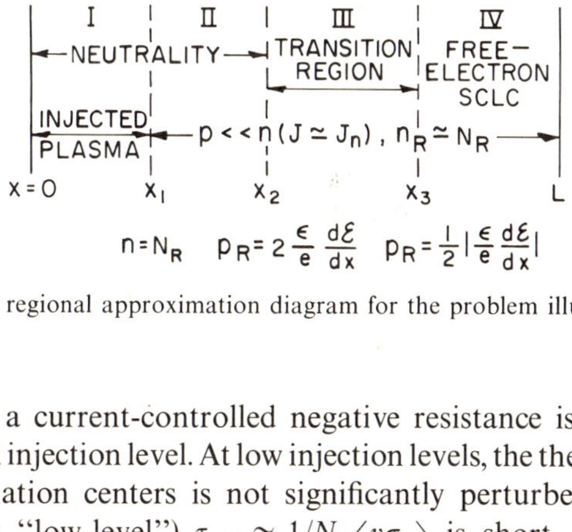

FIG. 24. Schematic regional approximation diagram for the problem illustrated by Fig. 23.

The origin of a current-controlled negative resistance is the increase of hole lifetime with injection level. At low injection levels, the thermal occupancy of the recombination centers is not significantly perturbed, and the hole lifetime ("l" for "low-level") $\tau_{p,1} \approx 1/N_R\langle v\sigma_p\rangle$ is short. Significant hole penetration into the insulator cannot occur until a voltage of magnitude $L^2/\mu_p\tau_{p,l}$ is reached, giving a hole transit time comparable to the low-level hole lifetime. This is a regime in which the relatively high-voltage region IV is dominant. At high injection levels, $n \approx p > N_R$, the occupancy of the recombination centers is inverted, i.e., changed to dominant hole occupancy, because $\sigma_p \gg \sigma_n$. The corresponding hole lifetime $\tau_h \approx 1/N_R\langle v\sigma_n\rangle \gg \tau_{p,1}$ ("h" for "high-level"; subscript p is now omitted because τ_h is the common lifetime for electrons and holes). Under this condition, hole penetration across the insulator occurs at substantially lower voltages—indeed, at a voltage as low as $L^2/\mu_p\tau_h$. A voltage dropping with increasing current defines a current-controlled negative resistance. The regime of decreasing voltage is a regime of increasing dominance of the relatively low-voltage region I. Finally, after region I has occupied the whole insulator, the current–voltage characteristic reverts to positive resistance, namely, that corresponding to the semiconductor square law (211) with N_R replacing $(n_0 - p_0)$. In effect, the insulator has been electronically transformed into an n-type semiconductor with an equivalent thermal density N_R of electrons.

We proceed with a more detailed discussion. The four regions are characterized as follows:

(1) *Region I* $(0 \leqslant x \leqslant x_1)$: $n > N_R$, $p_R \approx N_R$. The Poisson equation (252) is replaced by the neutrality condition

$$n = p + p_R, \tag{257}$$

or, taking $p_R \approx N_R$,

$$n = p + N_R. \tag{258}$$

Using Eq. (258) in (255), we can write

$$N_R \frac{d\mathscr{E}}{dx} = \frac{(a+1)n}{\mu_p\tau_h}, \qquad \frac{1}{\tau_h} = \langle v\sigma_n \rangle N_R, \tag{259}$$

where τ_h is the high-injection-level lifetime, which is the same for electrons and holes, namely, for $n \approx p \gg N_R$.

Using Eq. (258), Eq. (251) becomes

$$J = e\mu_n\mathscr{E}(a+1)n - e\mu_p N_R \mathscr{E}. \tag{260}$$

Substituting for $(a + 1)n$ from Eq. (260) into Eq. (259) gives the differential equation

$$\frac{\mathscr{E}}{\mathscr{E} + S}\, d\mathscr{E} = T\,dx, \qquad S = \frac{J}{e\mu_p N_R}, \qquad T = \frac{1}{\mu_n \tau_h}, \tag{261}$$

with solution

$$\mathscr{E} - S \ln \frac{\mathscr{E} + S}{S} = Tx \tag{262}$$

satisfying the boundary condition (256) at the anode, $x = 0$.

The right-hand end of region I is defined by the condition

$$x = x_1: \qquad n_1 = n(x_1) = N_R, \qquad \mathscr{E}_1 = \mathscr{E}(x_1) = \frac{J}{e\mu_n N_R} = aS. \tag{263}$$

The potential drop from 0 to x is

$$V(x) = \int_0^x \mathscr{E}\, dx = \int_0^{\mathscr{E}} \mathscr{E} \frac{dx}{d\mathscr{E}}\, d\mathscr{E},$$

or

$$V(x) = \frac{1}{T}\left[\frac{1}{2}\mathscr{E}^2 - S\mathscr{E} + S^2 \ln \frac{\mathscr{E} + S}{S}\right]. \tag{264}$$

At $x = x_1$ we have, using Eq. (263) in (262) and (264),

$$x_1 = \frac{S}{T}[a - \ln(1 + a)], \qquad V_{\mathrm{I}} = V(x_1) = \frac{S^2}{T}\left[\frac{1}{2}a^2 - a + \ln(1 + a)\right]. \tag{265}$$

(2) *Region II* ($x_1 \leqslant x \leqslant x_2$): $n < N_{\mathrm{R}}, p \ll n$. In this region, the recombination centers are still largely filled, since $n < N_{\mathrm{R}}$; $p \ll n$ because of the preferential capture of holes by the filled centers (since $\sigma_p \gg \sigma_n$). The neutrality condition (257) obtains, whence

$$n \approx p_{\mathrm{R}} \ll N_{\mathrm{R}}, \qquad n_{\mathrm{R}} \approx N_{\mathrm{R}}. \tag{266}$$

The current equation (251) simplifies to

$$J \approx e\mu_n n\mathscr{E}. \tag{267}$$

Using (266) in (254), we obtain

$$p \approx \frac{n^2\langle v\sigma_n\rangle}{N_{\mathrm{R}}\langle v\sigma_p\rangle}, \qquad p\mathscr{E} \approx \frac{J\langle v\sigma_n\rangle}{e\mu_n N_{\mathrm{R}}\langle v\sigma_p\rangle} n \tag{268}$$

where (267) has been used in obtaining the latter.

Insertion of (268) into (253) gives the important differential equation

$$-\frac{dn}{n^2} = \frac{eN_{\mathrm{R}}\langle v\sigma_p\rangle}{aJ}\, dx, \tag{269}$$

with solution

$$\frac{1}{n} = \frac{1}{N_{\mathrm{R}}} + \frac{eN_{\mathrm{R}}\langle v\sigma_p\rangle}{aJ}(x - x_1) \tag{270}$$

satisfying the boundary condition: $n = N_{\mathrm{R}}$ at $x = x_1$. The field intensity then is

$$\mathscr{E} \approx \frac{J}{e\mu_n N_{\mathrm{R}}} + \frac{N_{\mathrm{R}}\langle v\sigma_p\rangle}{\mu_p}(x - x_1), \tag{271}$$

where Eq. (267) has been used. Region II ends where the approximation of local neutrality runs out of self-consistency, that is, where the neglected space charge catches up to the terms retained in the original Poisson equation (252),

$$x = x_2: \qquad \frac{\varepsilon}{e}\left(\frac{d\mathscr{E}}{dx}\right)_2 \approx p_{R,2} = p_R(x_2) = n_2 = n(x_2). \tag{272}$$

From Eq. (271), the potential drop across region II, $V_{II} = \int_{x_1}^{x_2} \mathscr{E}\, dx$, is

$$V_{II} = \frac{J}{e\mu_n N_R}(x_2 - x_1) + \frac{N_R\langle v\sigma_p\rangle}{2\mu_p}(x_2 - x_1)^2. \tag{273}$$

(3) *Region III* ($x_2 \leqslant x \leqslant x_3$): $p \ll n \approx p_R \ll N_R$, $n_R \approx N_R$. The current equation for this region is (267).

Space charge is no longer negligible, so that the Poisson equation (252) obtains. Dropping p from this equation,

$$\frac{\varepsilon}{e}\frac{d\mathscr{E}}{dx} \approx p_R - n. \tag{274}$$

Although the space charge is indeed important, in that it is responsible for "turning the field around" in region III (the field increases monotonically in regions I and II, and must have a maximum in region III so that it can start its decrease toward 0 at the cathode, at $x = L$), it is nevertheless still true that $p_R \approx n$ in region III. Indeed, at the left-hand end of region III, $p_{R,2} = n_2$, and at the right-end of region III (see the discussion of region IV), $p_{R,3} = n_3/2$. Thus, so long as we are not dealing with the difference $n - p_R$, it is legitimate to replace n by p_R in studying region III. Since the derivation of Eq. (269) nowhere involved the difference $n - p_R$, it is legitimate to replace n by p_R in this equation:

$$-\frac{dp_R}{p_R^{\,2}} = \frac{eN_R\langle v\sigma_p\rangle}{aJ}\,dx, \tag{275}$$

which has the solution

$$\frac{1}{p_R(x)} - \frac{1}{p_{R,2}} = \frac{eN_R\langle v\sigma_p\rangle}{aJ}(x - x_2). \tag{276}$$

The spatial dependence of $\mathscr{E}$ is given by the solution to the differential equation (274), which can be written as $(\varepsilon/e)(d\mathscr{E}/dx) + J/e\mu_n\mathscr{E} - p_R = 0$, with p_R given by (276). We do not solve this differential equation, since our main interest is the potential drop across region III, $V_{III} = \int_{x_2}^{x_3} \mathscr{E}\, dx$. It suffices, instead, to use a simple interpolation scheme to approximate $\mathscr{E}(x)$, namely a quadratic expression, as discussed below.

(4) *Region IV* ($x_3 \leqslant x \leqslant L$): $p \ll p_R \ll n$, $n_R \approx N_R$. Only the injected free electrons are of significance in this region, so that the Poisson equation (274)

simplifies to

$$\frac{\varepsilon}{e}\frac{d\mathscr{E}}{dx} = -n. \tag{277}$$

The current equation is (267). The two equations (267) and (277) describe one-carrier, electron, trap-free, space-charge-limited current injection from the cathode at $x = L$. It is easy to check that the solution is

$$\mathscr{E}(x) = \left(\frac{2J}{\varepsilon\mu_n}\right)^{1/2}(L - x)^{1/2}, \qquad n(x) = \left(\frac{\varepsilon J}{2e^2\mu_n}\right)^{1/2}\frac{1}{(L - x)^{1/2}}. \tag{278}$$

As the distance from the cathode increases, $n(x)$ decreases as specified by (278). Finally, n gets small enough that it is comparable to the (neglected) trapped-hole density p_{R}. Thus, a reasonable criterion to fix the left-hand end of region IV is

$$x = x_3: \qquad p_{\mathrm{R},3} = p_{\mathrm{R}}(x_3) = \frac{n_3}{3} = \frac{n(x_3)}{2} = -\frac{\varepsilon}{2e}\left(\frac{d\mathscr{E}}{dx}\right)_3, \tag{279}$$

where $p_{\mathrm{R},3}$ is to be determined from (276) and n_3 from (278). For $x < x_3$, $p_{\mathrm{R}}(x)$ is large enough compared to $n(x)$ that its neglect in the Poisson equation (274) is no longer valid, and (277) is no longer a useful approximation.

For further mathematical discussion, it is convenient to switch to dimensionless variables:

$$u = \frac{\mathscr{E}}{\mathscr{E}^{**}}, \qquad w = \frac{x}{x^{**}}, \qquad v = \frac{V}{V^{**}};$$

$$\mathscr{E}^{**} = \frac{2aJ}{\varepsilon N_{\mathrm{R}}\langle v\sigma_p\rangle}, \qquad x^{**} = \frac{2a\mu_p J}{\varepsilon N_{\mathrm{R}}^2\langle v\sigma_p\rangle^2}, \tag{280}$$

$$V^{**} = \mathscr{E}^{**}x^{**} = \frac{4a^2\mu_p J^2}{\varepsilon^2 N_{\mathrm{R}}^3\langle v\sigma_p\rangle^3}.$$

The current–voltage characteristic will finally be given by a plot of $1/w_{\mathrm{c}}$ vs. $v_{\mathrm{c}}/w_{\mathrm{c}}^2$, since

$$\frac{1}{w_{\mathrm{c}}} = \frac{2a\mu_p}{\varepsilon N_{\mathrm{R}}^2\langle v\sigma_p\rangle^2 L}J, \qquad \frac{v_{\mathrm{c}}}{w_{\mathrm{c}}^2} = \frac{\mu_p}{N_{\mathrm{R}}\langle v\sigma_p\rangle L^2}V. \tag{281}$$

The separate regions are now characterized as follows:

Region I ($0 \leqslant w \leqslant w_1$). The characteristic differential equation (262) for this region becomes

$$\frac{u\,du}{u + A} = B\,dw; \qquad A = \frac{S}{\mathscr{E}^{**}} = \frac{\varepsilon\langle v\sigma_p\rangle}{2ea\mu_p}, \qquad B = \frac{Tx^{**}}{\mathscr{E}^{**}} = a\frac{\langle v\sigma_n\rangle}{\langle v\sigma_p\rangle}, \tag{282}$$

with S and T given in Eq. (261). The solution (262) becomes

$$u - A \ln \frac{u + A}{A} = Bw. \tag{283}$$

The right-end of region I is characterized by (263) and (265), rewritten here as

$$w = w_1: \qquad w_1 = (A/B)[a - \ln(1 + a)], \qquad u_1 = u(w_1) = aA, \tag{284}$$

and the dimensionless voltage is, from (265),

$$v_{\text{I}} = (A^2/B)[\tfrac{1}{2}a^2 - a + \ln(a + 1)]. \tag{285}$$

Region II ($w_1 \leqslant w \leqslant w_2$). The characteristic differential equation (269) for this region becomes

$$du/dw = 1, \tag{286}$$

and the solution (270) becomes

$$u - u_1 = w - w_1, \tag{287}$$

with u_1 and w_1 given in (284).

The right-hand end of region II, specified by (272), now becomes

$$u_2 = \tfrac{1}{2}, \qquad w_2 = \tfrac{1}{2} - aA + (A/B)[a - \ln(1 + a)]. \tag{288}$$

The dimensionless voltage drop across region II is, from (273),

$$v_{\text{II}} = \tfrac{1}{2}(u_2{}^2 - u_1{}^2) = \tfrac{1}{2}(\tfrac{1}{4} - a^2A^2). \tag{289}$$

Region III ($w_2 \leqslant w \leqslant w_3$). We first transcribe Eq. (276) into the dimensionless variables and then evaluate it at $w = w_2$:

$$\frac{1}{-\frac{1}{2}(du/dw)_3} - \frac{1}{(du/dw)_2} = 2(w_3 - w_2). \tag{290}$$

Using Eqs. (286) and (298) below, we can rewrite Eq. (290) as

$$4(w_c - w_3)^{1/2} = 1 + 2(w_3 - w_2). \tag{291}$$

As remarked above, we shall obtain $\mathscr{E}(x)$, i.e., $u(w)$, not by solving the dimensionless equivalent of (274), but by a simple interpolation scheme. Of a number of such schemes that might be used, we choose the following: Let $u = U_2(w)$ be the equation for the tangent to the curve $u = u(w)$ in region II at $w = w_2$; correspondingly, $u = U_3(w)$ is the tangent to the curve $u = u(w)$

in region IV at $w = w_3$. Then

$$U_2(w) = u_2 + (w - w_2)\left(\frac{du}{dw}\right)_2 = u_2 + (w - w_2),$$
$$U_3(w) = u_3 + (w - w_3)\left(\frac{du}{dw}\right)_3 = u_3 - \frac{w - w_3}{2(w_c - w_3)^{1/2}}, \tag{292}$$

where we have used Eqs. (286) and (298).

Further, let $\alpha(w)$ and $\beta(w)$ be two linear weighting functions:

$$\alpha(w) = \frac{w_3 - w}{w_3 - w_2}; \qquad \alpha(w_2) = 1, \quad \alpha(w_3) = 0;$$
$$\beta(w) = \frac{w - w_2}{w_3 - w_2}; \qquad \beta(w_2) = 0, \quad \beta(w_3) = 1. \tag{293}$$

Then the interpolation approximation is

$$u(w) = U_2(w)\alpha(w) + U_3(w)\beta(w), \tag{294}$$

or, substituting from Eqs. (292) and (293),

$$u(w) = (u_2 + w - w_2)\frac{w_3 - w}{w_3 - w_2} + \left[u_3 - \frac{w - w_3}{2(w_c - w_3)^{1/2}}\right]\left(\frac{w - w_2}{w_3 - w_2}\right), \tag{295}$$

satisfying $u(w_2) = u_2$ and $u(w_3) = u_3$, thereby assuring continuity of the $\mathscr{E}$-field across the transition planes $x = x_2$, $x = x_3$.

Integrating Eq. (295) we obtain for the voltage drop across region III

$$v_{\mathrm{III}} = \int_{w_2}^{w_3} u\,dw = \frac{1}{2}(u_2 + u_3)(w_3 - w_2) + \frac{1}{6}\left[1 + \frac{1}{2(w_c - w_3)^{1/2}}\right](w_3 - w_2)^2. \tag{296}$$

Region IV ($w_3 \leqslant w \leqslant w_c$). Equations (267) and (277) give the (dimensionless) equation

$$du^2/dw = -1, \tag{297}$$

with solution

$$u = (w_c - w)^{1/2}, \qquad du/dw = -1/2(w_c - w)^{1/2}, \tag{298}$$

being the dimensionless form of Eq. (278).

The voltage drop across region IV is given by

$$v_{\mathrm{IV}} = \int_{w_3}^{w_c} u\,dw = \tfrac{2}{3}(w_c - w_3)^{3/2}. \tag{299}$$

Collecting all of these results, the dimensionless current–voltage characteristic is given by:

$$\frac{1}{w_c} = \frac{1}{-1 - aA + (A/B)[a - \ln(1 + a)] + (y + 1)^2}, \qquad y = (w_c - w_3)^{1/2}, \tag{300}$$

$$\frac{v_c}{w_c^{\,2}} = \left(\frac{1}{w_c}\right)^2 \left\{\frac{1}{2}\left(\frac{1}{4} - a^2 A^2\right) + \frac{A^2}{B}\left[\frac{1}{2}a^2 - a + \ln(a + 1)\right]\right.$$
$$\left. + \frac{2}{3}y^3 + \frac{5}{3}y^2 + \frac{1}{4}y - \frac{1}{4} + \frac{1}{48y}\right\}. \tag{301}$$

In obtaining Eq. (300), we have noted that $w_c = w_2 + (w_3 - w_2) + (w_c - w_3)$ and have used Eqs. (288), (291), and (299). In obtaining Eq. (301), we have noted that $v_c = v_{\mathrm{I}} + v_{\mathrm{II}} + v_{\mathrm{III}} + v_{\mathrm{IV}}$ and have used Eqs. (299) ($v_{\mathrm{IV}} = 2y^3/3$), (296) [$v_{\mathrm{III}} = (5y^2/3) + (y/4) - (1/4) + (1/48y)$], (289), and (285).

As usual, current is not given as a direct, known function of voltage, but, instead, current and voltage are related through an auxiliary variable, in this case, $y = (w_c - w_3)^{1/2}$. In order to obtain the actual current–voltage characteristic, the auxiliary variable y must be eliminated by numerical computation. Such computations have been carried out for two prototype cases, with results exhibited in Fig. 25 by the two dashed curves. For both curves, the material parameters chosen were $\varepsilon/\varepsilon_0 = 12$, $\mu_n = \mu_p = 10^4$ cm^2/V-sec ($a = 1$) and $N_R = 10^{15}$ cm^{-3}. For the lower curve, $\langle v\sigma_n\rangle = 9 \times 10^{-10}$ cm^3/sec and $\langle v\sigma_p\rangle = 3 \times 10^{-6}$ cm^3/sec, giving $A = 10^{-3}$ and $B = 3 \times 10^{-4}$; for the upper curve, $\langle v\sigma_n\rangle = 10^{-9}$ cm^3/sec and $\langle v\sigma_p\rangle = 10^{-6}$, giving $A = 3 \times 10^{-4}$ and $B = 10^{-3}$. These are values such as might pertain to a silicon experiment at liquid-nitrogen temperature.

Some data are of interest concerning the relative importance of the various regions at different points in the current–voltage characteristic. On the lower curve, at the point labeled α, the electron space-charge region IV occupies 83% of the insulator and absorbs 81% of the voltage. The transition region III occupies essentially the rest of the insulator and takes up the remaining part of the voltage. Regions I and II are both negligible. At point β, the neutrality regions I and II are now in the picture; they take up 32% of the insulator width, although they absorb but 2% of the voltage. Region III takes 43% of the width and 72% of the voltage, region IV 25% of the width and 26% of the voltage. At point γ, regions I and II occupy 55% of the width and take 5% of the voltage, region III 34% of the width and 80% of the voltage, and region IV 11% of the width and 5% of the voltage. At point δ, the two neutrality regions I and II occupy 99% of the insulator width and

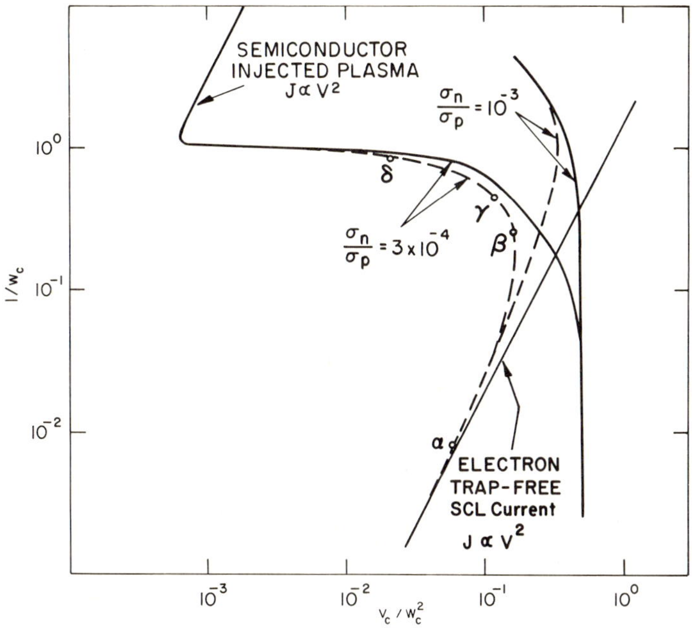

FIG. 25. Prototype universal current–voltage characteristics for double injection into an insulator with a single set of recombination centers completely filled in thermal equilibrium. Here $J \propto 1/w_c$ and $V \propto v_c/w_c^2$. The two cases calculated are $\sigma_n/\sigma_p = 10^{-3}$ and $\sigma_n/\sigma_p = 3 \times 10^{-4}$. The heavy solid lines are the characteristics obtained assuming neutrality; the dashed lines are those obtained including space charge.

98% of the voltage, the residual width and voltage being in the transition region III.

At low voltage, region IV occupies virtually the entire insulator. Therefore, there is negligible error in extending region IV up to the anode, $w = 0$. Taking $w_3 = 0$, $v_{IV} = v_c$ in (299), we obtain for the current–voltage characteristic at low voltages

$$\frac{1}{w_c} = \frac{9}{4}\left(\frac{v_c}{w_c^2}\right)^2 \rightarrow J = \frac{9}{8}\varepsilon\mu_n\frac{V^2}{L^3} \qquad (302)$$

which is shown as the straight line of slope 2 in Fig. 25.

An important critical current is the lowest current J_N at which the neutrality approximation (257) holds throughout the insulator, that is, at which regions I and II just fill the insulator. In Eq. (281), taking $w_c = w_2$ given by (288), we obtain

$$
\begin{aligned}
J_{\mathrm{N}} &= \frac{\varepsilon N_{\mathrm{R}}{}^2 \langle v\sigma_p \rangle^2 L}{2a\mu_p} \frac{1}{\frac{1}{2} - aA + (A/B)[a - \ln(1+a)]}, \\
V_{\mathrm{N}} &= \frac{N_{\mathrm{R}}\langle v\sigma_p\rangle L^2}{\mu_p} \frac{\frac{1}{2}(\frac{1}{4} - a^2A^2) + (A^2/B)[\frac{1}{2}a^2 - a + \ln(1+a)]}{\{\frac{1}{2} - aA + (A/B)[a - \ln(1+a)]\}^2},
\end{aligned}
\tag{303}
$$

where we have taken $v_{\mathrm{c}} = v_{\mathrm{I}} + v_{\mathrm{II}}$, with v_{I} given by Eq. (285) and v_{II} by Eq. (289). (Note that, strictly speaking, so long as we retain the boundary condition $\mathscr{E} = 0$ at $x = L$, region II cannot extend *right up to* the cathode; there *must* be regions III and IV in the insulator. However, for $J \gtrsim J_{\mathrm{N}}$, the latter two regions are completely insignificant. If, however, we let region II extend up to the cathode, because of the neutrality condition, we are left only with a first order differential equation, and the requirement $\mathscr{E} = 0$ at $x = 0$ exhausts the permitted boundary conditions. We drop the requirement that $\mathscr{E} = 0$ at $x = L$, for $J \gtrsim J_{\mathrm{N}}$, then have a picture that is both physically and mathematically self-consistent.)

A second important critical current is the lowest current J_{M} at which region I just fills the insulator. In Eq. (281), taking $w_{\mathrm{c}} = w_1$ given by Eq. (284), we obtain

$$
\begin{aligned}
J_{\mathrm{M}} &= h(a)\frac{eN_{\mathrm{R}}L}{\tau_{\mathrm{h}}}; \qquad \tau_{\mathrm{h}} = \frac{1}{N_{\mathrm{R}}\langle v\sigma_n\rangle}, \qquad h(a) = \frac{a}{a - \ln(1+a)}; \\
V_{\mathrm{M}} &= \frac{1}{g(a)}\frac{L^2}{\mu_p\tau_n}; \qquad g(a) = \frac{[a - \ln(1+a)]^2}{a[\frac{1}{2}a^2 - a + \ln(1+a)]};
\end{aligned}
\tag{304}
$$

where we have taken $v_{\mathrm{c}} = v_{\mathrm{I}}$ given by Eq. (285). The results (304) are precisely the same as obtained in a rigorous neutrality theory.[10]

From Eqs. (265), (273), and (304), it follows that in the current range $J_{\mathrm{N}} \leqslant J \leqslant J_{\mathrm{M}}$, the current–voltage characteristic can be written in the form

$$
J_{\mathrm{N}} \leqslant J \leqslant J_{\mathrm{M}}: \qquad V = V_{\mathrm{M}}\left[\left(\frac{J}{J_{\mathrm{M}}}\right)^2 + f(a)\left(\frac{J}{J_{\mathrm{M}}}\right)\left(1 - \frac{J}{J_{\mathrm{M}}}\right)\right] + V_{\mathrm{TH}}\left(1 - \frac{J}{J_{\mathrm{M}}}\right)^2
\tag{305}
$$

$$
f(a) = \frac{a[a - \ln(1+a)]}{\frac{1}{2}a^2 - a + \ln(1+a)}, \qquad V_{\mathrm{TH}} = \frac{L^2}{2\mu_p\tau_{p,1}}, \qquad \tau_{p,1} = \frac{1}{N_{\mathrm{R}}\langle v\sigma_p\rangle}.
$$

Although Eq. (305) is not valid for $J < J_{\mathrm{N}}$, if we imagined that it held down to arbitrarily small currents, we would find a voltage threshold for current flow, namely $V \to V_{\mathrm{TH}}$ as $J \to 0$. From Eqs. (305) and (304),

$$\frac{V_{\mathrm{TH}}}{V_{\mathrm{M}}} = \frac{g(a)}{2} \frac{\langle v\sigma_p \rangle}{\langle v\sigma_n \rangle}. \tag{306}$$

Since $g(a)$ decreases monotonically from $\frac{3}{8}$ at $a \ll 1$ to $1/a$ at $a \gg 1$, we see that unless a is very large, $V_{\mathrm{TH}} \gg V_{\mathrm{M}}$. In that case, (305) can be simplified still further to

$$J_{\mathrm{N}} \leqslant J \leqslant J_{\mathrm{M}}: \qquad V \approx V_{\mathrm{M}}\left(\frac{J}{J_{\mathrm{M}}}\right)^2 + V_{\mathrm{TH}}\left(1 - \frac{J}{J_{\mathrm{M}}}\right)^2. \tag{307}$$

In terms of the dimensionless variables, (304) and (305) correspond to

$$\left(\frac{1}{w_{\mathrm{c}}}\right)_{\mathrm{M}} = \frac{h(a)B}{aA}, \qquad \left(\frac{v_{\mathrm{c}}}{w_{\mathrm{c}}^2}\right)_{\mathrm{M}} = \frac{B}{ag(a)}; \qquad \left(\frac{v_{\mathrm{c}}}{w_{\mathrm{c}}^2}\right)_{\mathrm{TH}} = \frac{1}{2};$$
$$\frac{V}{V_{\mathrm{M}}} = \left(\frac{J}{J_{\mathrm{M}}}\right)^2 + f(a)\frac{J}{J_{\mathrm{M}}}\left(1 - \frac{J}{J_{\mathrm{M}}}\right) + \frac{ag(a)}{2B}\left(1 - \frac{J}{J_{\mathrm{M}}}\right)^2; \tag{308}$$
$$\frac{V}{V_{\mathrm{M}}} = \frac{(v_{\mathrm{c}}/w_{\mathrm{c}}^2)}{(v_{\mathrm{c}}/w_{\mathrm{c}}^2)_{\mathrm{M}}}, \qquad \frac{J}{J_{\mathrm{M}}} = \frac{(1/w_{\mathrm{c}})}{(1/w_{\mathrm{c}})_{\mathrm{M}}}.$$

The quantities J_{M} and V_{M} mark the high-current, low-voltage end of the negative-resistance regime. At higher currents, the current–voltage characteristic is given implicitly by the relations

$$J > J_{\mathrm{M}}: \qquad \mathscr{E}_{\mathrm{c}} - S\ln\frac{\mathscr{E}_{\mathrm{c}} + S}{S} = TL,$$
$$V = \frac{1}{T}\left[\frac{1}{2}\mathscr{E}_{\mathrm{c}}^2 - S\mathscr{E}_{\mathrm{c}} + S^2 \ln\frac{\mathscr{E}_{\mathrm{c}} + S}{S}\right], \tag{309}$$

with S and T given in Eq. (261). For $J/J_{\mathrm{M}} \gg 1$, $\mathscr{E}_{\mathrm{c}}/S \ll 1$, and expansion of the logarithm gives $\mathscr{E}_{\mathrm{c}}^2/2S \approx TL$ and $V \approx \mathscr{E}_{\mathrm{c}}^3/3TS$, which combine to give the semiconductor square law,

$$J \gg J_{\mathrm{M}}: \qquad J = \frac{9}{8} eN_{\mathrm{R}}\mu_n\mu_p\tau_{\mathrm{h}}\frac{V^2}{L^3}. \tag{310}$$

b. *Problem 2: Recombination Centers Partially Filled:* $\sigma_p \gg \sigma_n$

This problem is illustrated schematically by the energy-band diagram in Fig. 26. The only difference, in terms of the energy-band picture, from the previous problem illustrated in Fig. 23 is that here, the recombination centers are much closer to the Fermi level, so that they are only partially filled with electrons.

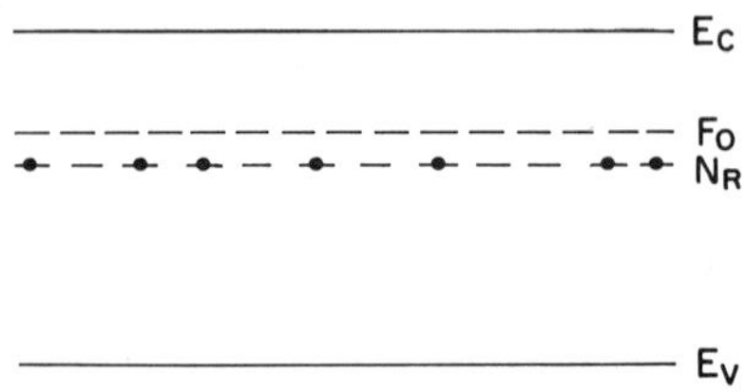

FIG. 26. Schematic energy-band diagram for the problem of double injection into an insulator with a single set of recombination centers partially filled in thermal equilibrium.

At high injection levels, where holes have penetrated deep into the insulator, this problem is essentially identical to the previous one. The striking difference in the two problems appears at low injection levels. In the previous problem, the injected electron space charge was necessarily free because, at low levels, there were no empty states in the gap to capture them. Thus, a trap-free, electron-space-charge square-law current flowed at low levels. In the present problem, there are empty states in the gap available to capture injected electrons. Consequently, there is a voltage threshold for current flow somewhat reminiscent of the voltage threshold V_{TFL} for current flow with one-carrier space-charge injection and deep trapping.[5] However, here, recombination processes are very much in the picture, even at low injection levels, and so the resemblance is not a deep one.

The equations characterizing this problem are the same as for the previous problem, (251) and (253)–(256), except that the Poisson equation (252) is here replaced by

$$\frac{\varepsilon}{e}\frac{d\mathscr{E}}{dx} = p + (p_R - p_{R,0}) - n, \tag{311}$$

where $p_{R,0}$ is the thermal equilibrium value of p_R.

Since this problem does not yield to exact analysis, we use the regional approximation method. Three regions are required as illustrated in Fig. 27.

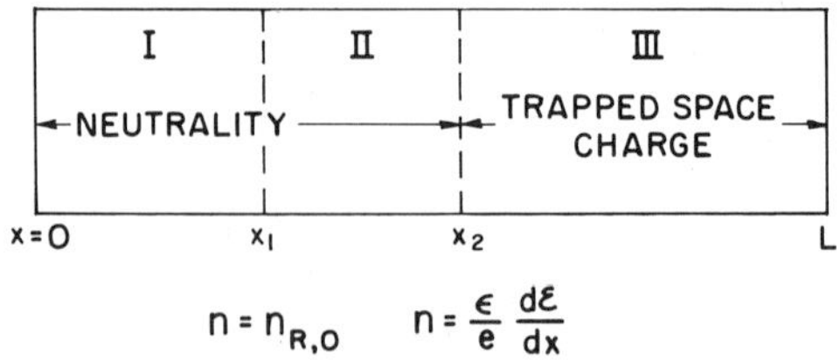

FIG. 27. Schematic regional approximation diagram for the problem illustrated by Fig. 26.

Regions I and II are very similar to the same regions of the preceding problem, Fig. 24. They are both regions in which local neutrality holds. In region I, the recombination centers have been largely depopulated of electrons, so that $p_R \approx N_R$, $n > n_{R,0}$. In region II, where $n < n_{R,0}$, the electron population of the recombination centers is not drastically changed from its thermal equilibrium value. Region III is a region dominated by space charge trapped in the recombination centers. Here, unlike the previous problem, we do not need a transition region between the neutrality-dominated regions I and II, and the trapped-space-charge-dominated region III.

We now proceed with a detailed discussion of the three regions.

(1) *Region I* $(0 \leqslant x \leqslant x_1)$: $n > n_{R,0}$, $p_R \approx N_R$. The Poisson equation (311) is replaced by the neutrality condition

$$n - p = N_R - p_{R,0} = n_{R,0}, \tag{312}$$

where we have replaced p_R by N_R.

Using Eq. (312) in (255), we can write

$$n_{R,0}\frac{d\mathscr{E}}{dx} = \frac{(a+1)n}{\mu_p \tau_h}, \qquad \tau_h = \frac{1}{\langle v\sigma_n \rangle N_R}, \tag{313}$$

where τ_h is the common high-injection-level lifetime for the injected electrons and holes, namely, the same as in Eq. (259).

Using Eq. (312) in (251), we obtain

$$J = e\mu_n \mathscr{E}(a+1)n - e\mu_p n_{R,0}\mathscr{E}. \tag{314}$$

Except for the replacement of N_R by $n_{R,0}$, Eqs. (313) and (314) are identical to (259) and (260), respectively. Thus, paralleling the results (261)–(265), we have the following:

The characteristic differential equation is

$$\frac{\mathscr{E}}{\mathscr{E} + S_0}\, d\mathscr{E} = T\, dx; \qquad S_0 = \frac{J}{e\mu_p n_{R,0}}, \qquad T = \frac{1}{\mu_n \tau_h}; \tag{315}$$

with solution

$$\mathscr{E} - S_0 \ln \frac{\mathscr{E} + S_0}{S_0} = Tx, \tag{316}$$

satisfying the boundary condition (256) at the anode, $x = 0$.

At the right-hand of region I,

$$x = x_1: \qquad n_1 = n(x_1) = n_{R,0}, \qquad \mathscr{E}_1 = \mathscr{E}(x_1) = \frac{J}{e\mu_n n_{R,0}} = aS_0. \tag{317}$$

The potential at position x is

$$V(x) = \frac{1}{T}\left[\frac{1}{2}\mathscr{E}^2 - S_0\mathscr{E} + S_0{}^2 \ln\frac{\mathscr{E} + S_0}{S_0}\right]. \tag{318}$$

At $x = x_1$, we have

$$x_1 = \frac{S_0}{T}[a - \ln(1 + a)], \qquad V_1 = V(x_1) = \frac{S_0{}^2}{T}\left[\frac{1}{2}a^2 - a + \ln(1 + a)\right]. \tag{319}$$

The parallelism of region I here with region I of problem 1 is obviously quite close.

(2) *Region II* $(x_1 \leqslant x \leqslant x_2)$: $n < n_{\mathrm{R},0}$, $p \ll n$. The parallelism of this region with region II of problem 1 is not quite as close as the parallelism of region I for the two problems.

The neutrality condition replacing the Poisson equation (311) is, dropping p,

$$n = p_{\mathrm{R}} - p_{\mathrm{R},0} = n_{\mathrm{R},0} - n_{\mathrm{R}}. \tag{320}$$

Replacing (314) is the simpler

$$J = e\mu_n n\mathscr{E}. \tag{321}$$

Noting that $(d/dx)(p_{\mathrm{R}}/n_{\mathrm{R}}) = (N_{\mathrm{R}}/n_{\mathrm{R}}{}^2)(dp_{\mathrm{R}}/dx)$ and $dp_{\mathrm{R}}/dx = dn/dx$, from Eq. (310), it follows that

$$\frac{dJ_p}{dx} = \frac{aJ\langle v\sigma_n\rangle N_{\mathrm{R}}}{\langle v\sigma_p\rangle n_{\mathrm{R}}{}^2}\frac{dn}{dx}. \tag{322}$$

From (253), $dJ_p/dx = -er = -en\langle v\sigma_n\rangle p_{\mathrm{R}}$. Combining this result with (322), we get

$$-\frac{dn}{n(n + p_{\mathrm{R},0})} = \frac{e\langle v\sigma_p\rangle n_{\mathrm{R}}{}^2}{aJN_{\mathrm{R}}}dx \approx \frac{e\langle v\sigma_p\rangle n_{\mathrm{R},0}^2}{aJN_{\mathrm{R}}}dx. \tag{323}$$

The replacement of n_{R} by $n_{\mathrm{R},0}$ in (323) is completely analogous to the replacement of n_{R} by N_{R} in region II of problem 1. Note that if $p_{\mathrm{R},0} = 0$, then Eq. (323) reduces to (269).

From Eq. (321), $dn = -J\,d\mathscr{E}/e\mu_n\mathscr{E}^2$, so that (323) can be rewritten as

$$\frac{d\mathscr{E}}{\mathscr{E} + (J/e\mu_n p_{\mathrm{R},0})} = \frac{e\langle v\sigma_p\rangle n_{\mathrm{R},0}^2 p_{\mathrm{R},0}}{aJN_{\mathrm{R}}}dx. \tag{324}$$

The right-hand end of region II is taken where the neglected space charge overtakes the retained terms in the Poisson equation (311),

$$x = x_2: \qquad \frac{\varepsilon}{e}\left(\frac{d\mathscr{E}}{dx}\right)_2 = n_2 = n(x_2). \tag{325}$$

(3) *Region III* ($x_2 \leqslant x \leqslant L$): $p \ll n \ll p_{\mathrm{R}} - p_{\mathrm{R},0}$. This region is dominated by trapped space charge, the appropriate Poisson equation being

$$\frac{\varepsilon}{e}\frac{d\mathscr{E}}{dx} = p_{\mathrm{R}} - p_{\mathrm{R},0}, \tag{326}$$

where n and p have been dropped from (311).

The current equation is (321).

Multiplying both sides of the recombination-rate equality $p\langle v\sigma_p\rangle n_{\mathrm{R}} = n\langle v\sigma_n\rangle p_{\mathrm{R}}$ [Eq. (254)] by $e\mu_p$ and using (321), we obtain

$$J_p \approx aJ\frac{\langle v\sigma_n\rangle}{\langle v\sigma_p\rangle}\frac{p_{\mathrm{R}}}{n_{\mathrm{R}}}. \tag{327}$$

Now, $(d/dx)(p_{\mathrm{R}}/n_{\mathrm{R}}) = (N_{\mathrm{R}}/n_{\mathrm{R}}{}^2)(dp_{\mathrm{R}}/dx) = (\varepsilon N_{\mathrm{R}}/en_{\mathrm{R}}{}^2)(d^2\mathscr{E}/dx^2)$, so that (326) gives

$$\frac{dJ_p}{dx} \approx \frac{\varepsilon aJ\langle v\sigma_n\rangle N_{\mathrm{R}}}{e\langle v\sigma_p\rangle n_{\mathrm{R}}{}^2}\frac{d^2\mathscr{E}}{dx^2}. \tag{328}$$

From the particle-conservation equation (253), we also have, using (321),

$$\frac{dJ_p}{dx} \approx -\frac{J\langle v\sigma_n\rangle p_{\mathrm{R}}}{\mathscr{E}\mu_n}. \tag{329}$$

Equations (328) and (329) together yield the characteristic differential equation

$$\mathscr{E}\frac{d^2\mathscr{E}}{dx^2} \approx -\frac{e\langle v\sigma_p\rangle n_{\mathrm{R}}{}^2 p_{\mathrm{R}}}{\varepsilon\mu_p N_{\mathrm{R}}} \approx -\frac{e\langle v\sigma_p\rangle n_{\mathrm{R},0}^2 p_{\mathrm{R},0}}{\varepsilon\mu_p N_{\mathrm{R}}}. \tag{330}$$

For further mathematical discussion, we switch to dimensionless variables:

$$u = \frac{\mathscr{E}}{\mathscr{E}^\ddagger}, \qquad w = \frac{x}{x^\ddagger}, \qquad v = \frac{V}{V^\ddagger};$$

$$\mathscr{E}^\ddagger = \frac{aN_{\mathrm{R}}J}{\{e\langle v\sigma_p\rangle n_{\mathrm{R},0}^2 p_{\mathrm{R},0}\varepsilon\mu_p N_{\mathrm{R}}\}^{1/2}}, \qquad x^\ddagger = \frac{aN_{\mathrm{R}}J}{e\langle v\sigma_p\rangle n_{\mathrm{R},0}^2 p_{\mathrm{R},0}}; \tag{331}$$

$$V^\ddagger = \mathscr{E}^\ddagger x^\ddagger = \frac{(aN_{\mathrm{R}}J)^2}{n_{\mathrm{R},0}^3\{e\langle v\sigma_p\rangle p_{\mathrm{R},0}\}^{3/2}\{\varepsilon\mu_p N_{\mathrm{R}}\}^{1/2}};$$

The current–voltage characteristic is given, as usual, by a plot of $1/w_c$ versus v_c/w_c^2:

$$\frac{1}{w_c} = \frac{aN_R}{e\langle v\sigma_p\rangle n_{R,0}^2 p_{R,0} L} J, \qquad \frac{v_c}{w_c^2} = \frac{1}{n_{R,0}^2 L}\left[\frac{\varepsilon\mu_p N_R}{e p_{R,0}\langle v\sigma_p\rangle}\right]^{1/2} V. \tag{332}$$

The separate regions are now characterized as follows:

Region I ($0 \leqslant w \leqslant w_1$). The characteristic differential equation (315) for this region becomes

$$\frac{u\,du}{u + C} = D\,dw; \qquad C = \left[\frac{\varepsilon\langle v\sigma_p\rangle p_{R,0}}{ea^2\mu_p N_R}\right]^{1/2},$$
$$D = \langle v\sigma_p\rangle N_R\left[\frac{a^2\varepsilon N_R}{e\langle v\sigma_p\rangle n_{R,0}^2 p_{R,0}\mu_p}\right]^{1/2}; \tag{333}$$

with the solution

$$u - C\ln\frac{u + C}{C} = Dw. \tag{334}$$

The right-hand end of region I is at w_1, corresponding to x_1 given by (319),

$$w_1 = \frac{C}{D}[a - \ln(1 + a)], \qquad u_1 = aC, \tag{335}$$

where (317) has also been used.

The dimensionless voltage drop across this region is

$$v_I = \frac{C^2}{D}\left[\frac{1}{2}a^2 - a + \ln(1 + a)\right], \tag{336}$$

corresponding to (319).

Region II ($w_1 \leqslant w \leqslant w_2$). The characteristic differential equation (324) for this region becomes

$$\frac{du}{u + E} = dw; \qquad E = \left\{\frac{\varepsilon\langle v\sigma_p\rangle n_{R,0}^2}{e\mu_p p_{R,0} N_R}\right\}^{1/2} \tag{337}$$

with solution

$$w - w_1 = \ln\frac{u + E}{u_1 + E}, \tag{338}$$

where u_1 and w_1 are given by (335).

The right-hand end of region II is specified by (325), which, using (337), can be written

$$u_2(u_2 + E) = 1 \qquad \text{with} \qquad u_2 = u(w_2). \tag{339}$$

Using Eq. (339) in (338), we get

$$w_2 - w_1 = \ln \frac{u_2 + E}{u_1 + E} = -\ln[u_2(u_1 + E)]. \tag{340}$$

The voltage across region II is, using Eq. (337),

$$v_{\mathrm{II}} = \int_{w_1}^{w_2} u\, dw = \int_{u_1}^{u_2} \frac{u}{u + E} du = u_2 - u_1 - E \ln \frac{u_2 + E}{u_1 + E}$$

$$= u_2 - u_1 + E \ln[u_2(u_1 + E)]. \tag{341}$$

Region III ($w_2 \leqslant e \leqslant w_c$). The characteristic differential equation (330) for this region becomes

$$u\, d^2u/dw^2 = -1. \tag{342}$$

We have encountered this equation previously in Section 3, Eq. (185) [the fact that (185) is an equation for u^2 and (342) an equation for u is not important]. Adopting the solution (186) to the current problem, we obtain for the solution to (342)

$$u = u_m \exp(-s^2), \qquad w - w_2 = \sqrt{2}\, u_m \int_{s_2}^{s} ds \exp(-s^2), \tag{343}$$

where u_m is the maximum value reached by u, namely, at $s = 0$.

From (343), at the left-hand end of region III, $u_2 = u_m \exp(-s_2{}^2)$. Since u increases monotonically with w in regions I and II, it must reach its maximum u_m inside region III. Thus, s_2 is necessarily negative, and can be written as $-|s_2|$. Further, the boundary condition $\mathscr{E} = 0$ at $x = L$, that is, $u = 0$ at $w = w_c$, corresponds to $s_c = s(w_c) = \infty$. Thus, taking $s = s_c$ in (343), we get

$$w_c - w_2 = \sqrt{2}\, u_2 (\exp s_2{}^2) \int_{-|s_2|}^{\infty} ds \exp(-s^2). \tag{344}$$

The voltage drop across region III is

$$v_{\mathrm{III}} = \int_{w_2}^{w_c} u\, dw = \int_{-|s_2|}^{\infty} u \frac{dw}{ds} ds = u_2{}^2 (\exp 2s_2{}^2) \int_{-\sqrt{2}|s_2|}^{\infty} ds \exp(-s^2). \tag{345}$$

Collecting all of the above results, the dimensionless current–voltage characteristic, $1/w_c$ versus $v_c/w_c{}^2$, is calculated from the following:

$$w_c = \sqrt{2}\, u_2 (\exp s_2{}^2) \int_{-|s_2|}^{\infty} ds\, (\exp -s^2) - [\ln u_2(aC + E)]$$

$$+ \frac{C}{D}[a - \ln(1 + a)], \tag{346}$$

where we have used Eqs. (335), (340) and (344) and the fact that $w_c = (w_c - w_2) + (w_2 - w_1) + w_1$, and

$$v_c = \frac{C^2}{D}\left[\frac{1}{2}a^2 - a + \ln(1+a)\right] + (u_2 - aC) + E \ln u_2(aC + E)$$
$$+ u_2{}^2(\exp 2s_2{}^2)\int_{-\sqrt{2}|s_2|}^{\infty} ds \exp(-s^2), \tag{347}$$

with

$$u_2 = \frac{E}{2}\left[-1 + \left(1 + \frac{4}{E^2}\right)^{1/2}\right], \tag{348}$$

where $v_c = v_{\mathrm{I}} + v_{\mathrm{II}} + v_{\mathrm{III}}$ and Eqs. (336), (341), (345), and (339) have been used.

The auxiliary variable linking w_c to v_c—namely, u_2—can be eliminated only through numerical computation. Calculations have been carried out for a prototype case, with results exhibited in Fig. 28 as the solid curve.

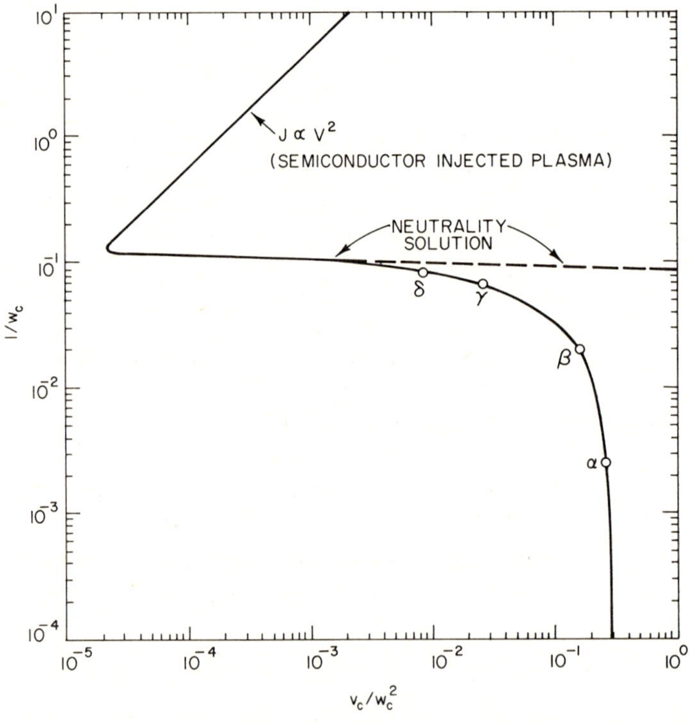

FIG. 28. Prototype universal current–voltage characteristic (solid line) for double injection into an insulator with a single set of recombination centers partially filled in thermal equilibrium.

The corresponding materials parameter values are $\varepsilon/\varepsilon_0 = 12$, $\mu_n = \mu_p = 10^4\ \text{cm}^2/\text{V-sec}$, $n_{\text{R},0} = p_{\text{R},0} = 5 \times 10^{14}\ \text{cm}^{-3}$, $\langle v\sigma_n\rangle = 10^{-9}\ \text{cm}^3/\text{sec}$ and $\langle v\sigma_p\rangle = 10^{-7}\ \text{cm}^3/\text{sec}$, giving $a = 1$, $C = E = 5.8 \times 10^{-3}$, and $D = 2.3 \times 10^{-4}$. At the point labeled α, region I occupies 2% of the insulator, region II 1%, and region III 97%, with essentially 100% of the applied voltage across region III. At point β, region I occupies 14% of the insulator, region II 7%, and region III 79%, with still essentially 100% of the voltage across region III. At point γ, region I occupies 48% of the insulator and absorbs 4% of the voltage, region II occupies 26% of the insulator and absorbs 14% of the voltage, and region III occupies 26% of the insulator and absorbs 82% of the voltage. At point δ, region I occupies 56% of the insulator and absorbs 8% of the voltage, region II occupies 29% of the insulator and absorbs 33% of the voltage, and region III occupies 15% of the insulator and absorbs 59% of the voltage.

An important critical current is the lowest current J_N at which the neutrality approximation $p + (p_\text{R} - p_{\text{R},0}) - n = 0$ holds throughout the solid, that is, at which regions I and II just fill the insulator. In (332), taking $w_\text{c} = w_2$ given by (340), we obtain

$$J_\text{N} = \frac{e\langle v\sigma_p\rangle n_{\text{R},0}^2 p_{\text{R},0} L}{aN_\text{R}} \frac{1}{(C/D)[a - \ln(1 + a)] - \ln[u_2(aC + E)]},$$

$$V_\text{N} = n_{\text{R},0} L^2 \left[\frac{ep_{\text{R},0}\langle v\sigma_p\rangle}{\varepsilon\mu_p N_\text{R}}\right]^{1/2} \times \frac{(C^2/D)[\frac{1}{2}a^2 - a + \ln(1 + a)] + u_2 - aC + E\ln[u_2(aC + E)]}{\{(C/D)[a - \ln(1 + a)] - \ln[u_2(aC + E)]\}^2}, \tag{349}$$

where we have used $v_\text{c} = v_\text{I} + v_\text{II}$, with v_I given by (336) and v_II by (341). The remarks which were made following Eq. (303) concerning the decrease in the number of permitted boundary conditions accompanying the neutrality assumption are equally valid in the present situation.

The low-voltage turnaround point, that is, the lowest current J_M at which region I just fills the insulator, is, from (319), given by

$$J_\text{M} = h(a)\frac{en_{\text{R},0}L}{\tau_\text{h}}, \qquad V_\text{M} = \frac{1}{g(a)}\frac{L^2}{\mu_p\tau_\text{h}}, \tag{350}$$

with τ_h given in (313). Note the similarity to (304). In the dimensionless plot of Fig. 28, J_M corresponds to $(1/w_\text{c})_\text{M} = 0.13$, and V_M to $(v_\text{c}/w_\text{c}^2)_\text{M} = 2.1 \times 10^{-4}$.

For currents between J_N and J_M, we have, from (338) and (341),

$$w_\text{c} = w_1 + \ln\frac{u_\text{c} + E}{u_1 + E}, \tag{351}$$

$$v_{\mathrm{II}} = \int_{u_1}^{u_c} \frac{u}{u+E}\,du = u_c - u_1 - E\ln\frac{u_c+E}{u_1+E}. \tag{352}$$

In this intermediate range of currents in the negative-resistance regime, we therefore obtain, from (332), for $J_{\mathrm{N}} \leqslant J \leqslant J_{\mathrm{M}}$,

$$J = \frac{e\langle v\sigma_p\rangle n_{\mathrm{R},0}^2 p_{\mathrm{R},0} L}{aN_{\mathrm{R}}} \frac{1}{(C/D)[a - \ln(1+a)] + \ln[(u_c+E)/(aC+E)]}$$

$$V = n_{\mathrm{R},0}L^2\left[\frac{ep_{\mathrm{R},0}\langle v\sigma_p\rangle}{\varepsilon\mu_p N_{\mathrm{R}}}\right]^{1/2} \tag{353}$$

$$\times\frac{(C^2/D)[\frac{1}{2}a^2 - a + \ln(1+a)] + u_c - aC - E\ln[(u_c+E)/(aC+E)]}{\{(C/D)[a-\ln(1+a)] + \ln[(u_c+E)/(aC+E)]\}^2}.$$

For currents exceeding J_{M}, the characteristic is given by (309) with S_0 replacing S, and, for $J \gg J_{\mathrm{M}}$, by (310) with $n_{\mathrm{R},0}$ replacing N_{R}.

If we attempt to extend the neutrality result (353) to low currents, we obtain a physically absurd result, namely, as $J \to 0$ (i.e., $u_c \to \infty$), $V \to \infty$. The limiting characteristic, as the current goes to 0, is properly studied simply by letting region III fill the entire insulator, that is, by taking $w_2 = 0$, $|s_2| = \infty$, $v_{\mathrm{III}} = v_c$ in (344) and (345), remembering that $u_2 \exp s_2{}^2 = u_{\mathrm{m}}$. This gives a threshold voltage for current flow, namely

$$\left(\frac{v_c}{w_c{}^2}\right)_{\mathrm{TH}} = \frac{1}{2\int_{-\infty}^{\infty} ds\exp(-s^2)} = \frac{1}{(4\pi)^{1/2}} = 0.28,$$

$$V_{\mathrm{TH}} = \left[\frac{e\langle v\sigma_p\rangle L^4 n_{\mathrm{R},0}^2 p_{\mathrm{R},0}}{4\pi\varepsilon\mu_p N_{\mathrm{R}}}\right]^{1/2}. \tag{354}$$

Ashley[22] has studied the threshold problem under more general conditions than obtain here. Under the more restricted conditions appropriate to the present problem, his result for V_{TH} is precisely (354). Note that we are assuming here that $V_{\mathrm{TFL}} = ep_{\mathrm{R},0}L^2/2\varepsilon > V_{\mathrm{TH}}$ as given by (354). Otherwise, before the onset of the double-injection current, there will flow a purely one-carrier space-charge-limited current of electrons, at the threshold V_{TLF}, which will completely fill the initially empty recombination centers. For $V > V_{\mathrm{TFL}}$, the problem would then be essentially identical to problem 1.[13]

[22] K. L. Ashley, Investigation of the effects of space charge on the conduction mechanism of double injection in semi-insulators. Ph.D. Thesis, Dept. Elec. Eng., Carnegie Inst. of Technol., Pittsburgh, Pennsylvania, 1963; see also K. L. Ashley and A. G. Milnes, *J. Appl. Phys.* **35**, 369 (1947).

IV. A Transistor Design Problem

The problems studied up to this point are basic, prototype problems involving idealized, simplified models. It is by the study of such models that the science of current injection has been built up. Applications to technology require the consideration of more complex models dictated by the particular applications in mind. We shall show that here, too, the regional approximation method can be a powerful tool to aid the design engineer.

The problem we consider is that of the planar $N^+P\nu N^+$ transistor structure illustrated in Fig. 29 (ν indicates a lightly doped N region). In earlier times,

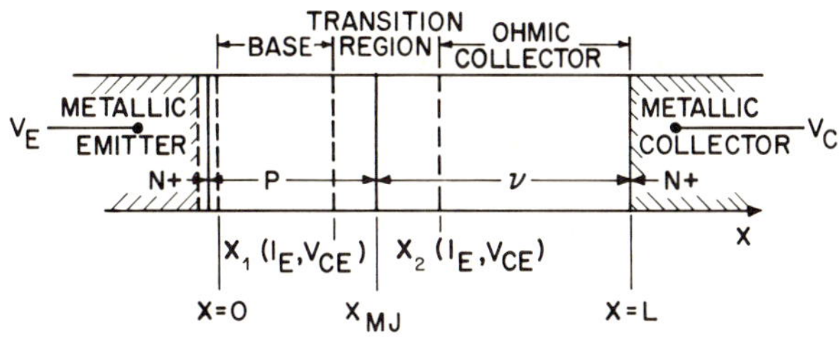

FIG. 29. Schematic regional approximation diagram for the one-dimensional transistor. Here, x_{MJ} is the location of the base–collector metallurgical junction.

the active region requiring study, namely, the base region, would be the fixed region contained between the emitter junction plane, at $x = 0$, and the collector junction plane, at $x = x_{MJ}$, both being defined by the metallurgical preparation of the structure. In actual fact, under low-voltage, high-current conditions, there can be substantial base widening, even to the point where the base reaches the metallic collector plane at $x = L$ in Fig. 29. The full equations describing transistor behavior have been studied by Gummel[23] on a computer. By self-consistent iteration, he obtained the solutions to the three second-order, nonlinear differential equations in the electrostatic potential and two quasi-Fermi potentials. The sheer complexity and purely numerical character of these solutions obviously limit the range of their applicability. Certainly, it is difficult to see how useful "rules of thumb" for design purposes and quick physical insight can evolve out of what might be described as "the total computer approach."

Here we adopt an intermediate approach. We break the problem up into three separate regions each dominated by separate physical considerations. This not only keeps the underlying physics clearly in view, but greatly simplifies the determining differential equations. In particular, we need deal

[23] H. K. Gummel, *IRE Trans. Electron Devices* **ED-11**, 455 (1964).

only with differential equations of first order. And, although, indeed, for many cases of overriding technological interest, we cannot solve these equations analytically, they posit an incomparably easier problem in digital computation than does the "total computer approach." In fact, the required storage capacity is limited enough to permit solution on a time-sharing basis.

The equations characterizing the transistor problem are the electron-flow equation

$$J_n = e\mu_n n\mathscr{E} + eD_n \frac{dn}{dx}; \tag{355}$$

the hole-flow equation

$$J_p = e\mu_p p\mathscr{E} - eD_p \frac{dp}{dx}; \tag{356}$$

the total-current equation

$$J = J_n + J_p = \text{const}; \tag{357}$$

the Poisson equation

$$\frac{\varepsilon}{e} \frac{d\mathscr{E}}{dx} = p - n + N(x); \tag{358}$$

where $eN(x)$ is the net ionic charge density and we have explicitly recognized the spatial dependence of this ionic charge;

the electrostatic potential equation

$$\Psi = -\int \mathscr{E}\, dx; \tag{359}$$

the electron particle-conservation equation

$$\frac{dJ_n}{dx} = -e(g - r); \tag{360}$$

and the hole particle-conservation equation

$$\frac{dJ_p}{dx} = e(g - r), \tag{361}$$

where g and r are the electron-hole generation and recombination rate densities, respectively.

For the case of the N–P–N transistor, the dominant carriers are electrons. The hole current J_p is therefore significantly less than either of its two components, that is, it is a small difference between two much larger currents.

Further, with operation at useful values of gain, neither the generation nor recombination of carriers significantly perturbs the current flow. Making these "classical" approximations, we can replace (355) by

$$J_n = e\mu_n n\mathscr{E} + eD_n\frac{dn}{dx} \approx J = \text{const} \tag{362}$$

and (356) by

$$\mathscr{E} \approx \frac{D_p}{\mu_p}\frac{dp/dx}{p}. \tag{363}$$

Equations (362) and (363), together with (358) and (359), constitute the simplified transistor equations. From these equations, given J, ψ, and the doping profile $N(x)$, the quantities $n, \mathscr{E}$, and the gain are determined. [In practice, J and a boundary condition on n are used to generate $n(x)$ and $\mathscr{E}(x)$, from which ψ is found from (359).]

Reduction of Eqs. (358), (362), and (363) to an equation in one variable yields a highly nonlinear, third-order differential equation. In contrast, through the use of the regional approximation method, only first-order differential equations will result, requiring substantially less computer time for solution and allowing rate-of-change calculations to be made using a slide rule. We proceed to our application of the regional approximation method.

There are three regions in the problem, as illustrated in Fig. 29. Region I, adjacent to the emitter, is the classical base region characterized by approximate, local neutrality. Region I terminates, at the plane x_1, where this neutrality approximation runs out of self-consistency, namely, where the neglected space charge $(\varepsilon/e)\, d\mathscr{E}/dx$ catches up with the separate components of charge in the base region. With this criterion, the base width is obviously not fixed, but varies with the current level. Herein, of course, lies the mathematical complexity of the problem.

We proceed to discuss the separate regions.

(1) *Region* I $(0 \leqslant x \leqslant x_1)$. This region is characterized by local neutrality, namely,

$$p - n + N(x) = 0 \tag{364}$$

replacing the Poisson equation (358).

Between this base region and the highly doped N^+ emitter is a depletion layer. The left-hand edge of region I, $x = 0$, is taken at the base edge of the depletion layer. The density $n(0)$ at $x = 0$ is known from junction theory, given a known emitter doping and voltage across the emitter–base depletion layer. This voltage, added to the voltage from 0 to L, gives the total transistor voltage.

A differential equation in n is obtained using (363) and (364) in (362), and the Einstein relations $D_n = \mu_n kT/e$ and $D_p = \mu_p kT/e$:

$$\frac{dn}{dx} = \frac{n[(dN/dx) + (J/eD_n)] - (J/eD_n)N}{2n - N}. \tag{365}$$

Equation (365) is an Abel equation of the second kind, and, with $N(x)$ specified, it is readily solved for $n(x)$ on a computer; $p(x)$ is then given by (364) and $\mathscr{E}(x)$ by (363).

The self-consistency condition on the neglected space charge defining the end of region I is

$$x = x_1: \qquad R(x_1) = R_1 = \frac{\varepsilon}{e}\frac{(d\mathscr{E}/dx)_1}{p_1} = 0.5. \tag{366}$$

It turns out that the final results are quite insensitive to the particular value of $R = R_1$ used to terminate the base, over the range $0.1 \leqslant R_1 \leqslant 1$. At the plane $x = 0$, R is order of magnitudes less than unity.

(2) *Region II* ($x_1 \leqslant x \leqslant x_2$). This region is characterized by the domination of drift current over diffusion current, so that (362) may be approximated by

$$J \approx e\mu_n n\mathscr{E}. \tag{367}$$

On the other hand, from the characterization (366) of the termination plane $x = x_1$, it is clear that space charge cannot be neglected beyond this plane. Thus, we must use the Poisson equation in place of the neutrality condition (364). However, since p is negligible in this region [which can be shown using (363) for p once $\mathscr{E}$ has been determined from (369)], (358) simplifies to

$$\frac{\varepsilon}{e}\frac{d\mathscr{E}}{dx} = -n + N(x). \tag{368}$$

Using (367) in (368), we obtain a differential equation in $\mathscr{E}$:

$$\frac{d\mathscr{E}}{dx} = -\frac{J}{\varepsilon\mu_n\mathscr{E}} + \frac{e}{\varepsilon}N(x). \tag{369}$$

Like (365), this is an Abel equation of the second kind, readily solvable by computer.

Note that the Eqs. (367)–(369) characterizing this region are identical, for the case $N(x) = \text{const}$ (positive, negative, or zero), to the equations characterizing one-carrier, space-charge-limited current theory (positive constant to the Ohm's-law–square-law transition problem, zero to the perfect-insulator problem, and negative constant to the trap-filled-limit problem) under homogeneous conditions, for which solutions are available. For voltage drops across this region exceeding a few $V_T = kT/e$, the neglect of the diffusion current in these problems has been justified by detailed studies.[18,24]

Region II ends, at plane x_2, where space charge is once again no longer important, that is,

$$x = x_2: \qquad R(x_2) = R_2 = \frac{\varepsilon}{e} \frac{|d\mathscr{E}/dx|_2}{n_2} = 0.5. \tag{370}$$

Note that region II will usually, but not always, contain the plane x_M of the metallurgical junction between the P and ν regions of the structure.

(3) *Region III* ($x_2 \leqslant x \leqslant L$). Region III is an ohmic region, that is, the space charge $(\varepsilon/e)(d\mathscr{E}/dx)$ can be dropped from (368), giving

$$n \approx N(x), \tag{371}$$

or, using (367),

$$J \approx e\mu_n N(x)\mathscr{E}, \tag{372}$$

which determines $\mathscr{E}(x)$.

Using the above theory, a computer calculation has been made for the prototype power transistor illustrated in Fig. 30, namely, one containing an exponential doping profile. Plots of electric field intensity versus position are

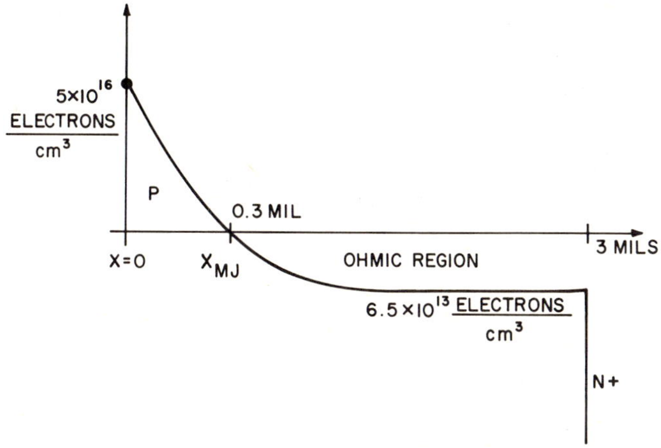

FIG. 30. An exponential doping profile for approximating a typical N–P–N transistor structure.

[24] M. A. Lampert and F. Edelman, *J. Appl. Phys.* **35**, 2971 (1964).

shown in Fig. 31 for operation at fixed current, $J = 4\ \text{A/cm}^2$, and at varying voltages ($1.5 < V_{CE} < 10$ V). First, note that all curves have approximately the same electric field at $x = 0$. This is due to the "built-in" field associated with a varying doping profile. At the right-hand end of each curve, we have a constant electric field due to the ohmic behavior (372).

The width of the ohmic region is critically affected by the voltage (at fixed current). With decreasing voltage, the base edge of the ohmic region moves toward the collector. Low values of voltage result in a "weak" sink condition for electrons in the region of the metallurgical junction x_{MJ}, leading to the base pushing into the collector. Eventually, as the voltage continues to drop, the base will widen until it reaches the N^+ collector. Base widening is a critical effect in transistor behavior. By control of the doping profile $N(x)$, which is the basis of all types of transistor design, base widening can be predicted and controlled in the design stage.

We now discuss in detail the shape of specific curves. For $V_{CE} = 10$ V, the effective base extends from x to 0 to x_1 to 5 μm. The electric field then rises

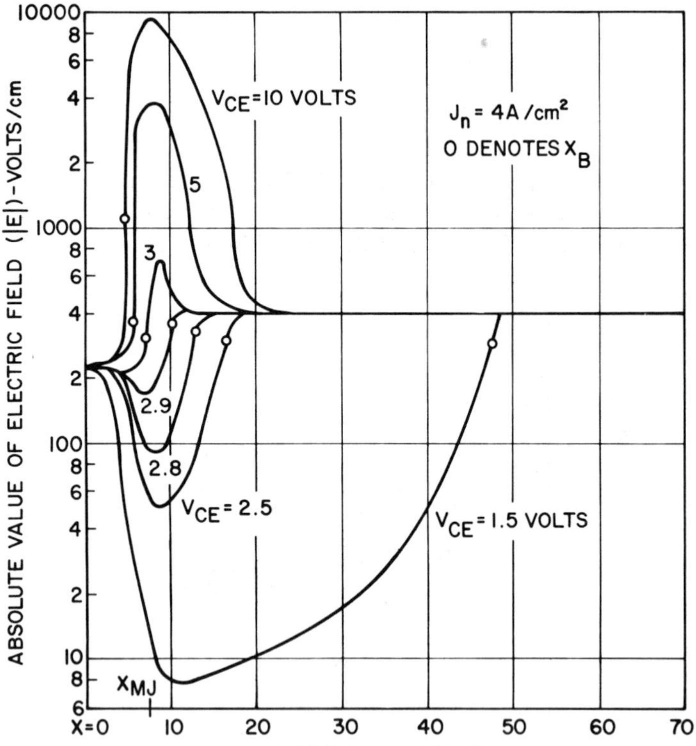

FIG. 31. Electric field as a function of position at fixed current and variable voltage.

sharply, peaking at approximately the metallurgical junction. The electric field thereafter decreases, reaching the ohmic field at approximately 24 μm. This type of behavior (i.e., high fields peaking at x_{MJ}) will be referred to as high-voltage or low-current behavior. It is to be noted that "high" or "low" is a relative matter. Thus, 10 V is high voltage at 4 A/cm^2, but may be considered low voltage at 40 A/cm^2. High or low voltage therefore depends on the current, and vice versa. The regional approach to transistor theory brings this out clearly.

At $V_{CE} = 5$ V the base edge has shifted to 6 μm. The position of the peak field has shifted from $x_{MJ} = 7.6$ to ~ 8.7 μm. Further decrease in V_{CE} to 3 V produces marked variation in the electric field profile. The base edge is now at 7.2 μm, close to x_{MJ}. The significant change in the field profile is the slope, $d\mathscr{E}/dx$ at $x = 0$. For $V_{CE} = 3$ V, this slope corresponds to net positive charge, whereas for $V_{CE} = 5$ and 10 V, the net charge at $x = 0$ is negative. A net positive charge front therefore develops at the origin as the base pushes toward the collector. For $V_{CE} = 5$ V, the net charge is negative, then positive, going through zero at x_{MJ}. For V_{CE} in the range 2.8–3.0 V, the net charge is in the sequence positive-negative-positive, and, for $V_{CE} = 2.5$ V, it is positive, then negative. With decreasing voltage, a positive front originates at the origin, followed by the disappearance of a positive front into the collector.

It is significant that a 0.1 V change in V_{CE}, namely, from 3.0 to 2.9 V, shifts x_1 by 3 μm in the vicinity of x_{MJ}, whereas a 2.0 V change (5 to 3 V) is required for a 1-μm shift to the left of x_{MJ}. The increased sensitivity in base widening at, and beyond, x_{MJ} leads one to the following condition for the onset of base widening: Base widening first becomes significant when the effective base crosses the base–collector metallurgical junction.

Having discussed the electric field profile at fixed current and varying voltage, we now consider conditions of fixed voltage and varying current. This is shown in Fig. 32 for $V_{CE} = 4$ V. The emitter current is varied from 200 mA (4 A/cm^2) to 5 A (100 A/cm^2).

As shown in Fig. 32, the electric field at $x = 0$ starts to depart from the built-in field at approximately 1 A (20 A/cm^2). The 4-V, 1-A curve corresponds to $n(0) = 4 \times 10^{15}$ cm^{-3}, which is roughly an order of magnitude below the background doping level, $N(0) = 5 \times 10^{16}$ cm^{-3}. For operation at 4 V, 5 A, $n(0) = 6.7 \times 10^{16}$ cm^{-3}, and $\mathscr{E}(0)$ has fallen from 250 V/cm to 65 V/cm.

At high currents ($V_{CE} = 4$ V), the base widens toward the collector. The large dip in the electric field profile is required to keep the area under the curve fixed (area is approximately constant for fixed V_{CE}) as the right-hand portion of the curve (ohmic region) rises with increasing current. At 5 A, we note that the ohmic region has almost disappeared. Under this condition, the base has widened almost to the end of the structure (76.2 μm).

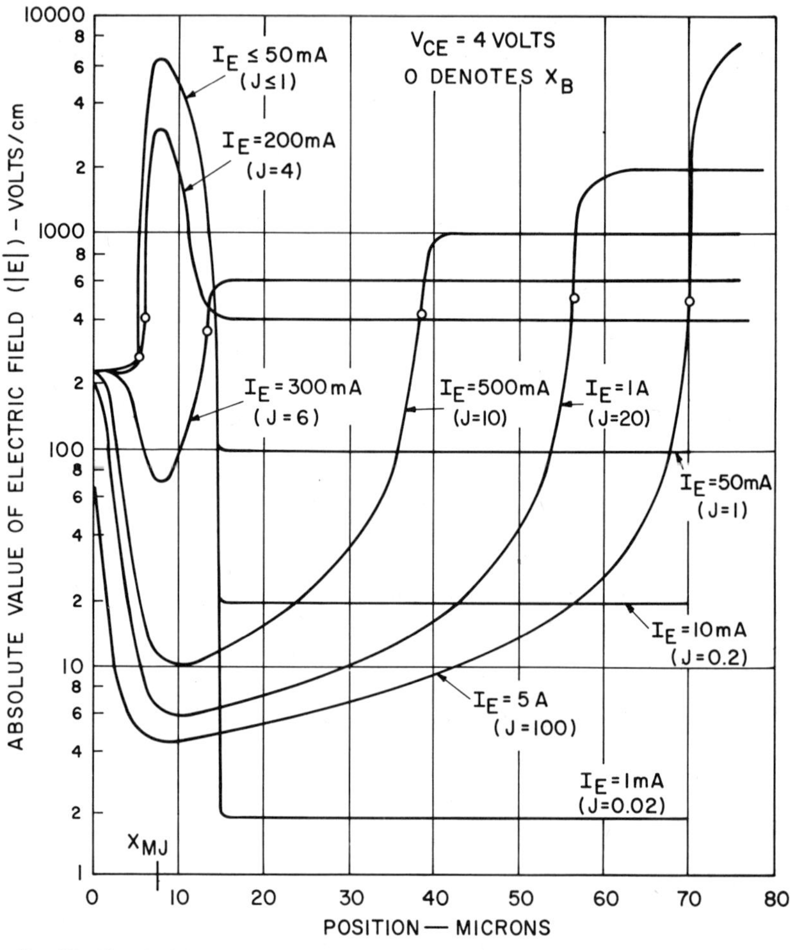

FIG. 32. Electric field as a function of position at fixed voltage and variable current.

The electron density as a function of position is shown in Fig. 33, for $V_{CE} = 4$ V and emitter currents from 200 mA to 5 A. At 200 mA, the electron density is monotonically decreasing, approaching very low values close to the metallurgical junction. Under this condition, the effective base termination is chosen where the electron concentration goes to zero (actually, just before $n = 0$). Zero electron density corresponds to a "perfect sink" condition. Choosing x_1 where $n = 0$ results in a sharp discontinuity in $n(x_1)$, as shown for the 200-mA curve. A small discontinuity in $n(x)$ will result at x_1 at each current, due to using continuity in electric field at x_1. Only for perfect sink conditions will the discontinuity be large.

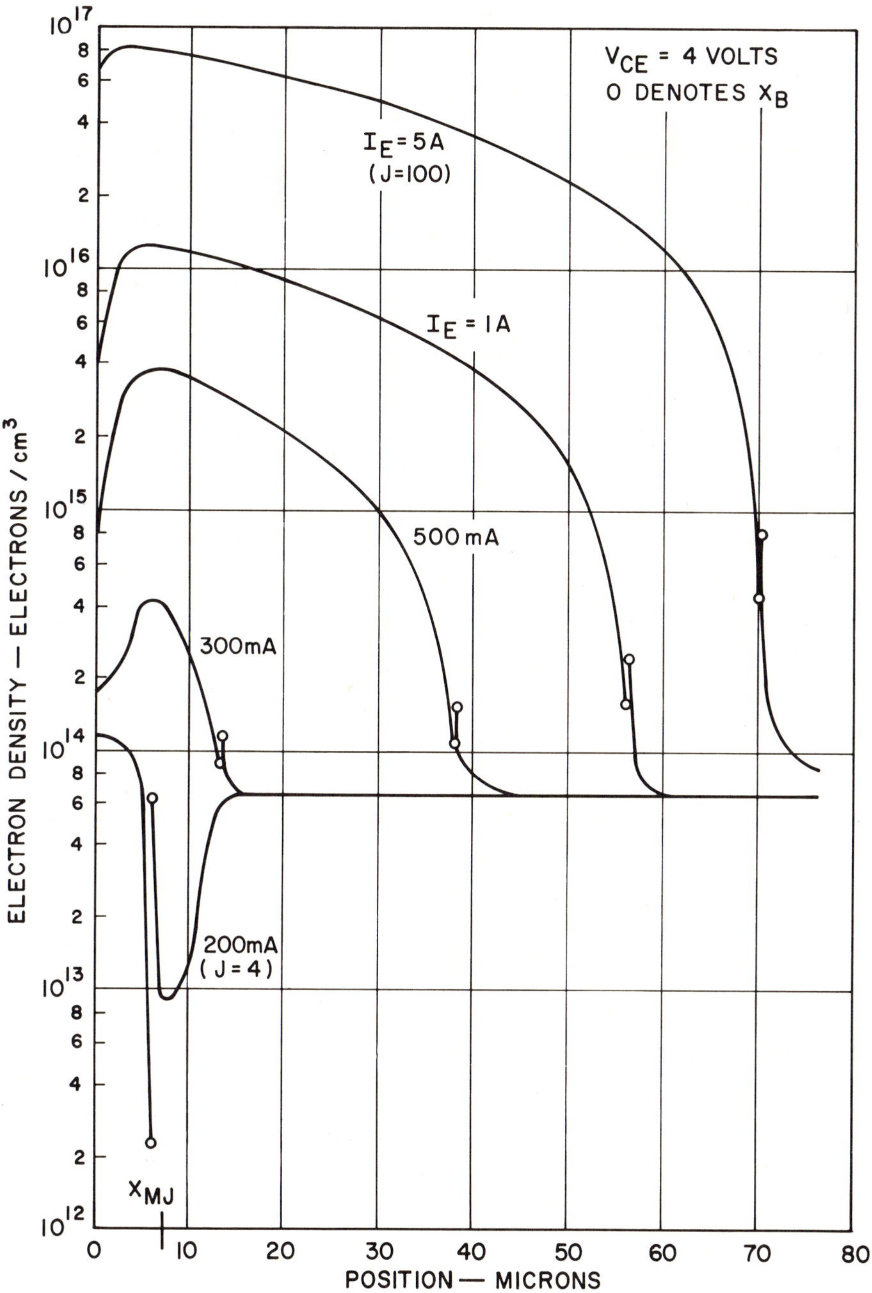

FIG. 33. Electron density as a function of position at fixed voltage and variable current.

For currents of 300 mA and above and $V_{CE} = 4$ V, the electron density has a positive slope at the origin. For each curve, the electron density reaches a peak and then monotonically decreases toward the value of the ohmic collector electron concentration (6.5×10^{13} cm^{-3}). The peak position shifts toward the origin with increasing current. Under the condition of increasing electron concentration at the origin, the drift and diffusion terms are in opposite direction. At the peak, only drift current is present, and beyond the peak, drift and diffusion are in the same direction.

Base widening is clearly evident in Fig. 33, as the position of x_1 increases with increasing current due to the large increase in total electron density with increasing current.

The above results indicate how useful the regional approximation method can be in exposing the underlying physical mechanisms operative in power transistor behavior.

Acknowledgments

A particular debt is owing to Dr. Albert Rose, the pioneer in the field of injection currents in insulators. The recognition that simplicity, insight, and usefulness in this field of problems can be achieved in exchange for total rigor and precise accuracy has marked all of his contributions and has particularly strongly influenced the senior author (MAL).

CHAPTER 2

Injection by Internal Photoemission

Richard Williams

I. General Ideas on Internal Photoemission

1. Early Use of the Concept

The process of the photoemission of electrons into vacuum has been intimately bound up with the development of atomic and solid-state physics. It has helped to establish numerous theoretical ideas, including the quantum nature of light, the concept of work function in metals, and the general nature of energy levels in solids. In addition, it has been a valuable experimental tool in the measurement of work functions and related properties. Incorporated into various photosensitive devices, it has been an integral part of countless experiments.

The essential feature of photoemission is the excitation of electrons in a solid by light and their subsequent injection into vacuum. Unexcited electrons in the solid cannot escape from the solid at room temperature because there is an energy barrier at the surface. Similar energy barriers often exist at the interface between two solids and may prevent the free passage of charge carriers across the interface. In analogy with the vacuum photoemission case, excited carriers can surmount the barrier and give a photoemission current through the interface. This concept is not very meaningful except in the context of the energy-band picture of solids. Thus, experiments on internal photoemission began only after the general features of the energy-band picture were well understood, some fifty years after the empirical experimental laws of photoemission into vacuum had been established. However, developments have been rapid, and most of the important features of vacuum photoemission now have their analog in internal photoemission. In addition, there are certain features of internal photoemission, such as the photoemission of holes, which have no analog in vacuum photoemission.

The concept of internal photoemission was introduced by Mott and Gurney[1] to interpret some photocurrent measurements of Gyulai.[2] The system studied was rock salt crystal containing particles of metallic sodium. The spectral response of the photocurrent was similar to the spectral response for photoemission of electrons from sodium into vacuum except that the whole curve was displaced to lower energies by approximately 0.5 eV. This was interpreted as photoemission of electrons from the sodium particles into the conduction band of the NaCl. The displacement of the curve to lower energies meant that the energy of an electron in the conduction band of NaCl was 0.5 eV lower than that of an electron in vacuum. This was in agreement with independent estimates of the energy of the conduction band made by Mott and Gurney. Similarly,[1] internal photoemission of electrons from silver particles into the conduction band of silver chloride was introduced to explain bleaching of the photographic latent image by infrared radiation.

The first direct measurement of internal photoemission appears to be the work of Gilleo[3] in 1953. He used electrodes of evaporated silver on single

[1] N. F. Mott and R. W. Gurney, "Electronic Processes in Ionic Crystals," 2nd ed., p. 73. Oxford Univ. Press, London and New York, 1950.

[2] Z. Gyulai, *Z. Physik* **35**, 411 (1926).

[3] M. A. Gilleo, *Phys. Rev.* **91**, 534 (1953).

crystals of KBr, NaCl, and AgCl. From measurements of the photocurrent as a function of the energy of the exciting light, he obtained a photoemission threshold energy of 4.3 eV for both KBr and NaCl. The threshold for photoemission from silver into vacuum is 4.8 eV. This indicates that the energy of the bottom of the conduction band lies 0.5 eV below vacuum, in agreement with the ideas of Mott and Gurney.[1] The photoexcited electrons were trapped in the alkali halide crystals and could be later released by F-band illumination, supporting the photoemission model for the effect. For silver on AgCl, a threshold energy of 1.1 eV was found.

With lower-band-gap materials, trapping effects are less severe and photoemission is easier to observe. Early results were obtained for copper and gold on cadmium sulfide.[4] These metals form blocking contacts to conducting n-type CdS. The resulting Schottky barrier has two features which are very desirable. First, the effective specimen thickness is the thickness of the Schottky barrier ($\sim 10^{-4}$ cm). A thin specimen minimizes trapping of injected carriers, and is difficult to obtain by other means. Second, in a Schottky barrier, there is a relatively high electric field at the metal–semiconductor interface, usually around 10^4 V/cm. This contributes to the efficient collection and measurement of injected carriers. With this system, photovoltaic currents were interpreted as internal photoemission. The accumulation of results on other systems has come rapidly, and will be taken up in the remainder of this chapter.

2. Areas of Application

a. Measurements of Barrier Heights and Energy Relations at Interfaces

The bulk of the work on internal photoemission thus far has been devoted to the determination of barrier heights. For metal–semiconductor systems, this method can be used together with measurements of capacitance and I–V characteristics to give highly reliable data for barrier heights. An extensive compilation of such data has been given by Mead[5] in a recent review article. This paper lists more than 60 barrier heights which have been determined by the photoemission method. The existence of a body of systematic data permits interesting conclusions[5] to be drawn on fundamental questions, such as the effects of surface states and of the chemical composition of the semiconductor on barrier heights.

b. Energy Losses of Excited Carriers in Metals

In photoemission, a clear experimental distinction can be made between electrons at the Fermi level and those having energies well above the Fermi

[4] R. Williams and R. H. Bube, *J. Appl. Phys.* **31**, 968 (1960).

[5] C. A. Mead, *Solid-State Electron.* **9**, 1023 (1966).

level. This is because the photoemission current is due solely to electrons having energies above the photoemission threshold. An electron excited above the threshold energy by light may lose some or all of its excitation energy before it reaches the metal–semiconductor interface. If it loses too much energy, it will not be collected and will make no contribution to the photoemission current. By using thin metal layers of different thicknesses, it is possible to excite the electrons at various distances from the interface.[6] Those electrons excited at greater distances from the interface are more attenuated by energy losses by the time they reach the interface, and the resulting quantum yield for photoemission is smaller. In this way, the mean free paths for energy loss by electrons in gold and several other metals have been determined. The experiments are sensitive mainly to electrons having energies near the photoemission threshold energy. By using different semiconductors with a given metal, a range of threshold energies can be covered and the energy losses can be determined over a range of energies. The photoemission method is a unique tool for measuring electron energy losses in metals for two reasons. First, it gives data for electrons over the entire range of excitation energies from 4 eV above the Fermi level down to 1 eV or less. Similar measurements involving photoemission into vacuum or secondary emission are restricted to electrons having energies greater than the vacuum work function of the metal, which is usually around 4 eV. The energy loss rates at these energies differ greatly from those at lower energies. Second, the internal-photoemission method permits the measurement of energy-loss properties for holes. These have not been measured by any other method.

(c) *Study of Transport and Other Properties of Insulators*

The study of electronic transport in insulators is becoming an increasingly important field. Since, in an insulator, the number of equilibrium free carriers is negligible, the carriers must either be excited in the material or injected from the contacts. If the band gap of the material is 2 or 3 eV, either of these methods may be used. In cadmium sulfide, for example, where the band gap is 2.5 eV, photoconductivity is readily excited by visible light, or, alternatively, electrons may be injected from ohmic indium contacts[7] in the dark. However, if the band gap is 8 eV, as in the case of SiO_2, both methods are extremely difficult. There seems to be no way to make ohmic contact to such a material with metal electrodes. Excitation of carriers in the bulk requires high-energy radiation with many accompanying experimental difficulties, not the least of which is the inevitable space-charge polarization when blocking contacts are used. In certain cases, such as the alkali halides, there are well-characterized impurity centers such as *F* centers which make the excitation problem easier,

[6] W. G. Spitzer, C. R. Crowell, and M. M. Atalla, *Phys. Rev. Letters* **8**, 57 (1962).

[7] R. W. Smith and A. Rose, *Phys. Rev.* **97**, 1531 (1955).

but the blocking-contact problem remains. It is probably such difficulties which have led to the preponderance of work on optical properties over that on electrical properties in the extensive literature on alkali halides, even though there are many interesting questions about electron transport in these materials.

Internal photoemission appears to be a general method for injecting carriers into insulators or high-band-gap semiconductors. In several cases, the difficulties described above can be overcome by this means. For example, SiO_2 is a classic insulator material for which knowledge of the electron-transport properties is interesting from many points of view. Until recently, almost nothing was known about the transport properties. By means of internal photoemission, either holes or electrons can be injected. The general features of transport and trapping for both carriers are now known and a Hall measurement of the electron mobility has been made. Details of this work will be discussed later.

By means of photoemission, it appears possible, in principle, to inject carriers from metals or low-band-gap semiconductors into almost any crystalline insulator. It is to be hoped that this can contribute to rapid expansion of existing knowledge in the field of electron transport in insulators.

II. Physics of Internal Photoemission

3. Contact between a Metal and a Semiconductor or Insulator

a. Ohmic Contacts

An ohmic contact[8] is one which supplies a reservoir of carriers to enter the material as needed. In the case of a resistor, the current is then limited by the bulk material. As the applied voltage is increased, more carriers must enter from the contacts per unit time, and the contact is able to provide these. The energy-band diagram for an ohmic contact of a metal to an *n*-type semiconductor is shown in Fig. 1. The curvature of the energy bands of the semiconductor near the interface indicates that electrons have spilled over from the metal into the semiconductor. The space charge of these electrons gives an energy barrier which provides the equilibrium condition, under which the metal always supplies just enough electrons to maintain the current flow through the crystal. Under various applied fields, the barrier adjusts to preserve this property. The main property of an ohmic contact for the present discussion is that it can always supply *more* carriers than the bulk material can carry under given applied voltage. Therefore, it can give no photoemission current. The number of electrons which the metal supplies to the

[8] A. Rose, "Concepts in Photoconductivity and Allied Problems," Chapter 8. Wiley, New York, 1963.

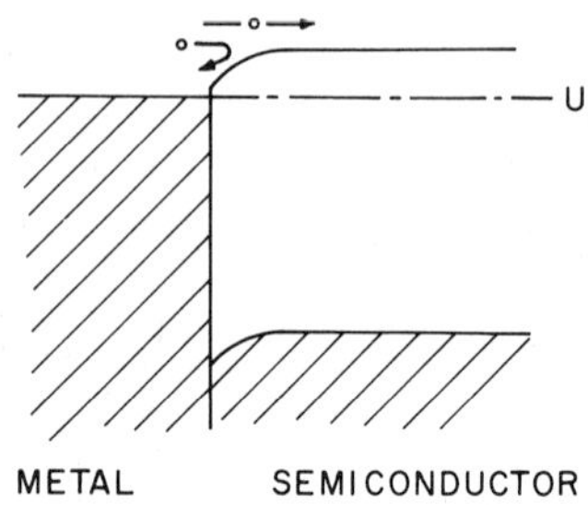

FIG. 1. Ohmic contact between a metal and an n-type semiconductor.

material is already more than the material can carry, and no increase of current results if still more are supplied by photoexcitation.

b. Blocking Contacts

A blocking contact[8] is one in which the semiconductor can carry away more carriers than the contact can supply. The current is then determined by the rate at which the contact can supply carriers. Blocking contacts and ohmic contacts are two extreme cases of the possible influence of contacts on current flow in solids. With blocking contacts, the current is completely determined by the maximum rate at which the contact can supply carriers, while with ohmic contacts, the current is completely independent of this rate. The energy-band diagram for a blocking contact is shown in Fig. 2. The main feature here is the energy barrier ϕ_B, defined as the energy difference between the Fermi level in the metal and the bottom of the conduction band in the semiconductor. A necessary, but not sufficient, condition for a blocking contact is $\phi_B \gg kT$. A simple kinetic argument[8] gives the condition for a blocking contact. This happens when the electrons in the semiconductor are carried away as fast as they arrive at the surface. We consider a very thin slice

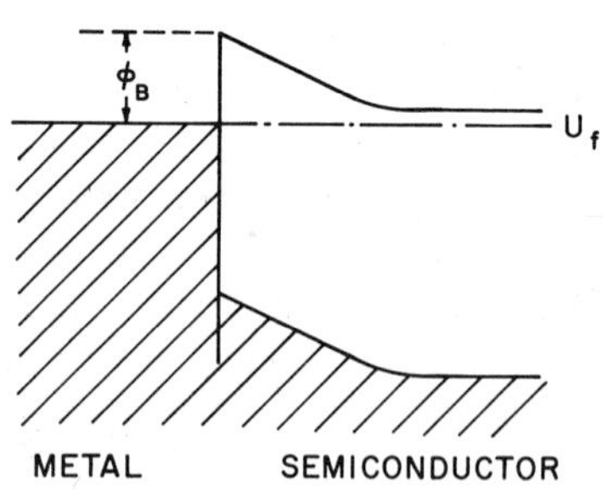

FIG. 2. Blocking contact for electrons between a metal and an n-type semiconductor. Here, U_f is the Fermi level.

of the semiconductor at the interface with the metal and assume that the density of n_0 of electrons in the conduction band is the same as the density at the same energy in the adjoining metal. If the thermal velocity of electrons at this energy is v, then the current to the surface is $n_0ev/4$. If the surface field in the semiconductor is E and the electron mobility is μ, then, within the semiconductor, they move away from the surface at the rate $n_0e\mu E$. The current saturates at the field

$$E_{sat} = v/4\mu \quad \text{V/cm}. \tag{1}$$

In typical semiconductors, the saturation field will be from 10^4 to 10^5 V/cm. A surface field of this magnitude is ordinarily present in a Schottky barrier when the ionized donor concentration in the semiconductor is greater than 10^{14} cm^{-3}. This condition is ordinarily fulfilled. Typically, the Schottky barrier is 10^{-4} cm thick or less and ϕ_B is 1 eV. Thus, even with no applied voltage, the surface field can be greater than E_{sat}. With an insulator, the saturation condition may be difficult to achieve. There is then no Schottky barrier, and, if the crystal is thick, a high applied voltage is required.

In measurements of photoemission, it is particularly desirable to work above the saturation field. Here, the collection efficiency for photoemitted carriers is at its maximum and it is easiest to interpret the observed effects of light.

4. Thermionic Emission versus Photoemission

The behavior of a metal–semiconductor blocking contact is remarkably similar to that found at the interface between metal and vacuum. Thermionic emission of electrons from metal to semiconductor[9] follows very closely the Richardson equation for thermionic emission of electrons into vacuum:

$$j_{th} = AT^2 \exp(-\phi_B/kT), \tag{2}$$

where A is a constant having the value 120 A/cm^2-deg^2 and ϕ_B is, for the vacuum case, the vacuum work function of the metal. The equation has been tested for several metal-semiconductor systems and fits quite well. To observe the thermionic emission, ordinarily, a small reverse bias is applied to the metal–semiconductor diode to saturate the emission current. The main difference from the vacuum case is that the metal–semiconductor barrier height may be arbitrarily small. Values around 1 eV are typical. Thus, the thermionic emission current at room temperature is quite large. For room temperature, Eq. (2) has the approximate form:

$$j_{th} = 10^{7-17\phi_B} \quad \text{A/cm}^2. \tag{3}$$

[9] H. K. Henisch, "Rectifying Semiconductor Contacts," Chapter 7. Oxford Univ. Press, London and New York, 1957.

This gives, for example, $j_{th} = 10^{-10}$ A/cm^2 for a barrier of 1.0 eV and 1 A/cm^2 for a barrier of 0.41 eV. General experimental agreement with the Richardson equation has been found for several systems. Kahng[10] studied gold on *n*-type silicon. Data were satisfactorily fit by the Richardson equation with values of 10–70 A/cm^2-deg^2 for the constant A. Similar data were obtained by Goodman[11] for contacts of evaporated gold on *n*-type cadmium sulfide. There was somewhat more scatter for the barrier heights with this system, and A was not explicitly determined, but here again the data are in agreement with the value 120 A/cm^2-deg^2. A question arises in applying the Richardson equation, since A contains the electron effective mass m^*,

$$A = 4\pi m^* e k^2 / h^3 . \tag{4}$$

Here, k is the Boltzmann constant, e the electronic charge, and h is Planck's constant. For a metal–semiconductor system, the potential barrier which the electron must surmount in order to escape into the semiconductor lies within the semiconductor, just inside the surface. Crowell[12] has taken this into account in a derivation of the thermionic emission equation which includes the structure of the energy bands of the semiconductor. The correction is somewhat more complicated than is suggested by Eq. (4), mostly due to the anisotropy of the bands. The theoretical value of A is changed by about 20% for germanium and by about a factor of two for silicon. The theoretical predictions are in good agreement with experimental data[12] for W–Si and Au–GaAs diodes.

Photoemission is always observed against a background of thermionic emission. Photoexcited carriers in a metal are indistinguishable from thermally excited carriers having the same energy. (This statement might be subject to qualification if an optical transition creates an anisotropic velocity distribution of excited electrons.) The effect of light is to increase the concentration of electrons with energies high enough to surmount the energy barrier at the metal–semiconductor interface. These then move in all directions, some toward the surface. Those not losing too much energy on the way will get over the barrier and be measured as a photocurrent. This picture of the increase in current over the barrier on illumination is very clear-cut in cases where the barrier is high; 1 eV at room temperature, for example. Where the barrier is lower, the situation is not so clear. Here, a small temperature rise in a metal electrode due to light absorption can increase the thermionic emission current, and it is difficult to distinguish experimentally between an increase in current due to this factor and one due to photoemission.

[10] D. Kahng, *Solid-State Electron.* **6**, 281 (1963).

[11] A. M. Goodman, *J. Appl. Phys.* **35**, 573 (1964).

[12] C. R. Crowell, *Solid-State Electron.* **8**, 395 (1965).

To illustrate the above, we consider the temperature coefficient of the thermionic emission current. Differentiation of Eq. (2) gives

$$dj_{th}/dT = AT^2[(2/T) + (\phi_B/kT^2)]\exp(-\phi_B/kT)$$
$$= (j_{th}/T)[2 + (\phi_B/kT)]. \qquad (5)$$

If the photon flux reaching the metal is 10^{15} quanta/cm^2-sec, then, since the quantum yield for photoemission is not likely to be higher than 0.1, the maximum photoemission current would be around 10^{-5} A/cm^2. We compare this with the increase in j, due to thermionic emission, which would be caused by a temperature rise ΔT. A small temperature rise is inevitable when a metal surface is illuminated. Some light is always absorbed in transitions which do not excite electrons to energies high enough to get over the barrier, and the excitation energy is then converted into heat. If ϕ_B is 1.0 eV, then a ΔT of 1°C at room temperature will increase j_{th} by $\Delta j_{th} \approx 10^{-11}$ A/cm^2. This is clearly small in comparison with 10^{-5} A/cm^2, and the photoemission current is easily distinguishable from Δj_{th} which would be caused by the assumed value of ΔT. The situation is quite different when ϕ_B is smaller. For $\phi_B = 0.41$ eV, $j_{th} = 1.0$ A/cm^2. In this case, even if ΔT is as small as 10^{-4}°C, Δj_{th} is 0.6×10^{-5} A/cm^2. Thus, for a barrier of this magnitude, a very small temperature rise in the metal due to light absorption would already give enhanced thermionic emission comparable to the maximum photoemission current which could be expected. There would be no way to determine the true origin of the observed Δj without further experiments, such as measurements of response time or spectral response. Equation (5) is not strictly applicable to the latter case which has been examined. When the increase in the thermionic current Δj_{th} corresponds to an electron flow comparable to the photon flux incident on the surface, there is a cooling effect. This is due to the flow of excited carriers from the metal into the semiconductor and has not been taken into account in the above discussion. This does not affect the qualitative conclusion which has been drawn here. To summarize this argument, when ϕ_B/kT is 40 or larger, it is easy to distinguish photoemission currents from effects due to enhanced thermionic emission. When ϕ_B/kT is 20 or less, this is no longer the case.

5. Schottky Barrier

When a metal makes blocking contact to a semiconductor, the simplest situation which can arise is that a Schottky barrier is established in the semiconductor. With an n-type semiconductor, for example, there is a separation of charges in which a sheet of negative charge is located in the plane of the interface. A corresponding positive charge is inside the semiconductor, located on the ionized donor centers within the barrier. The energy-band diagram for the barrier is ordinarily drawn as in Fig. 2. Here, the scale of distance is such that the potential appears to undergo a sharp discontinuity at the interface. In this case, an electron in the metal having energy above the bottom of the semiconductor conduction band would, if moving in the right direction, be free to enter the semiconductor. All would do so except for a fraction returned to the metal by quantum-mechanical reflection. For many purposes, this is an adequate model of the interface and is sufficient to interpret the general features of photoemission.

Several additional features of photoemission can be understood by considering the changes in Fig. 2 which are introduced when the distance scale is expanded.[13–15] This is illustrated by Fig. 3, which is adapted from a figure by Crowell and Sze.[15] The potential discontinuity of Fig. 2 is replaced

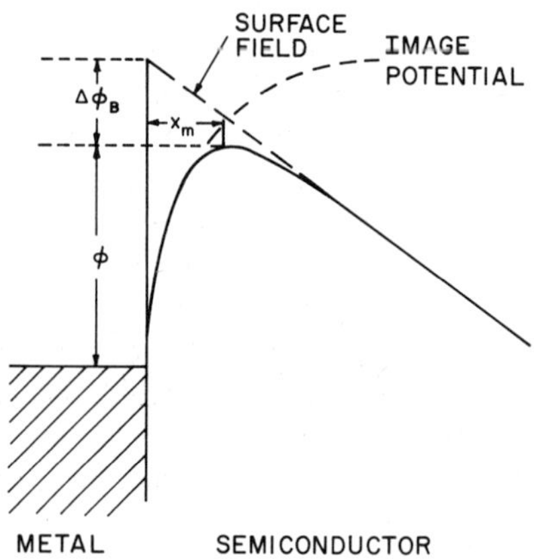

FIG. 3. Energy-band relations at a metal–semiconductor blocking contact, showing behavior near the interface. The potential in the semiconductor is determined by the superposition of two potentials: that due to the surface field, and that due to electron image forces, expressing the mutal attraction between an electron and a nearby metal.

by a more realistic potential which changes gradually over a distance of order 60 Å. The shape of the potential near the maximum is determined by superposition of two separate potential curves: (1) the potential determined by the surface field in the Schottky barrier, and (2) the image force potential experienced by an electron leaving the metal. There are two important consequences of this. First, the barrier height is a function of the voltage applied to the barrier. This is schematically shown in Fig. 3. The barrier

[13] C. R. Crowell and S. M. Sze, *Solid-State Electron.* **8**, 979 (1965).

[14] S. M. Sze, C. R. Crowell, and D. Kahng, *J. Appl. Phys.* **35**, 2534 (1964).

[15] C. R. Crowell and S. M. Sze, *Solid-State Electron.* **9**, 1035 (1966).

height is lowered from what it would be at the surface in the absence of the image field to the value of the potential maximum obtained by superposition of the two potential curves. Quantitatively, the lowering of the barrier is given by the equation[14]

$$\Delta\phi_B = -e(eE_{esu}/\varepsilon)^{1/2}. \tag{6}$$

This equation is in electrostatic units; here E_{esu} is the field in esu and ε is the optical dielectric constant. In practical units, this becomes

$$\Delta\phi_B = 3.8 \times 10^{-4}(E/\varepsilon)^{1/2} \quad \text{eV}. \tag{7}$$

Here, $\Delta\phi_B$ and E are in eV and V/cm, respectively. Thus, for a typical case where E is 10^5 V/cm and ε is 10, ϕ_B is lower by 0.038 eV. This effect has been quantitatively demonstrated by Sze *et al.*[14] for barriers of gold on *n*-type silicon. Similarly, both Goodman[16] and Mead *et al.*[16a] have measured the lowering of the barrier by high electric fields in the interfaces SiO_2–Si and SiO_2–metal. These are not Schottky barriers, strictly speaking, since there are not ionized donors in the volume of the insulator. Nonetheless, Eq. (6) is still valid.

The second effect due to the finite thickness of the potential step at the interface is that on electron scattering. A maximum in the curve of potential energy versus distance lies at a distance x_m from the interface,

$$x_m = (e/4\varepsilon E_{esu})^{1/2}. \tag{8}$$

With the field E in V/cm, this becomes

$$x_m = 1.9 \times 10^{-4}(\varepsilon E)^{-1/2} \quad \text{cm}. \tag{9}$$

Typically, E is 10^4 V/cm and ε is around 10, so that x_m is 60 Å and the potential maximum lies inside the semiconductor at this distance from the interface. In photoemission, an electron scattered within this distance is likely to return to the metal and not be photoemitted, undergoing the equivalent of surface recombination. It may be scattered either by absorption of a phonon or by emission of a phonon, depending on its energy and the temperature.

Generally speaking, high temperature and low electron energy favor phonon absorption, while low temperature and high electron energy favor phonon emission. The problem of collection efficiency under these conditions has been analyzed in detail by Crowell and Sze.[13,15] They treated the case

[16] A. M. Goodman, *Phys. Rev.* **144**, 588 (1966).

[16a] C. A. Mead, E. H. Snow, and B. E. Deal, *Appl. Phys. Letters* **9**, 53 (1966).

where the electrons were excited to the required energy thermally (thermionic emission), and also the case where excited electrons were injected from another barrier contact under forward bias. The results are relevant to photoemission, though they cannot be transferred directly, since photoexcited carriers are likely to have an energy distribution quite different from that of thermally excited carriers. For electrons with sufficient energy, emission of optical phonons is important in representative cases. For example, hot-electron data show that the mean free path for optical phonon emission is 76 Å in Si and 58 Å in GaAs. Thus, scattering is likely within the distance x_m from the surface. As the field is increased, the value of x_m decreases. Scattering back into the metal is decreased and collection efficiency increases. By taking account of both absorption and emission of phonons and averaging over a Boltzmann distribution of electrons, Crowell and Sze[15] obtained the appropriate average probability f_p that an electron with energy greater than ϕ_B will be collected without being scattered back into the metal. In Fig. 4,

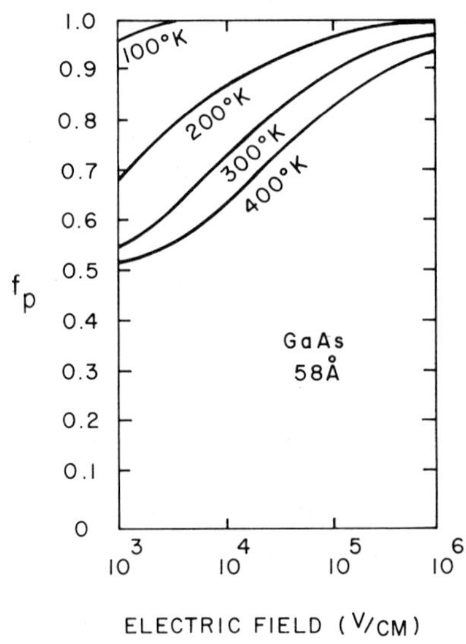

FIG. 4. Calculated[15] emission probability f_p averaged over a Maxwellian distribution of electrons incident on the potential energy maximum of a GaAs Schottky barrier. It is plotted as a function of electric field at the metal–semiconductor interface. The calculation takes account of the effects of phonon scattering.

this is shown as a function of the surface electric field at several different temperatures for a GaAs Schottky barrier. It is seen that as the field is increased from 10^3 V/cm to 10^6 V/cm, the collection efficiency increases from

55% to 95%. In this case, the effect of back scattering on the collection efficiency is not drastic, since, even at low fields, the efficiency is not cut down by more than a factor of two. It may be anticipated, however, that the effect will be more serious in experiments on photoemission into insulators, because the mobility is usually small in insulators. This implies a small mean free path and an increased probability for back scattering. For example, if the mobility is 10 cm^2/V-sec, the mean free path for scattering is about 6 Å. Thus, an electron would be scattered several times before reaching the potential energy maximum at the distance x_m from the interface. It appears likely that the generally low quantum yields for photoemission from metals into insulators[17,18] are due, in part at least, to this cause. Detailed calculations of the effect for this case have not been done, though a similar problem involving the effect of recombination on the quantum yield for photoconductivity has been treated by Kepler and Coppage.[19]

An understanding of the general properties of Schottky barriers is important to an understanding of their usefulness in studying photoemission. The single feature which contributes most to their usefulness is the fact that they provide a thin layer of material $\sim 10^{-4}$ cm thick which is nearly depleted of free carriers, permitting observation of photocurrents with a minimum of interference from dark currents and trapping effects. The barrier thickness d is ordinarily determined from measurements of the capacitance and application of the formula for a parallel-plate capacitor of unit area: $C = \varepsilon/4\pi d$. In practical units, this is (with d in cm)

$$C\,(\mathrm{pF/cm^2}) = 0.088\varepsilon/d. \tag{10}$$

This refers to the small-signal capacitance, in which the ac measuring signal used is small compared with the barrier height. The total charge in the barrier, which is a measure of the integral capacitance over a range of voltage equal to the barrier height, is twice that which would be inferred from Eq. (10).

The barrier height can also be derived from capacitance data and compared with the same quantity measured by photoemission methods. The relation between C, applied reverse-bias voltage V, diffusion voltage V_d, and the concentration N of ionized donors in the barrier is[9]

$$\frac{1}{C^2} = \frac{8\pi(V_d + V)}{e\varepsilon N}. \tag{11}$$

[17] R. Williams, *Phys. Rev.* **140**, 569 (1965).

[18] R. Williams and J. Dresner, *J. Chem. Phys.* **46**, 2133 (1967).

[19] R. G. Kepler and F. N. Coppage, *Phys. Rev.* **151**, 610 (1966).

In practical units, with C in $\mathrm{pF/cm^2}$, V_d and V in volts, and N in $\mathrm{cm^{-3}}$, this becomes

$$\frac{1}{C^2} = \frac{1.4 \times 10^8 (V_d + V)}{\varepsilon N}. \tag{12}$$

Again the formula refers to small-signal capacitance, where the ac measuring voltage is small compared to V_d and V. Measurements of $1/C^2$ as a function of V may be extrapolated to give V_d. In the simplest case, V_d is smaller than ϕ_B by the energy difference V_f between the Fermi level and the majority-carrier band edge in the bulk. Experimental tests of this model will be discussed later. More-complicated relations have been derived for the case where there is a thin layer of insulator between metal and semiconductor.[20]

6. Insulator without Schottky Barrier

The energy-band relations at a metal–insulator interface are similar to those shown for the metal–semiconductor interface in Figs. 1, 2, and 3. In the simplest case, there is no space charge in the volume of the insulator. Under applied voltage, a uniform field extends through the entire thickness. Contacts may be either ohmic or blocking, as in the case of semiconductors. Generally speaking, the higher the band gap of a material, the harder it is to make ohmic contacts. Often, it is necessary to use inconvenient electrode materials such as reactive electrolyte solutions.[21] In materials with very high band gaps, such as alkali halides and SiO_2, apparently all electrode materials make blocking contact for fields high enough to satisfy the conditions given by Eq. (1). To achieve high fields and good collection efficiency, it is desirable to work with thin samples of material. This is easily done in cases where the insulator can be grown as a thin layer on a substrate of metal or low-band-gap semiconductor. Such a system which has been extensively investigated is that of SiO_2 obtained by thermal growth on single-crystal silicon.[22] The oxide is atomically bonded to the silicon and provides the kind of intimate interface required for meaningful results. Thicknesses from 5×10^{-6} to 10^{-3} cm of oxide have been studied.

A typical experimental cell is shown in Fig. 5. This is, in effect, a thin layer of the insulator with two electrodes, one of silicon and the other of any desired metal. The oxide will sustain a field greater than 10^{-6} V/cm. This is more than enough to satisfy Eq. (1) for electrons, for which μ is about $20\ \mathrm{cm^2/V\text{-}sec}$. A similar method is used to study Al_2O_3 which is also readily available in the form of thin layers. Here, special problems arise because the layers are under 100 Å in thickness. Photoemission from both electrodes is

[20] C. R. Crowell, H. B. Shore, and E. E. LaBate, *J. Appl. Phys.* **36**, 3843 (1965).

[21] P. Mark and W. Helfrich, *J. Appl. Phys.* **33**, 205 (1962).

[22] M. M. Atalla, E. Tannenbaum, and E. J. Scheiber, *Bell System. Tech. J.* **38**, 749 (1959).

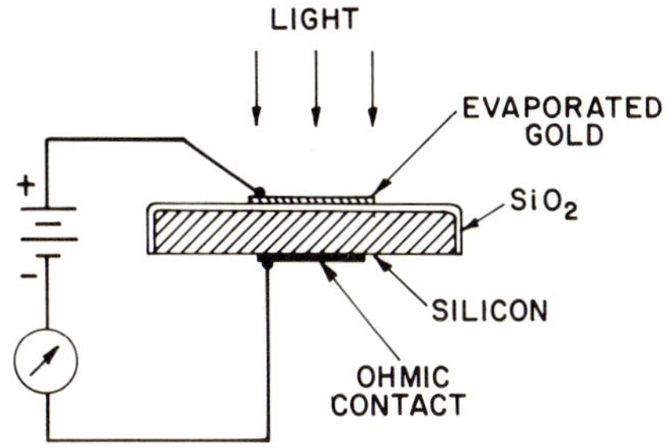

FIG. 5. Typical experimental cell for investigating photoemission of electrons into SiO_2.

significant under most conditions and special analysis is necessary to interpret experiments.[23,23a] Relatively thick insulator specimens have been used for photoemission measurements,[3,18] but there do not appear to be many insulators available at this time which have trap concentrations low enough to draw steady currents without polarization.

7. Photoemission as a Contact-Controlled Current

The type of photoemission current easiest to interpret is the saturated current discussed in connection with Eq. (1). Here, the current is independent of applied voltage and depends only on the supply of carriers with the proper energy. This in turn, is proportional to the light intensity. If the spectral response of the photocurrent varies with the electrode metal, it is then very likely due to photoemission of carriers excited in the metal. If there is no variation of spectral response as the electrode metal is changed, then the photocurrent may still be due to photoemission with the surface barrier determined by surface states, and, thus, independent of the metal. It is usually possible in this case, by simple experiments,[4] to determine whether the excited carriers arise in the metal or in the semiconductor itself. In the latter case, the process would not be photoemission, but photoconductivity.

If the field is low, or if, for any other reason, the I–V curve is not saturated, then results are more difficult to interpret, since the photocurrent may not be directly proportional to light intensity, and this distorts the spectral response curve. In an extreme case, at low fields, the reservoir of photoexcited carriers from the metal may act as an ohmic contact. This would give a photocurrent independent of light intensity over a limited range. It could be either ohmic or space-charge limited, depending on the injection level.

[23] A. Braunstein, M. Braunstein, G. S. Picus, and C. A. Mead, *Phys. Rev. Letters* **14**, 219 (1965).
[23a] F. Schuermeyer and J. A. Crawford, *Appl. Phys. Letters* **9**, 317 (1966).

8. Spectral Response of Photoemission Current

A characteristic property of photoemission currents is the form of the spectral response curve. For most systems which have been studied, the dependence of the quantum yield Y on the energy $h\nu$ of the exciting light quantum has the simple form

$$Y = C(h\nu - h\nu_0)^2, \qquad \text{when} \qquad (h\nu - h\nu_0) \gtrsim 6kT. \tag{13}$$

Here, $h\nu_0$ is the threshold energy, equal to ϕ_B, and C is a constant. In the determination of barrier heights by the photoemission method, it is important to have an understanding of the dependence of Y on energy, since the threshold is obtained by extrapolation of data for energies above $h\nu_0$. In addition, the characteristic dependence is evidence that an observed photocurrent may be due to photoemission. The dependence shown in Eq. (13) should be found only for the case where both the emitter and the "collector" of the photoexcited carriers have wide energy bands. "Wide" here means more than 0.5 eV or so, and, by this definition, most common metals, semiconductors, and insulators have wide bands.

There are certain materials in which the energy bands are narrow, such as anthracene, where they are about 0.02 eV in width.[24,25] Another case where the energy band may be effectively narrow is that in which electrons are photoemitted from a degenerate n-type semiconductor.[26] There is a certain range of photon energies in which electrons are emitted only from the conduction band and not from the valence band. Since the electrons in the conduction band are concentrated in a range of energies of order kT at the bottom of the conduction band, the effective width of the band which is the source of the electrons is kT. For narrow bands, the dependence of quantum yield on photon energy differs from Eq. (13) and gives information on the energy-band structure of the materials involved. Simplified derivations of the spectral response for several important cases are given below.

a. Wide Bands

This case is the same as that treated by Fowler for photoemission from metals into vacuum. It is discussed in detail by Hughes and DuBridge.[27] A simplified version is presented here in which certain features, such as the temperature dependence, are neglected.

[24] O. H. LeBlanc, Jr., *J. Chem. Phys.* **35**, 1275 (1961).

[25] R. Sibley, J. Jortner, S. A. Rice, and M. T. Vala, *J. Chem. Phys.* **42**, 733 (1965).

[26] A. M. Goodman, *Phys. Rev.* **152**, 785 (1966).

[27] A. L. Hughes and L. A. DuBridge, "Photoelectric Phenomena," p. 243. McGraw-Hill, New York, 1932.

It is assumed throughout that the electrons excited by the light have no preferred direction of motion, i.e., the momentum distribution function is spatially isotropic. The basic reason for the variation of quantum yield with photon energy for energies above threshold is that an electron can escape only if it has a component of momentum normal to the surface which is greater than p_0, where

$$p_0^2 = 2m^* U_0. \tag{14}$$

Here, U_0 is the total kinetic energy of the electron measured from the bottom of the conduction band. If it is moving in a direction exactly normal to the surface, any electron with $p > p_0$ will escape. Any electron moving parallel to the surface will not escape, regardless of how much energy it may have. The geometric construction in Fig. 6 shows a cone of velocities or momenta

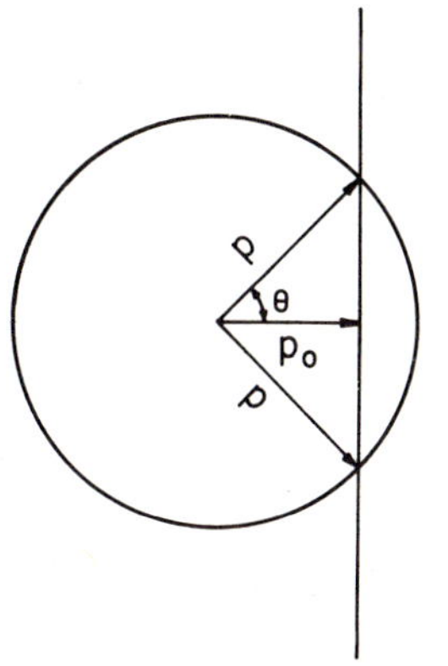

FIG. 6. Construction of cone of velocities which determines the fraction of electrons with total momentum p in an isotropic distribution, which have a component p_0 normal to an interface.

drawn so that any electron moving in a direction lying outside the cone will not be able to escape. It is seen that an electron with $p > p_0$ will have a normal component of momentum greater than p_0 whenever its angle with the normal is equal to or less than θ. This defines a cone with vertex angle θ, where $\cos\theta = p_0/p$. The fraction $f(p)$ of all electrons having momentum p which are moving in directions inside the escape cone is simply the ratio of the surface area of the sphere included within the cone to the total surface area. For an isotropic momentum distribution,

$$f(p) = \tfrac{1}{2}(1 - p_0/p). \tag{15}$$

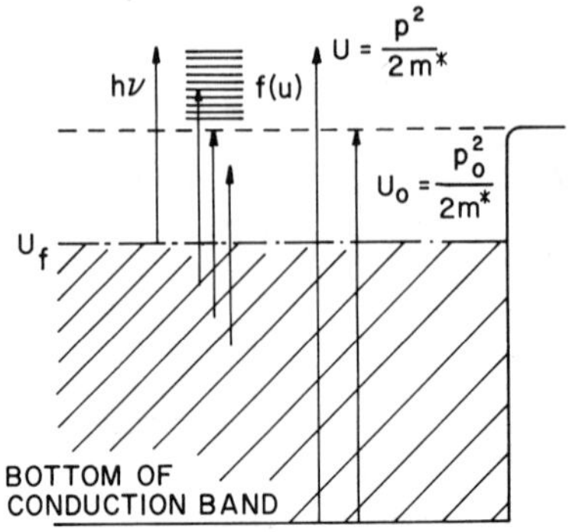

FIG. 7. Illustration of terminology used for derivation of the photoemission spectral response curve.

The relevant energy relations can be understood from Fig. 7. This shows the simplest model for the conduction band of a metal. With momenta measured from the bottom of the conduction band, p_0 and p are indicated by arrows. A photon with energy $h\nu > h\nu_0$ can excite electrons into a range of final states extending from the Fermi level up to an energy $h\nu$ above the Fermi level. (It is assumed that all energy-conserving transitions are allowed.) All those having a normal component of momentum greater than p_0 will escape, while none of those with normal component of momentum less than p_0 will escape. Further assumptions are: (1) there are no collisions of electrons or energy losses before they reach the surface; (2) the density of states near the Fermi level is independent of energy and the intensities of optical transitions are simply proportional to the densities of initial states. We consider only those electrons with $U > U_0$ which can escape.

Excitation produces a distribution spatially isotropic and uniformly distributed in energy over the range from U_0 to U. For any energy within this range, there is a corresponding escape probability which may be written either as a function of momenta or of energy. As a function of energy, Eq. (15) becomes

$$f(U) = \frac{1}{2}\left(1 - \frac{\sqrt{U_0}}{\sqrt{U}}\right) = \frac{1}{2}\,\frac{\sqrt{U} - \sqrt{U_0}}{\sqrt{U}} = \frac{U - U_0}{2[U + (U_0 U)^{1/2}]}. \tag{16}$$

The last form of Eq. (16) contains a numerator which is directly proportional to $(U - U_0)$ and a denominator which is slowly varying. Typical conditions for measurement of photoemission from a metal into a semiconductor are as follows: U_f is 8 eV above the bottom of the conduction band; ϕ_B is 1 eV; the measurements are carried out over a range of photon energies extending from the threshold energy to $\frac{1}{2}$ eV higher; the variation of quantum yield over this range is one to two orders of magnitude. For this case, U_0 is 9 eV, and the highest value of U is 9.5 eV. Over this range, the denominator varies by about 4%. Thus, the variation of one or two orders of magnitude in the quantum yield must be due to variations in the numerator. It is a reasonable approximation, then, that

$$f(U) \approx C'(U - U_0), \qquad C' = \text{const}. \tag{17}$$

From Fig. 7 and our assumptions, we see that for each energy interval dU above U_0, it is equally probable that absorption of a photon with energy $h\nu$ will excite the electron into the energy interval. The escape probability is then $f(U)$ and the probability that a carrier will be excited into dU and also escape is proportional to $f(U)\,dU$. The photoemission probability is then

$$\begin{aligned} Y(U) &= \int_{U_0}^{U} f(U)\,dU = C'\int_{U_0}^{U} (U - U_0)\,dU \\ &= \tfrac{1}{2}C'(U - U_0)^2. \end{aligned} \tag{18}$$

Since $U - U_0 = h\nu - h\nu_0$, Eq. (18) is equivalent to Eq. (13). This is commonly used in the form

$$\sqrt{Y} = C(h\nu - h\nu_0). \tag{19}$$

Considering the rather drastic approximations which go into it, it is remarkable how often and how well this equation fits experimental data (see, for example, Spitzer *et al.*[6] and Goodman[11]). It even fits reasonably well in cases where the assumed density of states in the absorbing material is clearly wrong. An example of this is the photoemission of electrons[17] excited from the filled valence band of silicon into the conduction band of SiO_2. Clearly, the density of states at the top of the valence band cannot be independent of energy, but it is found empirically that Eq. (19) fits the data reasonably well. There is always a small tail in the plot of quantum-yield data near the threshold which departs from Eq. (19). This is due to a finite concentration of thermally excited electrons up to an energy several kT above the Fermi level. This is clearly shown by Goodman.[11] It is considered in the original Fowler theory, and, where it is important, requires the data to be plotted in a different way.[27] In summary, Eq. (19) has a broad empirical usefulness in getting objective values for ϕ_B by extrapolation of quantum-yield data for energies above the threshold.

b. Narrowband Emitter

A narrowband emitter is shown in Fig. 8. This was found experimentally by Goodman.[26] In degenerate *n*-type silicon, there is a high density of electrons concentrated into an energy range lying within a few kT of the bottom of the conduction band. There is also a higher concentration of electrons in the

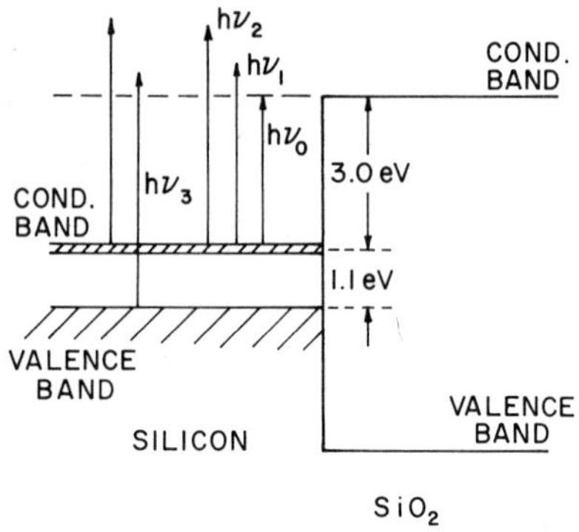

FIG. 8. The various transitions possible in photoemission of electrons from *n*-type degenerate silicon into SiO_2. This is a case of a narrowband emitter.

valence band of the silicon which begins 1.1 eV below the bottom of the conduction band. Photons with energies $h\nu_0$, $h\nu_1$, and $h\nu_2$ can excite electrons from the conduction band of the silicon into the conduction band of the SiO_2. At higher energies, such as $h\nu_3$, electrons can be excited into the conduction band of the SiO_2 from either the conduction band or the valence band of the silicon. Taking, for example, $h\nu_0$, it can be seen that when the excitation is from a narrow band, the excited electrons are not distributed over a wide range of energies as they were in Fig. 7. They are distributed over a narrow range of energies no wider than the band from which they originate. As long as this applies, then the escape probability is the same for all electrons excited by a single frequency such as $h\nu_1$. Thus, for the case of a narrowband emitter, $f(U)$ is given by Eq. (17), and, since the excited electrons all have nearly the same energy, no integration of the equation is required. The dependence of quantum yield on frequency is simply

$$Y = C(h\nu - h\nu_0). \tag{20}$$

This dependence has been found experimentally for the Si–SiO_2 system over an energy range of about 0.5 eV. At higher energies, the absorption from the valence band becomes dominant, and the data for this range fit Eq. (19).

c. Narrowband "Collector"

We shall use the term "collector" to designate the semiconductor or insulator into which electrons go in the process of photoemission. When this has narrow energy bands, one may anticipate a very unusual photoemission spectrum. The reason for this may be understood from the energy-band diagram of Fig. 9. Here, the collector, which may be a narrowband insulator such as anthracene, is shown on the right. The emitter is a normal metal with

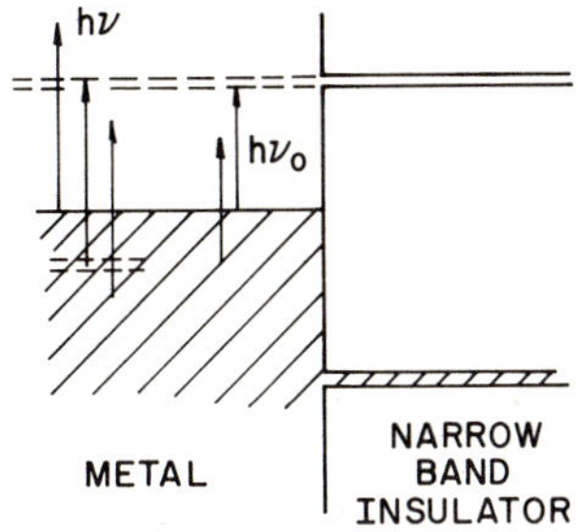

FIG. 9. Energy-band diagram for photoemission from a metal into a narrowband insulator such as anthracene. Various transitions, including those leading to photoemission, are labeled. This is a case of a narrowband "collector."

wide energy bands. The general assumptions about densities of states in the metal are the same as for the previous cases. For photon energies less than $h\nu_0$, no electrons will be collected. At the threshold energy $h\nu_0$, the small fraction of the excited electrons which are moving in the right direction will be able to enter the narrow conduction band of the insulator. For energies above threshold, such as $h\nu$, the excited electrons will be able to enter the conduction band only if they have an energy lying within the narrow range covered by the conduction band. This range is calculated to be about 0.02 eV or less for anthracene.[24,25]

The meaning of the concept of narrow bands is that only those electrons lying within the appropriate narrow range of energies are capable of motion within the crystal. Thus, only those electrons excited from the initial states with energies between the dashed lines drawn in the metal can enter the insulator. Of those excited from this energy interval by a quantum of energy $h\nu$, only the small fraction going in the right direction will be able to enter. Since the density of states in the metal is assumed to be independent of energy, the quantum yield for energies above the threshold is small and nearly independent of the energy of the exciting light. This is illustrated in Fig. 10. Whether

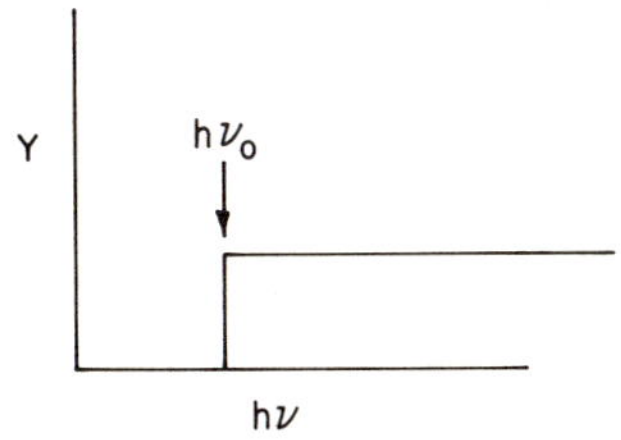

FIG. 10. Quantum yield Y for photoemission of electrons from a metal into a narrowband insulator having a single band, as a function of photon energy. The threshold energy $h\nu_0$ is indicated.

the quantum yield for energies above $h\nu_0$ remains constant as drawn, or actually decreases, depends on more-detailed assumptions, such as whether or not the metal is thick enough to absorb all the light which is incident. Some data which may be an illustration of this case have been obtained for photoemission of holes from metals into anthracene.[18] To interpret these data, it was necessary to invoke the concept[25] that there are several narrow bands, separated by intervals corresponding to the molecular vibration frequencies, which, in anthracene, are of order 0.2 eV. This is illustrated in Fig. 11, along with the photoemission spectrum (Fig. 12) which would be anticipated by applying the above ideas to the many-band case. If this is, indeed, the correct interpretation of the structure in the photoemission spectra reported by Williams and Dresner,[18] it promises to be a useful tool for examining the structure of narrowband materials.

It is possible that both emitter and collector could have narrow bands. In this case, the photoemission spectrum would simply be a narrow band at the threshold frequency. Participation of many bands would lead to something

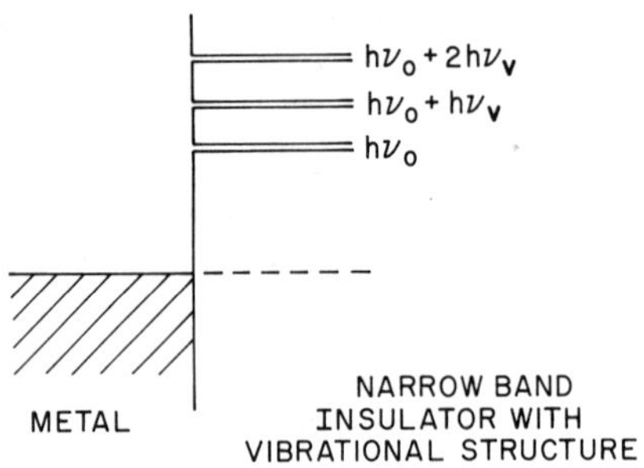

FIG. 11. Energy-band diagram for the interface between a metal and a narrowband insulator. The effect of vibrations, leading to a many-band structure, is included.

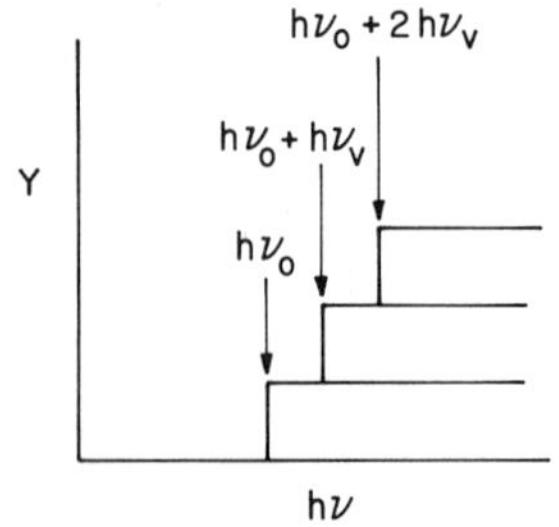

FIG. 12. Quantum yield versus photon energy for photoemission from a metal into a narrow-band insulator having several narrow bands equally spaced by a dominant vibrational energy $h\nu_v$.

resembling vibronic bands in molecular spectra. Experiments corresponding to this case have not been reported.

The spectral response has been discussed here with reference to electrons, but all arguments apply equally well to holes.

9. Determination of Barrier Heights

Perhaps the most common application of internal photoemission is in the measurement of energy-band relationships at the interface between two solids. The quantum yield for photoemission Y is measured as a function of photon energy over a range of energies and extrapolated to obtain $h\nu_0$, which is equal to ϕ_B. Ordinarily, Eq. (19) is used, or, alternatively, the Fowler plot.[27] The theoretical relations refer to the quantum yield with respect to absorbed photons. Experimental data are commonly plotted with reference to incident light rather than absorbed light because it is experimentally simpler to do. This appears to be satisfactory in many cases, especially where the reflectivity of the metal does not vary strongly with energy.

A typical plot for the determination of a threshold is shown in Fig. 13, taken from Goodman.[11] This is for an electrode of evaporated gold on n-type CdS. The deviation of experimental points from the line at low energies is due to the effect of thermally excited electrons at energies above the Fermi level. It is to be expected when data are plotted in this way, since the theory [Eq. (19)] does not consider thermally excited electrons.

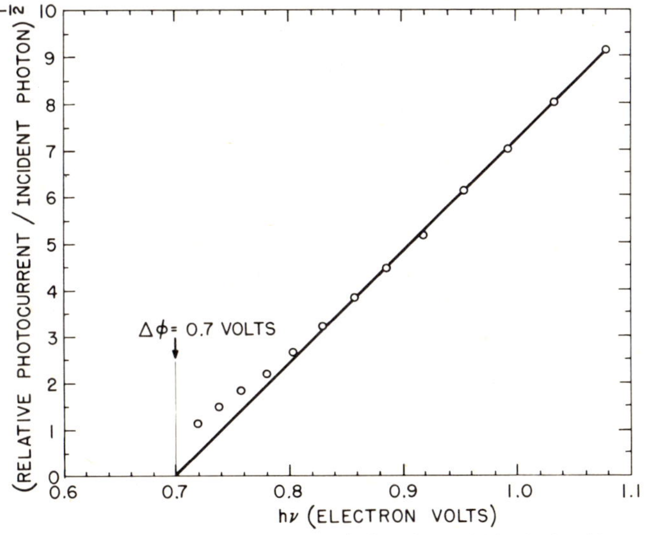

Fig. 13. Typical plot showing how the photoemission threshold is obtained by extrapolation of data for energies near the threshold. (Taken from Goodman.[11]) Evaporated gold on n-type CdS.

III. Experimental Results for Barrier Heights

10. Alkali Halides

As mentioned in the first section, the earliest controlled experiments on internal photoemission were those of Gilleo,[3] who measured the photoemission of electrons from silver into KBr and NaCl in 1953. These remain, to this date, the only such experiments which have been done with alkali halides. Silver electrodes were evaporated onto cleavage plates having the dimensions 18 × 10 × 1 mm. The other electrode was a nickel screen making capacitive contact to the opposite face of the crystal. When the silver electrode was negative, a photocurrent was obtained. For KBr, the plot of photocurrent versus photon energy fit a Fowler plot with $\phi_B = 4.3$ eV. With NaCl, only the threshold wavelength was measured, but this gave the same value for the threshold. It was concluded that the photocurrent was not due to photoconductivity of the KBr crystal, because it disappeared when the polarity of the applied voltage was reversed. Further evidence that electrons were being photoemitted into the crystal from the metal was provided by the fact that as the current flowed, trapped electrons appeared in the crystal. These could be then excited by *F*-band illumination and the corresponding current measured. Since the work function of silver is 4.7 eV, these results place the conduction bands of KBr and NaCl about 0.4 eV below the vacuum level (energy of an electron at rest in vacuum). This in reasonable agreement with earlier estimates by Mott and Gurney[1] showing that the electron affinity of NaCl should be very small. It is also in agreement with the fact that it has apparently not been possible to make ohmic metal contacts to alkali halides at room temperature. With any metal, the work function difference is simply too large to inject electrons.

11. Si, Ge, Diamond, SiC

Silicon has been widely used in photoemission experiments because well-behaved and reproducible Schottky barriers can be obtained with several different metals. Extensive results have been obtained for electrodes of evaporated gold on *n*-type Si. Spitzer *et al.*[6] found a good straight line when the square root of the quantum yield for photoemission was plotted against the photon energy. They obtained a threshold of 0.79 ± 0.01 eV at room temperature for Au evaporated onto silicon surfaces freshly cleaved in vacuum. In further experiments, Crowell *et al.*[28] measured the temperature dependence of the threshold of gold on *n*-type Si and found it to be the same as the temperature dependence of the band gap of silicon. This implies that the Fermi level at the interface is pinned in relation to the valence band edge.

[28] C. R. Crowell, S. M. Sze, and W. G. Spitzer, *Appl. Phys. Letters* **4**, 91 (1964).

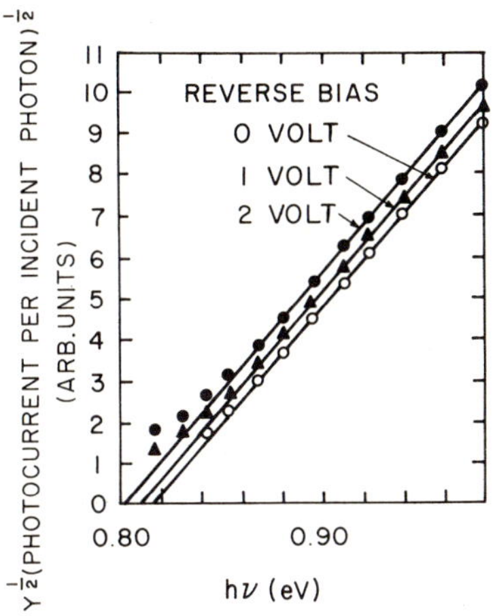

FIG. 14. Photothreshold plot for a diode of gold on *n*-type silicon (0.2 ohm-cm). The Schottky lowering of the barrier is shown by the variation of the threshold with applied voltage. (Taken from Sze *et al.*[14])

The lowering of barrier height by the image force field was discussed above and is summarized by Eq. (7). This effect was studied for the Au–Si barrier by Sze *et al.*[14] Figure 14 reproduces their photothreshold plot for three values of reverse bias voltage. It is clear that the threshold decreases with increasing bias voltage as one would expect. It is not obvious, *a priori*, what value should be used for the dielectric constant in Eq. (7) in order to compare theory with experiment. This is because an electron moves through the region near the barrier maximum in a very short time and consideration of the frequency dependence of the dielectric constant is important. It was found that the data in Fig. 14 could be fit quantitatively to the equation with a value of 12 for the dielectric constant. This is in good agreement with the known infrared dielectric constant for silicon. It was shown that the transit time of an electron through the relevant part of the barrier is comparable to the period of the infrared radiation used to measure the dielectric constant. It should be noted, however, that the general problem of the tunneling time of an electron is actually quite involved, and has been discussed by Thomber *et al.*[29]

Early evidence for photoemission of electrons from tin into *n*-type germanium was reported by Mahlman.[30] There was a photoresponse for wavelengths out to 2.5 μm. It was shown that this was not due to absorption in the germanium, and it was interpreted as photoemission. The effect was found at liquid-nitrogen and liquid-helium temperatures, but not at room

[29] K. K. Thomber, T. C. McGill, and C. A. Mead, *J. Appl. Phys.* **38**, 2384 (1967).

[30] G. W. Mahlman, *Phys. Rev. Letters* 7, 408 (1961).

temperature. This may be an example of the difficulty of observing photoemission when the barrier is low, discussed above in connection with Eq. (5). Measurements of the barrier heights were made for evaporated Au and Al on cleaved germanium specimens by Mead and Spitzer.[31] They reported values of 0.45 eV for Au and 0.48 eV for Al. Since the barriers are nearly the same and the work functions for the two metals differ by about 1 eV, this is a case where the barrier height must be determined by surface states.

Mead and Spitzer[31] report further barrier-height measurements for SiC and diamond. For photoemission of electrons from Au and Al into SiC, the barriers are 1.95 and 2.0 eV, respectively. For photoemission of holes from Au into *p*-type diamond, the barrier is 1.35 eV.

12. II–VI MATERIALS

a. General Results and Effects of Surface States

Barrier-height measurements have been made on several II–VI materials. The most extensive work has been done with CdS.[4,5,11,31,32] A summary of data is given by Mead.[5] Cadmium sulfide is representative of materials in which the barrier height is not determined by surface states. There is a linear relation between ϕ_B and the electronegativity of the metal for barriers on *n*-type CdS. This was seen by Goodman[11] and by Mead.[5] Similar results have been obtained for ZnS. An especially instructive comparison has been made by Mead[5] and is reproduced in Fig. 15. Here, ϕ_B is plotted against the electronegativity of the metal for various metals on ZnS, where the barrier is *not* controlled by surface states, and also for various metals on GaAs, where ϕ_B *is* controlled by surface states. The part played by surface states is clearly evident.

In CdSe, the barrier is controlled by surface states. Mead[33] has made barrier-height measurements on mixed crystals of CdS_xSe_{1-x} over the entire composition range. The results show a continuous transition in behavior from the pure selenide, where surface states determine the barrier height, to the pure sulfide, where they do not. Results are understandable on the basis of general discussions of surface states which have been given by Shockley[34] and, more recently, by Levine and Mark.[35] In moderately ionic materials, such as CdS, the surface states tend to split into two groups, one of which lies in a band of energies near the conduction band edge, and the other near the valence band edge. It is only when the surface states have energies near that

[31] C. A. Mead and W. G. Spitzer, *Phys. Rev.* **134**, A713 (1964).

[32] A. M. Goodman, *Surface Sci.* **1**, 54 (1964).

[33] C. A. Mead, *Appl. Phys. Letters* **6**, 103 (1965).

[34] W. Shockley, *Phys. Rev.* **56**, 317 (1939).

[35] J. D. Levine and P. Mark, *Phys. Rev.* **144**, 751 (1966).

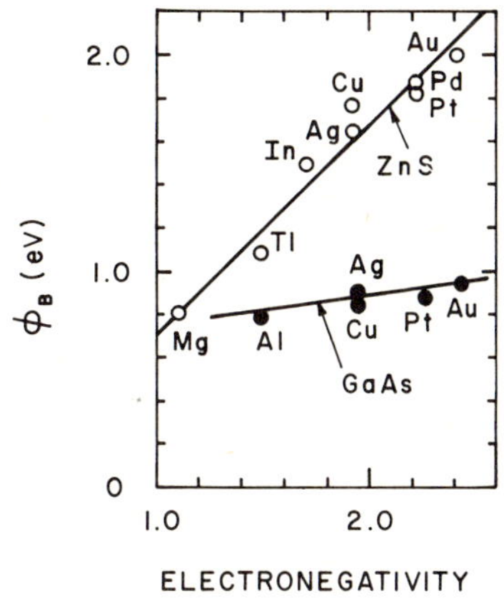

FIG. 15. Plots of barrier heights versus electronegativities for various metals on ZnS and on GaAs. For GaAs, ϕ_B is nearly the same for all metals, indicating that it is determined by surface states. For ZnS, ϕ_B varies widely from one metal to another and is related to the electronegativity, indicating that surface states are not important. (Taken from Mead.[5])

of the metal Fermi level that they determine the barrier heights. This is apparently what happens in the less-ionic materials such as CdSe and the group IV semiconductors, where the states lie near the middle of the band gap. In this case, any movement up or down of the Fermi level requires the filling of a high concentration of surface states, with the result that the Fermi level is pinned somewhere in the middle of the energy range covered by the surface states.

Effects of surface preparation methods on the barrier height have been described by Mead and Spitzer.[31] For materials such as CdS where the barrier is strongly dependent on the metal work function, the barrier height is strongly influenced by surface preparation. This is clearly seen on comparing results for electroplated metals[32] with those for metals evaporated onto freshly cleaved crystals.[31] For groups IV and III–V materials, barrier heights were not strongly dependent on the surface preparation.

13. III–V Materials

With III–V compounds, it is possible, in several cases, to make both *n*- and *p*-type material. This permits the observation of photoemission of both electrons and holes. The first observation of the photoemission of holes was that from tin into *p*-type GaAs.[36] The energy-band relations for a Schottky barrier in a *p*-type semiconductor are shown in Fig. 16. The surface field is such as to collect holes from the metal, but not electrons. Requirements to

[36] R. Williams, *Phys. Rev. Letters* **8**, 402 (1962).

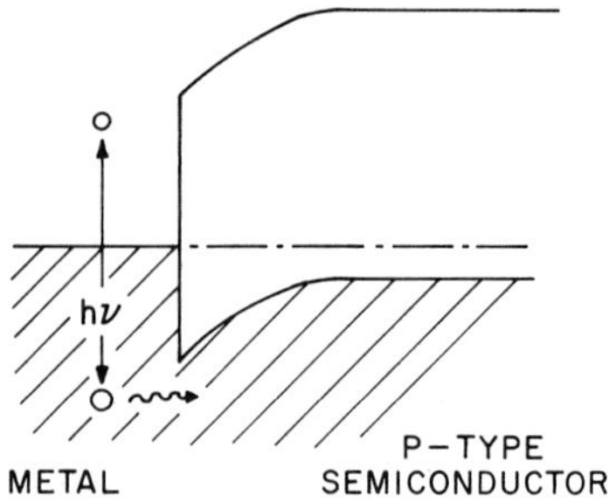

FIG 16. Barrier at the interface between a metal and a *p*-type semiconductor, showing the conditions required for collection of photoemitted holes.

observe a photocurrent of holes are the same as those for electrons. Excited holes must travel from their point of origin to the interface without losing their excitation energy, and the factors determining the spectral response of the photocurrent are the same as for electrons. The observed behavior is, in general, identical with that for photoemission of electrons.

With GaAs and GaP, both hole and electron photoemission have been observed in each case. It is of interest to compare the sum of the threshold energies for holes and electrons with the known band gaps of the semiconductors. With GaP, the threshold for photoemission of electrons from Au into *n*-type material is 1.3 eV[37] and that for photoemission of holes from Au into *p*-type material is 0.72 eV.[38] The sum is 2.02 eV and may be compared with the band gap for room temperature,[39] which is 2.18 eV. Similarly, for GaAs, the electron threshold is 0.90 eV and the hole threshold is 0.42 eV.[31] The sum (1.32 eV) is to be compared with the value[39] of 1.38 eV for the band gap of GaAs. The agreement is good enough, in each case, to suggest this as a method for determining band gaps where other methods are not available. In well-behaved semiconductors, it is easier to measure the band gap by other means. With insulators, however, the other means are not always available or easy to interpret. In such cases, the method for measuring thresholds for electron and hole photoemission and taking the sum may be the most practical method of determining the band gap.[18,40]

The lowest photoemission threshold which has been reported is apparently that for Au on *n*-type InSb. This was measured[31] at 77°K using both vacuum-cleaved and chemically prepared surfaces. In both cases, a threshold of 0.17 eV was found. This indicates that infrared response must be found out

[37] M. Cowley and H. Heffner, *J. Appl. Phys.* **35**, 255 (1964).

[38] H. G. White and R. A. Logan, *J. Appl. Phys.* **34**, 1990 (1963).

[39] J. J. Tietjen and J. A. Amick, *J. Electrochem. Soc.* **113**, 724 (1966).

[40] A. M. Goodman, *Phys. Rev.* **152**, 780 (1966).

to about 7 μm, with possibly useful applications for infrared detection. With *p*-type InSb, no threshold was measured, since the contact appeared ohmic. This illustrates the principle discussed earlier, that photoemission is difficult to detect when the threshold is low. In this case, the threshold for hole photoemission should be about 0.07 eV, since the band gap of InSb at 77°K is 0.24 eV and the threshold for electron emission is 0.17 eV. Hence, the ratio of the threshold energy to kT is about ten to one. Difficulties arise when this ratio falls below twenty to one.

14. Silicon Dioxide

This insulator has been extensively studied by photoemission. Material of excellent quality is available because of its use in the MOS field-effect transistor. It is grown on single crystals of silicon by placing them in an atmosphere containing oxygen, either free or combined, at temperatures around 1200°C. The result is a layer of SiO_2, chemically bonded to the silicon crystal, which may have any thickness up to 10 μm or so. It is vitreous,[41] rather than crystalline, and has the same chemical composition as fused quartz. The silicon crystal serves as one electrode, and a second electrode of metal is evaporated on the outer surface. This gives a thin layer of SiO_2 between two electrodes as shown in Fig. 5.

The earliest results for this system[17] were for photoemission of electrons from the silicon into the oxide. A threshold of 4.25 eV was found with both *n*- and *p*-type silicon. Comparable values have been reported by others. Goodman[16] found values of 4.19 and 4.31 eV, depending on the oxide thickness, while Deal *et al.*[42] found 4.35 eV, independent of the orientation of the silicon crystal.

Since the applied fields can be high, Schottky lowering of the barrier is important in this system.[16,42] Shifts up to 0.3 eV and more are found in fields somewhat above 10^6 V/cm. To fit the data to Eq. (7), it was necessary to use the value 2.15 for the dielectric constant, rather than 3.8, which is the measured low-frequency dielectric constant for SiO_2. Apparently, the electron passes through the potential maximum near the interface in a time too short for the low-frequency dielectric constant to be applicable.

Goodman[40] has also reported a threshold for photoemission of holes from silicon into SiO_2. The value is 4.92 eV. It should be remembered that the threshold for electrons gives the energy difference between the conduction band in the oxide and the valence band of the silicon, while the threshold for holes gives the difference between the valence band in the oxide and the conduction band in the silicon.

[41] M. M. Atalla, E. Tannenbaum, and E. J. Scheibner, *Bell. System Tech. J.* **38**, 749 (1959).

[42] B. E. Deal, E. H. Snow, and C. A. Mead, *J. Phys. Chem. Solids* **27**, 1873 (1966).

We can use the above data, together with data on photoemission of electrons from silicon into vacuum,[43] and data on the optical absorption of fused quartz,[44] to construct a self-consistent energy-band diagram[17] for the Si–SiO_2 interface. This is shown in Fig. 17. The fact that the conduction band is only 0.9 eV below the vacuum level indicates that it will probably not be possible to make ohmic contacts to this material in the ordinary way and that study of its transport properties will be possible only by the use of photoemission or other special injection techniques. Evidence for the trapping level shown will be given below.

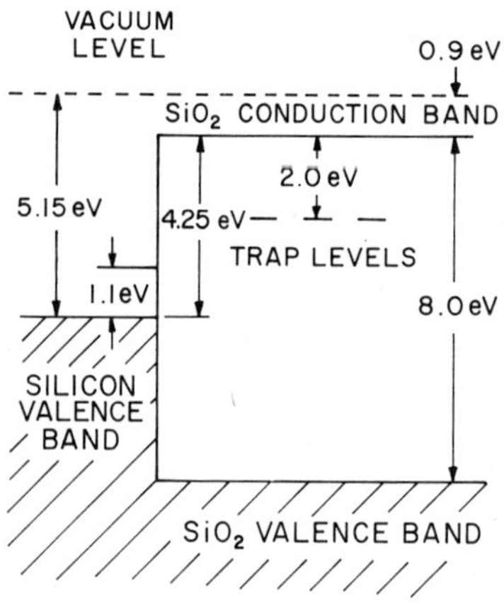

FIG. 17. Energy relations at the interface between silicon and SiO_2, as determined from photoemission experiments and other data. Band bending in the silicon is often present, but has been omitted for simplicity. (Taken from Williams.[17])

In ordinary silicon, photoemission of electrons takes electrons from the valence band of the silicon to the conduction band of the oxide. Even if the material is *n*-type, the density of states in the conduction band which are occupied by electrons is too small to make any significant contribution to the photoemission. This is no longer true if degenerate *n*-type silicon is used, since the density of occupied states in the conduction band may now be comparable with the density of occupied states over a similar energy range in the valence band. In this way, an additional contribution to the photoemission arises from transitions which take an electron from the conduction band of the silicon to the conduction band of the oxide. The new process should have a threshold smaller than the normal threshold by 1.1 eV, which is the

[43] G. W. Gobeli and F. G. Allen, *Phys. Rev.* **127**, 141 (1962).

[44] W. Groth and H. V. Weyssenhof, *Z. Naturforsch.* **11a**, 165 (1956).

band gap of silicon. In addition, the dependence of quantum yield on photon energy should be that given in Eq. (20), rather than the usual relation of Eq. (19). This is because the degenerate silicon is a narrowband emitter for those electrons which are excited from the conduction band. The electrons are concentrated in a narrow range of energies near the bottom of the conduction band. For electrons coming from the valence band, it is still a wideband emitter. This is clearly shown in the data of Goodman,[26] reproduced in Fig. 18. This shows the square root of the quantum yield for photoemission of electrons from degenerate n-type silicon into SiO_2 as a function of photon energy. The two thresholds are shown, along with the different dependences of quantum yield on photon energy for the two different regions.

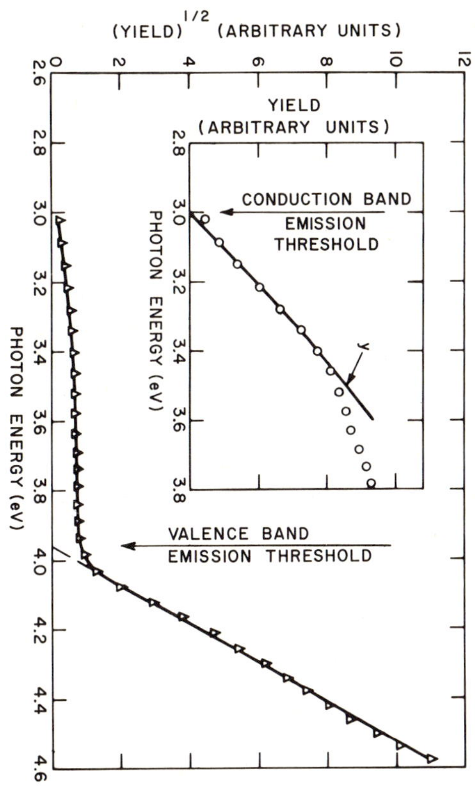

FIG. 18. Square root of quantum yield versus photon energy for photoemission of electrons from degenerate n-type silicon into SiO_2. Insert shows a linear plot of quantum yield versus photon energy for the low-energy part of the curve. (Taken from Goodman.[26])

A similar effect can be produced by a treatment[45] which causes mobile positive ions to drift under an applied voltage and accumulate near the silicon–SiO_2 interface. This produces a strong electric field which bends the bands in the silicon enough to produce *n*-type degeneracy near the surface in silicon, which was not originally degenerate. As a result, after this treatment, there is a new low-energy threshold, similar to that in Fig. 18, which apparently arises from transitions due to electrons in the conduction band.

Thresholds have been reported for photoemission of electrons[16a,42,46] from nine different metals into SiO_2. The values vary widely from one metal to another, ranging from 2.25 eV for magnesium to 4.30 eV for platinum. They are approximately in the same relative order as the metal work functions or electronegativities. Lowering of the barrier by high electric fields has been studied for electrodes of aluminum on 550-Å-thick layers of SiO_2. The zero-field threshold is 3.2 eV, but in fields of 4×10^6 V/cm, it drops to 2.5 eV, an enormous shift. There are significant departures from the field dependence given by Eq. (7). At high fields, there is an additional lowering not predicted by the equation. This additional effect is attributed by Mead *et al.*[16a] to penetration of the electric field into the metal electrode. The differences in barrier heights from one metal to another, as measured by photoemission, were in agreement with the same quantities determined by capacitance measurements.[42]

15. Other Insulators

Extensive use has been made of photoemission methods to study barrier heights and shapes for very thin layers of metal oxides sandwiched between two metal electrodes.[16a,23,47–56] Most work has been done with aluminum oxide, but results are also available[50] for oxides of niobium, tantalum, and titanium. The aluminum oxide is grown by first evaporating a layer of aluminum on glass, then growing a layer of Al_2O_3, typically to a thickness 20–40 Å, by oxidation in air, a discharge, or anodically. A second electrode of

[45] R. Williams, *J. Appl. Phys.* **37**, 1491 (1966).
[46] A. M. Goodman and J. J. O'Neill, Jr., *J. Appl. Phys.* **37**, 3850 (1966).
[47] G. Lucovsky, C. J. Repper, and M. E. Lasser, *Bull. Am. Phys. Soc.* **7**, 399 (1962).
[48] K. L. Chopra and L. C. Bobb, *Proc. IEEE* **51**, 1784 (1963).
[49] D. V. Geppert, *J. Appl. Phys.* **35**, 2151 (1964).
[50] K. L. Chopra, *Solid-State Electron.* **8**, 715 (1965).
[51] K. W. Shepard, *J. Appl. Phys.* **36**, 796 (1965).
[52] A. I. Braunstein, M. Braunstein, and G. S. Picus, *Phys. Rev. Letters* **15**, 956 (1965).
[53] O. L. Nelson and D. E. Anderson, *J. Appl. Phys.* **37**, 77 (1966).
[54] F. L. Schuermeyer and J. A. Crawford, *Appl. Phys. Letters* **9**, 317 (1966).
[55] F. L. Schuermeyer, *J. Appl. Phys.* **37**, 1998 (1966).
[56] G. W. Lewicki and C. A. Mead, *Appl. Phys. Letters* **8**, 98 (1966).

aluminum or another metal is evaporated on top of this. The barriers are not of the same height at the inner and outer faces of the oxide, even if both electrodes are of aluminum. Tunneling currents cannot be neglected, and the barrier height and shape are strongly affected by image forces and the Schottky effect. Light is generally absorbed about equally in both top and bottom electrodes and the applied voltage is small, often less than 1 V. Thus, electrons excited in both metal electrodes come through the oxide simultaneously from opposite directions. The main problem for this system has been to sort out what is going on and to obtain numerical data for the barriers at the inner and outer faces of the oxide.

Figure 19 shows an energy-band diagram for the metal–insulator–metal sandwich with and without a bias voltage applied. We shall describe the

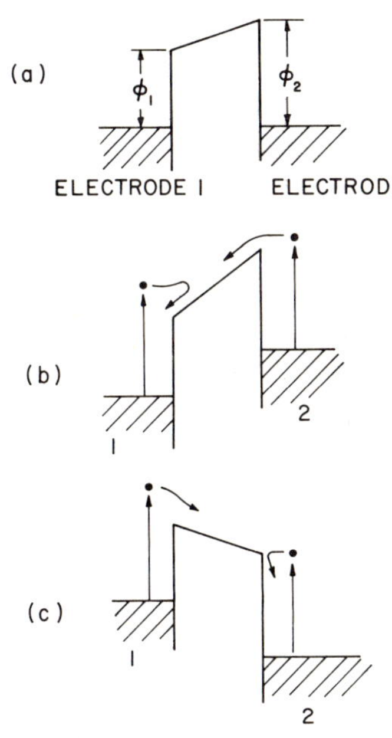

FIG. 19. Energy-band diagram for a thin insulator between two metals, such as Al–Al_2O_3–Al. Different values of ϕ are usually found at the two interfaces, even though the electrodes are of the same metal. (a) No applied voltage; (b) electrode 2 negative; (c) electrode 1 negative.

work of Braunstein *et al.*[23] on the Al–Al_2O_3–Al system, which is representative of much of the work done on systems of this type. An important assumption is that an electron which enters the oxide from the metal loses any energy it may have above the bottom of the conduction band in a distance which is short compared to the thickness of the oxide layer. All electrons in the oxide move at the bottom of the conduction band. Consider first the case shown in Fig. 19(b), where a negative bias is applied to electrode 2, which has the higher barrier. For photon energies greater than ϕ_2, electrons are excited in both electrodes. Those originating in electrode 2 can go over the barrier or tunnel through near the top and pass through the oxide, in which case, they are measured as a current in the external circuit. Electrons originating in electrode 1 may enter the oxide, but lose their excess energy and cannot move against the field. They return to electrode 1, making no contribution to the current. A plot of the square root of the photoresponse versus the photon energy gives ϕ_2. For Fig. 19(c), where electrode 1 is negative, the same arguments show that only the electrons from electrode 1 are collected, and ϕ_1 is measured. With no applied voltage, the built-in field favors electrons from electrode 2, but tunneling near the top of the barrier is more important. The values obtained in this way for ϕ_1 and ϕ_2 must be corrected for Schottky lowering of the barrier, which is significant even with no applied voltage. The result is $\phi_1 = 1.49$ eV and $\phi_2 = 1.92$ eV. This is a slightly oversimplified description of the experimental work presented by Braunstein *et al.*[23] and the reader is referred to the original article for details.

Generally, the barriers in the metal–Al_2O_3 system are high, running up to 2.5 eV or more, and vary from one metal to another. Further detail is provided[52] by measuring the barrier height as a function of applied voltage. Thin-film systems of this general kind present a remarkable opportunity to study insulators in which the significant dimension is about the same as that of a molecule of moderate size and to observe the effects of electric fields up to 10^7 V/cm.

Photoemission of holes from several metals into anthracene has been reported.[18] Threshold values vary from 1.17 eV for gold to 1.97 eV for magnesium. It should be possible, in principle, to measure thresholds for hole and electron photoemission for the same metal, and, thereby, to obtain the band gap. This is a controversial quantity for anthracene, not easily measured by other means. Electron transport is always more difficult to observe in this material than hole transport, but this difficulty will probably be overcome in the future. As discussed above, anthracene should be a narrowband collector, and this leads to unusual features in the photoemission spectral response curve. Some experimental evidence suggesting that this is the case, is shown in Fig. 20, taken from Williams and Dresner.[10] There is a periodic structure in the spectral response curve for photoemission of holes from aluminum into anthracene which appears to be real. This is enhanced by plotting the derivative of the photocurrent with respect to photon energy as a function of photon energy, as also done in Fig. 20. The peaks are com-

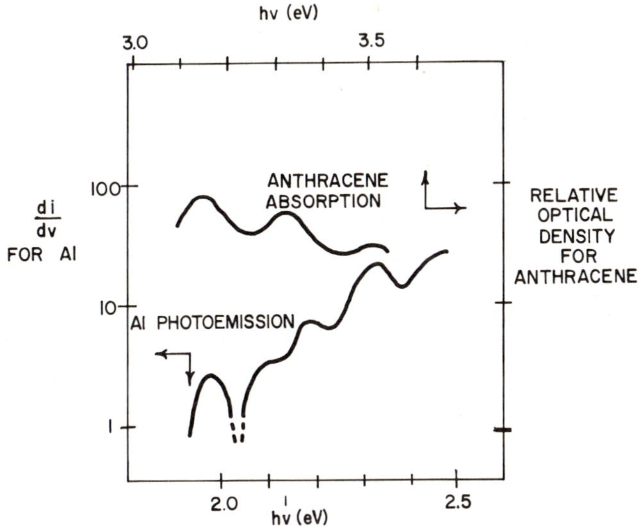

FIG. 20. Plot of the derivative of the photoemission current per incident photon with respect to photon frequency (di/dv) as a function of photon frequency $h\nu$. The data are for photoemission of holes from aluminum into anthracene. The absorption spectrum of crystalline anthracene is shown for comparison. Both curves are for room temperature. (Taken from Williams and Dresner.[18])

parable in width and spacing to those in the optical absorption spectrum of crystalline anthracene for the same temperature. The structure in the optical absorption spectrum is a typical example of the effect of superposing molecular vibrations on an electronic transition, and the peaks are separated by a dominant vibrational frequency. The expected effect on the photoemission spectrum of narrow bands separated in energy by the energy of a dominant vibrational mode is shown in Fig. 12. The derivative of this would be a series of evenly spaced lines corresponding to the peaks in Fig. 20. If the observed structure is, in fact, due to the narrowband properties of anthracene, then the photoemission spectrum will be a valuable tool for investigation of the band structure.

Several groups have been able to observe photoemission of electrons from alkali metals into anthracene.[56a–56d] All have observed structure in the

[56a] J. Dresner, *Phys. Rev. Letters* **21**, 356 (1968).

[56b] A. Many, J. Levinson, and I. Teucher, *Phys. Rev. Letters* **20**, 1161; **21**, 57 (1968).

[56c] G. Vaubel and H. Baessler, *Phys. Status Solidi* **26**, 599 (1968).

[56d] H. Baessler and G. Vaubel, *Solid State Commun.* **6**, 97 (1968).

spectral response curve and attributed this to corresponding features in the energy-band structure of the anthracene. There are still conflicting interpretations of the data, and, for brevity, we describe only the interpretation of Dresner,[56a] whose article is the most recent of those cited. According to this model, the photoemission takes place mainly from a narrow band of surface states (presumed to be anthracene negative ions formed by reaction of the crystal with the alkali metal) into a narrow conduction band lying 0.79 eV above. In addition, there is a broader conduction band 0.3 eV wide lying 0.05 eV above the narrow band, and photoemission of electrons into this band also is observed. This, then, is a case where both emitter and collector are narrow band materials, with additional complications due to the close-lying broad band. Vaubel and Baessler[56e] have combined data on the threshold values for photoemission of electrons and holes into anthracene to obtain the value of 3.72 eV for the band gap. Lakatos and Mort[56f] measured the spectral response curve for photoemission of holes from metals into the organic polymer poly-N-vinyl-carbazole. They obtained threshold values for Au, Cu, and Al, as well as evidence for narrowband structure of the polymer.

IV. Electron and Hole Energy Losses in Metals

16. Review of Results on Au, Ag, Cu, Pd and Al

The transport of hot electrons in metals was very poorly understood until rather recently. A brief review of the status in 1962 was given by Crowell *et al.*[57] It was originally believed that in photoemission into vacuum, the excited electrons could come only from the surface layer of atoms in the metal. Among the earliest convincing evidence that electrons excited deeper in the volume could be photoemitted was the work of Mayer and Thomas[58] on alkali metals. They concluded that *L*, the electron attenuation length, was ~1000 Å for electrons about 2.2 eV above the Fermi level in potassium. Work of Gobeli and Allen[43] on photoemission from silicon into vacuum indicates that the photoemitted electrons are produced between 10 and 100 Å from the surface. It turns out that *L* is strongly dependent on electron energy above the Fermi level. Much larger values of *L* are obtained at low energies. Experiments which measure the escape of electrons into vacuum can study only those electrons having energies greater than the vacuum work function. By using internal photoemission as a method for separating excited electrons

[56e] G. Vaubel and H. Baessler, *Phys. Letters* **27A**, 328 (1968).

[56f] A. I. Lakatos and J. Mort, *Phys. Rev. Letters* **21**, 1444 (1968).

[57] C. R. Crowell, W. G. Spitzer, L. E. Howarth, and E. E. LaBate, *Phys. Rev.* **127**, 2006 (1963).

[58] H. Mayer and H. Thomas, *Z. Physik* **147**, 419 (1957).

from those at the Fermi level, it is possible to cover an important range of energies not otherwise accessible. In addition, the behavior of hot holes can be studied.

By the utilization of an experimental method of Spitzer *et al.*,[6] consistent data on electron and hole attenuation lengths have now been accumulated and interpreted for several metals over a wide range of energies.[57,59–62] This experiment may be understood from Fig. 21. A metal film of thickness x is evaporated onto a semiconductor, which may be either n- or p-type,

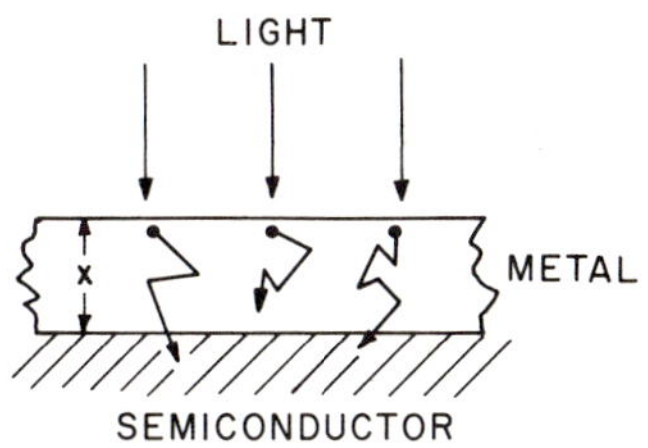

FIG. 21. Illustration of the arrangement used to determine the mean free path for energy loss by hot electrons from photoemission data. Light is absorbed near the upper surface of the metal, and the quantum yield for photoemission is determined as a function of the thickness x of the metal.

depending on whether electrons or holes are to be studied. In practice, x is varied from about 100 to 1000 Å and it is shown (or assumed) that x is much larger than the absorption depth of the light. Thus, excited electrons effectively originate at the outer surface of the metal. For photons which excite electrons to energies just above the barrier, it is found that the quantum yield for photoemission, $Y(x)$, varies according to the equation

$$Y(x) = Ae^{-x/L}. \tag{21}$$

Here, A is a constant, and the equation serves to define the attenuation length L. Measurement of Y as a function of x gives L for electrons or holes having energies near the top of the barrier. Going to a different semiconductor gives a barrier of a different height. Using photons of different energies with a given barrier also gives some control over the energy of the electrons observed.

[59] C. R. Crowell, W. G. Spitzer, and H. G. White, *Appl. Phys. Letters* **1**, 3 (1962).

[60] L. B. Leder, M. E. Lasser, and D. C. Rudolph, *Appl. Phys. Letters* **5**, 215 (1964).

[61] R. Stuart, F. Wooten, and W. E. Spicer, *Phys. Rev.* **135**, 495 (1964).

[62] S. M. Sze, J. L. Moll, and T. Sugano, *Solid-State Electron.* **7**, 509 (1964).

The metal most thoroughly studied has been gold. Here, L for electrons is found[57,62] to vary from 700 Å at 1 eV to 200 Å at 2 eV and 70 Å at 5 eV. (Smaller values of L were reported by Leder *et al.*[60]) Other values for electrons are: Ag, $L = 440$ Å; Cu, 50–200 Å; Pd, 170 Å; all near 1 eV. For holes[59,60]: Au, $L = 550$ Å; Al, $L = 50$ Å; both near 1 eV. Data have been analyzed by Stuart *et al.*[61] using Monte Carlo calculations to determine the values of electron–electron scattering length l_e and electron–phonon scattering length l_p that are required to fit the data. For the case of gold, with electrons 0.9 eV above the Fermi level, $l_e > 4000$ Å and $l_p \approx 400$ Å. Thus, the electron–electron scattering length is much greater than the attenuation length. A consistent body of data and interpretation have now been built up on the basis of the internal photoemission technique.

17. Hot-Electron Devices and Mean Free Paths

The rather long mean free paths for transport of hot electrons in metals indicated by the above measurements have suggested several devices. A triode consisting of a thin layer of metal sandwiched between two semiconductors has been proposed by Rose.[63] This would consist of two Schottky barriers, back to back, joined to a metal layer of order 100 Å thick. Separate connections to the metal and the two n-type semiconductors would be the three connections of the triode in which the metal layer would be the analog of the grid. Forward bias between one semiconductor and the metal would inject hot electrons into the metal. These would migrate through the metal and be collected by the other semiconductor, at reverse bias with respect to the metal. Possible advantages for high-frequency operation were described by Rose. This structure has been extensively examined by Crowell and Sze[13] with special attention given to the system with gold sandwiched between two crystals of n-type silicon. As an example of the collection efficiency to be expected, they find that at room temperature, 68% of the electrons injected by the emitter will be received by the collector.

It appears likely that hot electrons could also be injected into vacuum by replacing the collector in the above structure by a fractional monolayer of cesium, which would lower the work function of the metal on this face to about 1.4 eV. Electrons injected from a Schottky barrier with $\phi_B > 1.4$ eV could then traverse the metal layer and escape into vacuum. An experiment which successfully demonstrated part of the requirements for a cold cathode of this type has been described by Lifshitz and Musatov.[64] Actual emission into vacuum has recently been achieved by Williams and Wronski.[64a]

[63] A. Rose, U.S. Patent 3250966 (10 May 1966).

[64] T. M. Lifshitz and A. L. Musatov, *Zh. Eksperim. i Teor. Fiz. Pis'ma v Redaktsiyu* **4**, 295 (1966) [*English Transl.*: *JETP Letters* **4**, 199 (1966)].

[64a] R. Williams and C. Wronski, *Appl. Phys. Letters* **13**, 231 (1963).

Use of photoemission over low metal–semiconductor barriers would appear to be a possible source for a fast infrared detector. The general result that L increases with decreasing energy above Fermi level means that one might anticipate higher quantum efficiencies than in devices based on vacuum photoemission, where the excited electrons are necessarily more than 1 eV above the Fermi level. There appear to be no published references to this application.

V. Transport and Trapping in Insulators

18. Trapping in SiO_2

a. Electron Trapping

Injection of electrons into SiO_2 by photoemission allows the study of transport and trapping properties. The results reported by Williams[17] serve to illustrate the kind of information which can be obtained. Considering that the material is vitreous rather than crystalline, the trapping effects were remarkably small. With an applied field of 10^6 V/cm, the *schubweg* was much greater than the layer thickness, which was about 1 μm. Some trapping was evident, however, and the trapped carriers could later be released by visible light not energetic enough to excite electrons into the oxide from the electrode. Since the contact was completely blocking for the injection of electrons under these conditions, the current due to motion of carriers excited by the visible light died out with time. In the dark, the trapped electrons were stable against thermal ionization for hours at room temperature. This indicates they they were at energy levels at least 1 eV below the conduction band edge.

The current due to trapped carriers excited by visible light is shown in Fig. 22. This is a typical example of a polarizing current with blocking contacts. The lower part of the figure shows the stability of the trapped carriers against thermal ionization. The area under the curves gives the concentration of filled traps. When all are filled, this is the total concentration of traps, and was about 3×10^{14} cm^{-3} for this oxide. This is a very low trap concentration, and indicates the high degree of material perfection which can be achieved in SiO_2. Highly purified single-crystal materials often have much higher trap concentrations. By measuring the spectral response of the initial current due to the emptying of traps, an estimate of their depth is obtained. The spectral response is shown in Fig. 23 and indicates that the traps lie at energies 2 eV or more below the conduction band edge. Measurements of the kinetics of trapping gave the capture cross section S of the traps for electrons. This was 1.3×10^{-12} cm^2 and indicates[8] that the trap is a positively charged coulomb attractive center. Use of an argument due to

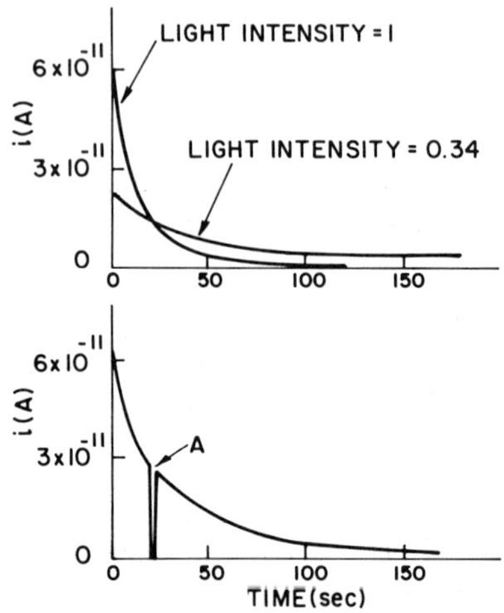

FIG. 22. Current due to excitation by visible light of trapped carriers in SiO_2 which were introduced into the oxide by photoemission from the silicon electrode. The photoemission requires ultraviolet light. The upper curve shows how the current decreases with time as the fixed number of trapped carriers are excited and move out of the oxide. The lower curve shows the thermal stability of trapped carriers at room temperature. At point *A*, the illumination was stopped for 90 min and then resumed. The voltage was left on during this time. Resumption of the current without change in magnitude indicates that no carriers were thermally excited out of the traps during this interval. (Taken from Williams.[17])

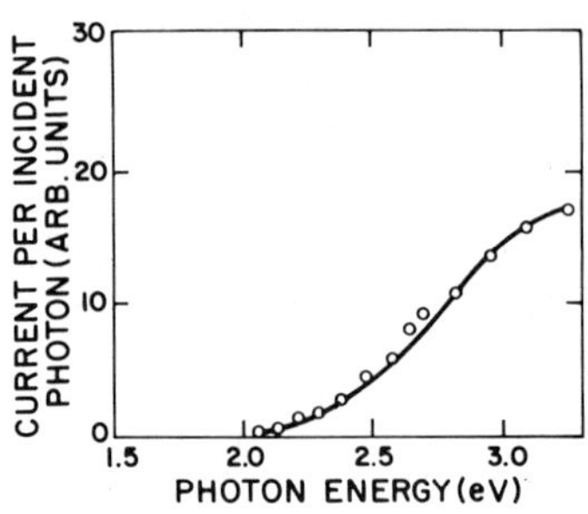

FIG. 23. Excitation spectrum for the trapped electrons in SiO_2. (Taken from Williams.[17])

Crandall[65] gave an estimate of the electron mobility in the oxide from the magnitude of *S*. The value of μ obtained by this argument depends on whether the trapping center is singly or doubly charged, which was not known in this case. The values are $\mu = 34\ \text{cm}^2/\text{V-sec}$ assuming a singly charged center

[65] R. S. Crandall, *Phys. Rev.* **138**, 1242 (1965).

and $\mu = 17\ \text{cm}^2/\text{V-sec}$ assuming a doubly charged center. These are surprisingly large for a noncrystalline material. This has been confirmed by direct measurements to be described below. Further information on trapping was obtained by Goodman.[16] By measuring photocurrent versus applied voltage, the product of mobility μ_n and lifetime τ_n was obtained. The order of magnitude was $\mu_n\tau_n \approx 10^{-9}\ \text{cm}^2/\text{V}$. This, combined with the estimate for mobility given above, shows that the electrons are free for about 10^{-10} sec before being trapped. Similar data were reported[40] for holes introduced into the oxide by photoemission. Here, the product $\mu_p\tau_p$ was less than $10^{-10}\ \text{cm}^2/\text{V}$, showing that, in a given field, holes are trapped within a much shorter distance from their point of origin than are electrons.

b. Hole Trapping and Radiation Damage

The fact that holes are trapped much more readily than electrons in SiO_2 has a very important effect on the behavior of devices containing this material in the presence of high-energy radiation. Zaininger[66,67] measured changes in metal–SiO_2–silicon (MOS) capacitors exposed to high-energy electrons in the range 0.1–1 MeV. The results are also applicable to MOS transistors. Changes in the capacitance produced by the radiation were found to be due to accumulation of positive space charge in the oxide near the SiO_2–Si interface. The devices were remarkably sensitive to the high-energy electron bombardment. Measurable effects were found for values of total electron flux well under $10^{12}\ \text{cm}^{-2}$. The capacitor could be restored to its original condition before electron bombardment by illumination with ultraviolet light. Zaininger proposed that the initial step in the radiation damage was the creation of hole–electron pairs in the oxide by the high-energy electrons. Following this, the free electron was able to escape into the silicon, but the hole was quickly trapped. The result was positive space charge in the oxide. This has a drastic effect on device characteristics. Subsequent illumination with UV light brings electrons back into the oxide from the silicon electrode and restores the oxide to its original condition. It should be noted that if neither holes nor electrons were trapped, this kind of damage would not arise. Also, if both holes and electrons were trapped equally, it would not arise.

19. Mobility in SiO_2

A method has been developed by Goodman[68] for measuring the Hall mobility of electrons injected into SiO_2 by photoemission from a silicon substrate. This method promises to be of considerable usefulness for deter-

[66] K. H. Zaininger, *Appl. Phys. Letters* **8**, 140 (1966).

[67] K. H. Zaininger, *IEEE Trans. Nucl. Sci.* **NS-13**, 237 (1966).

[68] A. M. Goodman, *Phys. Rev.* **164**, 1145 (1967).

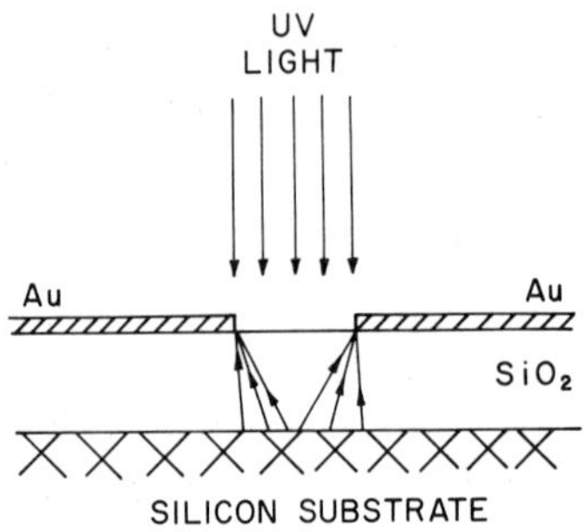

FIG. 24. Electrode geometry for Hall measurement of electrons in SiO_2. The electrons are introduced into the oxide by photoemission from the silicon substrate. The magnetic field is applied normal to the plane of the page. (Taken from Goodman.[68])

mining mobilities in insulators. The principle may be understood from the electrode configuration shown in Fig. 24. The SiO_2 layers were from 3.5 to 6.6 μm thick on substrates of single-crystal silicon. On the outer face of the oxide were evaporated two metal electrodes with a spacing comparable to the thickness of the oxide. On illumination as shown, electrons are photoemitted from the silicon and travel through the oxide along the electric field lines, indicated by the arrows. In the absence of a magnetic field, this is symmetrical with respect to the two electrodes, and equal numbers of electrons go to each. A magnetic field is applied normal to the plane of the page. This distorts the electron trajectories so that more now go to one electrode than to the other. This gives rise to a potential difference and current flow between the electrodes. Either current flow or potential difference can be measured and related to the Hall mobility of the electrons in the oxide. Using this three-electrode method, the mobility of electrons was determined. The average value was $\mu_e = 29\ \text{cm}^2/\text{V-sec}$. This value is in good agreement with the estimate made from the cross section for trapping by a coulomb attractive center.[17] This geometry for the Hall measurement is especially well suited to insulators.

20. Other Insulators

The mean free path for energy attenuation by hot electrons in thin Al_2O_3 films was studied by Braunstein *et al.*[69] They used oxide layers 40 Å thick, sandwiched between two aluminum electrodes. In this case, on illumination, electrons are excited in both electrodes and pass into the oxide. The current was analyzed in detail using a model in which one of the parameters was the energy attenuation length λ for the electrons. They assumed it to have the

[69] A. Braunstein, M. Braunstein, and G. S. Picus, *Phys. Rev. Letters* **15**, 956 (1965).

form $\lambda(E) = \lambda_0/(E - E_c)$, where $(E - E_c)$ is the excess energy of the electron above the bottom of conduction band, in eV. Their data could be fit with the value $\lambda_0 = 76$ Å. Thus, $\lambda(E) = 76/(E - E_c)$ in angstroms. In this way, detailed information on transport and energy losses of hot electrons is obtained by analysis of photoemission experiments.

CHAPTER 3

Current Filament Formation

Allen M. Barnett

I. Introduction

The double-injection device, which exhibits current-controlled negative resistance, will be discussed in this chapter. Specifically, studies that show evidence of two-carrier space-charge-limited filaments will be reviewed. A current filament is defined as the nonuniform distribution of current density in a sample with a uniform electric field, as shown in Fig. 1.

An attempt has been made to trace the history of filamentary double-injection devices. This historical review is complicated by the fact that current-controlled negative resistance in double-injection devices was reported as early as 1954,[1] and was followed by numerous analytical and experimental

[1] W. W. Tyler, *Phys. Rev.* **96**, 226 (1954).

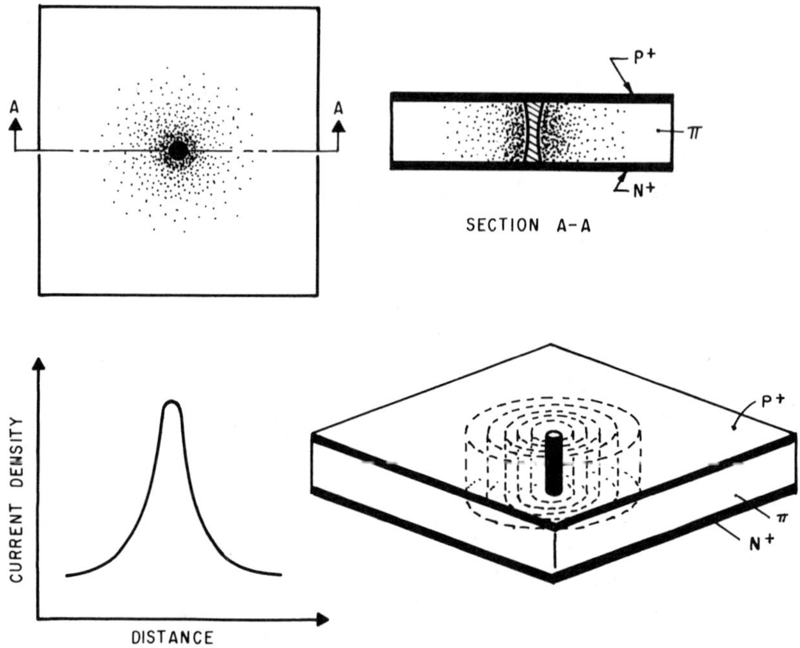

FIG. 1. Drawing of a current filament in a P^+–π–N^+ device.

studies. The results of the principal analytical study of double-injection negative resistance appeared in 1962.[2] The first experimental observation and study of space-charge-limited current filaments was reported in 1966.[3]

In this review, this early work will be described by defining as filamentary all *P–I–N* devices that exhibit the voltage–current characteristics that are indigenous to the filamentary-device characteristic.

The voltage–current characteristic of filamentary double-injection devices is unique. The characteristic curve is shown in Fig. 2 (the voltage drop across the junction is ignored). At low levels the characteristic is ohmic, followed by a region of one-carrier space-charge-limited current, with $I \propto V^2$. The voltage increases to a threshold voltage V_{TH}, and increases in current beyond this point mark the onset of negative resistance and two-carrier space-charge-limited current. The voltage then decreases along a path described by the load line to a minimum voltage, V_M. Now, as the current is increased, the voltage remains nearly constant until it reaches the region of a high-current power law, where $I \propto V^{1.5-2.0}$. The filament has been analytically

[2] M. A. Lampert, *Phys. Rev.* **125**, 126 (1962).

[3] A: M. Barnett and A. G. Milnes, *J. Appl. Phys.* **37**, 4215 (1966).

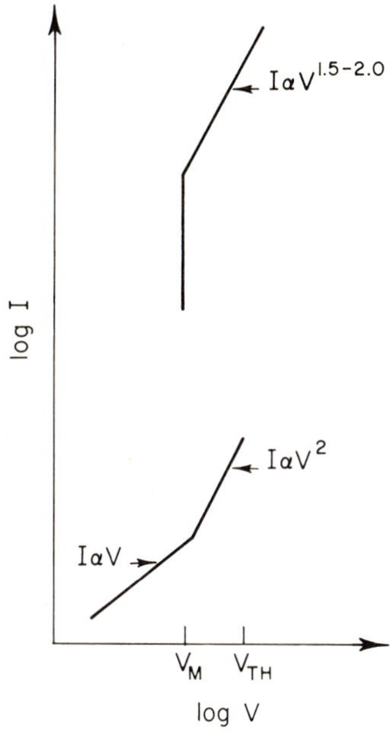

FIG. 2. Characteristic curve of filamentary double-injection device.

described in the negative resistance region.[4,5] Stable filaments have been observed in the region where current increases at constant voltage[6] and in the high-current power-law region.[3]

The theory of Lampert,[2] which describes the voltage–current characteristic below the threshold voltage, assumes a uniform current distribution throughout. Therefore, beyond the threshold voltage, this theory predicts a high-current square-law region, $I \propto V^2$, governed by two-carrier space-charge-limited current equations. This uniform-current-distribution characteristic curve has been observed for long *P–I–N* devices.[7]

In this chapter, a model for postbreakdown filaments is developed. The simplified Lampert theory is reviewed, and the observations that led to the development of the filamentary model will be analyzed. The early experimental and analytical filamentary work will also be analyzed. In addition, a new model, which analytically predicts filaments using the starting conditions of the threshold voltage predicted by the simplified Lampert theory, will be presented.

[4] B. K. Ridley, *Proc. Phys. Soc.* (*London*) **82**, 954 (1963).

[5] H. C. Bowers and A. M. Barnett. To be published; also see Sect. 6.

[6] A. M. Barnett and H. A. Jensen, *Appl. Phys. Letters* **12**, 341 (1968).

[7] J. W. Mayer, R. Baron, and O. J. Marsh, *Appl. Phys. Letters* **6**, 38 (1965).

1. Historical Review of Filamentary Device Studies

The first report of a current-controlled negative-resistance characteristic in filamentary devices was made by Tyler in 1954.[1] His studies were made on iron-doped germanium at liquid-nitrogen temperatures. Figure 3 shows Tyler's prebreakdown and postbreakdown characteristic curves. The same effect was observed by Lebedev *et al.*[8] in 1957 in gold-doped germanium

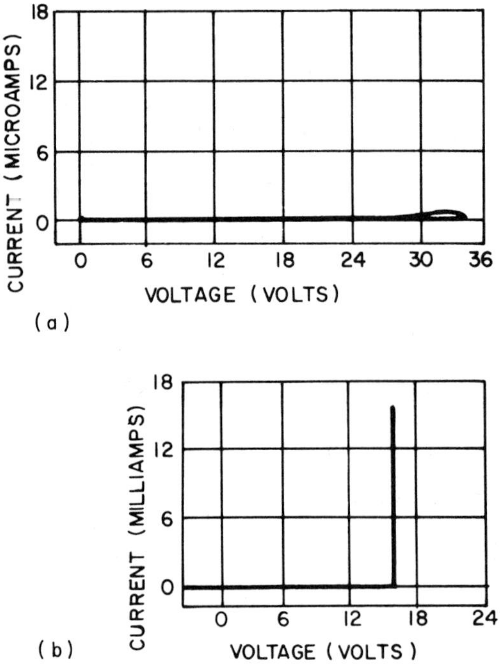

Fig. 3. (a) Prebreakdown and (b) postbreakdown oscillograms for double injection in iron-doped germanium. (After Tyler.[1])

diodes at temperatures ranging from 77°K to 180°K. Figure 4 shows the negative-resistance curve that they observed, which is typical for the devices studied in this chapter. Later, Stafeev[9] described the negative-resistance effect by the modulation of the diffusion length by injected carriers.

[8] A. A. Lebedev, V. I. Stafeev, and V. M. Tuchkevich, *Zh. Tekhn. Fiz.* **26**, 2131 (1956) [*English Transl.*: *Soviet Phys. Tech. Phys.* **1**, 2071 (1957)].

[9] V. I. Stafeev, *Fiz. Tverd. Tela.* **1**, 841 (1959) [*English Transl.*: *Soviet Phys.-Solid State* **1**, 763 (1959)].

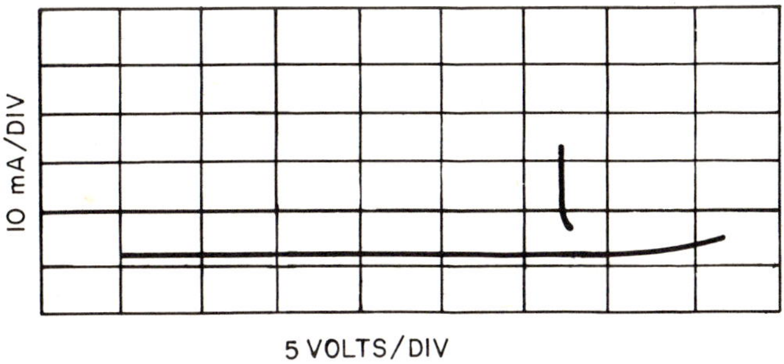

FIG. 4. Typical negative-resistance oscillogram for double-injection device. (After Lebedev *et al.*[8])

This negative-resistance effect was next reported in plastically deformed germanium *p–n* junctions in 1961[10] and 1962.[11]

Melngailis and Rediker[12] observed negative resistance in InSb $N^+–P–P^+$ diodes. They observed that the breakdown point seemed to depend on the density of injected carriers and not on the critical value of current density or electric field. They also reported large changes due to magnetic field effects and developed a series of devices based on these effects.[13] Also, in 1962, Holonyak *et al.*[14,15] observed double injection with negative resistance in GaAs, Si, and Ge. All of these devices operated at room temperature. Holonyak also reported that the threshold voltage for these devices was reduced in the presence of light.

Lampert[2] developed a model for the negative-resistance effect in 1962. Subsequently, he reviewed injection currents in insulators[16,16a]; his most recent work appears in Chapter 1 of this volume.

Pearson *et al.*[17,17a] reported current controlled negative resistance for a series of semiconductor glasses including As-Te-I, As-Tl-Se, and V-P-O. Recently he has described the operation of these devices in the high con-

[10] O. Ryuzan, *J. Phys. Soc. Japan* **16**, 2177 (1961).

[11] L. J. Van Ruyven and W. H. Th. Adriaens, *Phys. Letters* **3**, 109 (1962).

[12] I. Melngailis and R. H. Rediker, *J. Appl. Phys.* **33**, 1892 (1962).

[13] I. Melngailis and R. H. Rediker, *Proc. IRE* **50**, 2428 (1962).

[14] N. Holonyak, Jr., *Proc. IRE* **50**, 2421 (1962).

[15] N. Holonyak, Jr., S. W. Ing, Jr., R. C. Thomas, and S. F. Bevacqua, *Phys. Rev. Letters* **8**, 426 (1962).

[16] M. A. Lampert, *Proc. IRE* **50**, 1781 (1962).

[16a] M. A. Lampert, *Rept. Progr. Phys.* **27**, 329 (1964).

[17] A. D. Pearson, J. F. Dewald, W. R. Northover, and W. F. Peck, Jr., "Advances in Glass Technology," p. 375. Plenum Press, New York, 1962.

[17a] J. F. Dewald, A. D. Pearson, W. R. Northover, and W. F. Peck, Jr., *J. Electrochem. Soc.* **109**, 243C (1962).

ductance region as being caused by a current filament.[17b] (Most of the effects observed in glass-type devices are not included in this review.)

An experimental study of current filaments was reported by Melngailis and Milnes[18] in 1962. These filaments were not due to two-carrier space-charge-limited current, but were caused by impact ionization of shallow impurities in compensated germanium at 4.2°K. A current–controlled negative resistance is also a characteristic of this effect. (A review of impurity impact ionization is not included.) The filaments were observed by detecting the transverse conductance in a sample with the electric field applied in the longitudinal direction. A *p–n* junction was built around a cylindrical sample and as the junction was reverse-biased, the depletion layer was extended into the filament. The model for this filament was conceived as having a uniform current density, because the free carriers were of only one sign and were bound to the ionized impurity atoms by coulombic attraction. The smallest filament observed had a diameter of 10^{-3} cm at 40 μA, which is consistent with recent low-level studies of double-injection current filaments in GaAs.[6] The observation of current filaments associated with impurity impact ionization was noted during the course of the development of a generalized theory of filaments for current-controlled negative-resistance devices by Ridley[4] in 1963. This theory, which is based on thermodynamic considerations, is discussed in greater detail in Section 3.

In 1963, Keating[19] extended the model of Lampert to include the effect of partially filled recombination centers. He noted that one must show extra care in the interpretation of the experimental threshold voltage V_{TH}. Holonyak and Bevacqua[20] reported oscillations in the prebreakdown and post-breakdown regions, and a secondary postbreakdown region in cobalt-compensated silicon and GaAs. Ing and Jensen[21] developed a light-emitting diode on the same GaAs substrate as the negative-resistance device. They were able to lower the device threshold voltage with the application of light from the light-emitting diode.

Among the novel achievements made in 1964 was the observation of pronounced negative resistance in CdS at 4.2°K by Litton and Reynolds.[22] They considered the negative-resistance effect that they observed as most nearly following the double-injection model of Lampert.[2] Also, in 1964, Weiser and Levitt[23] reported negative resistance effects in GaAs *P–I–N*

[17b] A. D. Pearson and C. E. Miller, *Appl. Phys. Letters* **14**, 280 (1969).

[18] I. Melngailis and A. G. Milnes, *J. Appl. Phys.* **33**, 995 (1962).

[19] P. N. Keating, *Solid State Commun.* **1**, 210 (1963).

[20] N. Holonyak, Jr. and S. F. Bevacqua, *Appl. Phys. Letters* **2**, 71 (1963).

[21] S. W. Ing, Jr. and H. A. Jensen, *Proc. IEEE* **51**, 852 (1963).

[22] C. W. Litton and D. C. Reynolds, *Phys. Rev.* **133**, A536 (1964).

[23] K. Weiser and R. S. Levitt, *J. Appl. Phys.* **35**, 2431 (1964).

devices. The observation of a reduction in the threshold voltage in the presence of external light led to the development of a new theoretical model by Dumke.[24] Dumke's model was based on the mechanism of reabsorption of recombination radiation and is discussed in Section 3. Ing *et al.*[25] also reported negative resistance in GaAs. They observed a small spot of light through the p^+ junction, but were unable to separate the effect of this small spot of light from the limited geometry caused by the alloyed n^+ junction.

Keating[26] extended his earlier work to consider arbitrary thermal equilibrium, reoccupation of recombination centers, and shallow trapping of both carriers. At this time, the studies of gold-doped germanium at reduced temperatures were also extended.[27] Ashley and Milnes[28] discussed, analytically and experimentally, the effects of space charge trapped on the recombination centers for gold-doped germanium at 77°K.

In 1965, Shohno[29] observed the negative resistance effect in the forward direction of high resistivity (2×10^4 ohm-cm) silicon at room temperature. He also observed current oscillations, in the forward direction, of longer (0.2–0.6 mm) devices.[30] Mayer *et al.* wrote several experimental and theoretical studies of double injection in silicon; this work is discussed in Chapter 4 of this volume. Only one of their papers reports the observation of negative resistance, which occurred in long (2.0–8.1 mm) diodes at −40°C to lower temperatures.[7] This report of double injection is unusual because there is no indication of filamentary conduction, and the devices correspond to the early Lampert model. This work will be discussed in Section 12. Weiser[31] extended the experimental study of GaAs *P–I–N* devices. Wagener and Milnes[32] studied double injection in deep-level impurity-doped silicon diodes of varying length as a function of temperature.

Additional studies of negative-resistance effects in silicon devices at reduced temperatures were reported in 1966.[33,34] Deshpande[35] also observed double injection in long silicon structures. Negative resistance was observed

[24] W. P. Dumke, *in* "Radiative Recombination in Semiconductors" (*7th Intern. Conf.*), p. 611. Dunod, Paris and Academic Press, New York, 1965.

[25] S. W. Ing, Jr., H. A. Jensen, and B. Stern, *Appl. Phys. Letters* **4**, 162 (1964).

[26] P. N. Keating, *Phys. Rev.* **135**, A1407 (1964).

[27] V. P. Sondaevskii and V. I. Stafeev, *Fiz. Tverd. Tela* **6**, 80 (1964) [*English Transl.*: *Soviet Phys.–Solid State* **6**, 63 (1964)].

[28] K. L. Ashley and A. G. Milnes, *J. Appl. Phys.* **35**, 369 (1964).

[29] K. Shohno, *Japan. J. Appl. Phys.* **4**, 114 (1965).

[30] K. Shohno, *Japan. J. Appl. Phys.* **4**, 699 (1965).

[31] K. Weiser, *IBM J. Res. Develop.* **9**, 315 (1965).

[32] J. L. Wagener and A. G. Milnes, *Solid-State Electron.* **8**, 495 (1965).

[33] K. Shohno, *Japan. J. Appl. Phys.* **5**, 358 (1966).

[34] J. M. Brown and A. G. Jordan, *J. Appl. Phys.* **37**, 337 (1966).

[35] R. Y. Deshpande, *Solid-State Electron.* **9**, 265 (1966).

at 100°K, but the voltage–current characteristic curve shows no sign of filaments. Pilkuhn and Rupprecht[36] reported negative resistance for Cr-doped GaAs *P–I–N* devices at room temperature. Shohno[37] was able to develop third-terminal gate control for silicon devices.

Crowder *et al.*[38] reported negative resistance in ZnTe. They consider their devices to be exhibiting a combination of the avalanche injection mechanism and, possibly, the photoconductivity model of Dumke.[24] It is interesting to note that, at 77°K, they observed that the light emission "from underneath the negative electrode comes from a small spot, so that local current densities may be orders of magnitude higher." This report of a bright spot of light, in addition to the current-controlled negative-resistance characteristic curve, is taken as indirect evidence for the existence of current filaments.

The first analytical reports of the observation of current filaments in semi-insulating silicon were published in 1966. Barnett[39] and Barnett and Milnes[3,40] examined these current filaments by photographing them with their recombination radiation. A model was developed that predicted the current rise, at nearly constant voltage, in the postbreakdown region and in the high-current power-law region. This model also discussed the radial distribution of current density in the filament. Photographs of these filaments were taken in the high-current power-law region, and the analytical and experimental results were in relatively good agreement.[40a]

In 1967, negative resistance was reported in GaAs devices[41] formed by diffusing zinc into *n*-type material. This author believes that the insulator region was formed during the zinc diffusion. These experimental results will be discussed in a later section. Oscillations were again reported in silicon devices,[42,43] as was negative resistance in silicon and germanium devices irradiated with 3-MeV electrons, neutrons, and gamma radiation.[44] Negative resistance was also observed at room temperature for germanium-[45] and gold-doped silicon devices.[46] The gold-doped silicon work mentions the

[36] M. Pilkuhn and H. Rupprecht, *J. Appl. Phys.* **37**, 3621 (1966).

[37] K. Shohno, *Japan. J. Appl. Phys.* **5**, 414 (1966).

[38] B. L. Crowder, F. F. Morehead, and P. R. Wagner, *Appl. Phys. Letters* **8**, 148 (1966).

[39] A. M. Barnett, Ph.D. Thesis, Carnegie Inst. of Technol., Pittsburgh, Pennsylvania, 1966.

[40] A. M. Barnett and A. G. Milnes, *IEEE Trans. Electron Devices* **ED-13**, 816 (1966).

[40a] This work will be discussed in various sections throughout this chapter.

[41] D. Meyerhofer, A. S. Keizer, and H. Nelson, *J. Appl. Phys.* **38**, 111 (1967).

[42] B. G. Streetman, M. M. Blouke, and N. Holonyak, Jr., *Appl. Phys. Letters* **11**, 200 (1967).

[43] J. S. Moore, N. Holonyak, Jr., and M. D. Sirkis, *Solid-State Electron.* **10**, 823 (1967).

[44] F. M. Berkovskii and R. S. Kasymova, *Fiz. Tverd. Tela* **8**, 1985 (1966) [*English Transl.*: *Soviet Phys.–Solid State* **8**, 1580 (1967)].

[45] F. Driedonks, R. J. J. Zijlstra, and C. Th. J. Alkemade, *Appl. Phys. Letters* **11**, 318 (1967).

[46] I. V. Varlomov, I. A. Sondaevskaya, and V. P. Sondaevskii, *Fiz. Tekhn. Poluprov.* **1**, 452 (1967) [*English Transl.*: *Soviet Phys.–Semicond.* **1**, 375 (1967)].

possibility of a current filament broadening with increasing current until it is limited by the device geometry, similar to the impurity impact ionization work of Melngailis and Milnes.[18] Gerhard and Jensen[47] reported negative resistance and light ouptut in $GaAs_xP_{1-x}$. The small spot of light observed in these studies was attributed to the possibility of filament formation. Aven and Garwacki[48] reported light emission and negative resistance in $ZnSe_xTe_{1-x}$ at 77°K. The voltage–current characteristic looks filamentary below 60 mA. Above 60 mA, the voltage increases as the current remains constant.

The current-controlled negative-resistance effect has also been observed in evaporated silicon films.[49] Recombination radiation photographs revealed small spots of light in the postbreakdown region which were attributed to current filaments. The observation of the spot of light led the authors to describe the negative-resistance effect as double-injection breakdown. The observation of current filaments in GaAs was also reported in 1967.[50] These GaAs filaments were observed in the low-current postbreakdown region, where current increases with constant voltage, and it was also the first report of filaments in a direct-energy-gap material.

Silicon devices with negative resistance were observed with threshold voltages as high as 600 V: the postbreakdown region was described as current proportional to V^8 from 1 to 50 mA.[51] This characteristic curve resembles the filamentary double-injection device curve. Schibli and Milnes[52] reported negative resistance in semi-insulating silicon devices, at liquid nitrogen temperatures, containing a density gradient of deep impurities. Computed analytical solutions and experimental results show that the injection of holes, majority carriers, at the contact near the lower concentration of deep acceptors causes a higher threshold voltage than in devices with the injecting contact on the opposite face. Epstein[53] also observed negative resistance in GaP devices. These devices emitted light: the light wavelength is dominated by impurity trapping in the prebreakdown region. Band-to-band recombination radiation was observed in the postbreakdown region. These studies will be examined in greater detail in Section 4.

At the time of this writing, several new papers have appeared. Double injection in InSb has been described,[54] and a model describing the negative-

[47] G. C. Gerhard and H. A. Jensen, *Appl. Phys. Letters* **10**, 333 (1967).

[48] M. Aven and W. Garwacki, *J. Appl. Phys.* **38**, 2302 (1967).

[49] M. Braunstein, A. I. Braunstein, and R. Zuleeg, *Appl. Phys. Letters* **10**, 313 (1967).

[50] A. M. Barnett and H. A. Jensen, *Bull. Am. Phys. Soc.* **12**, 405 (1967).

[51] J. Afanasjevs and J. E. Nordman, *Proc. IEEE* **55**, 2043 (1967).

[52] E. Schibli and A. G. Milnes, *Solid-State Electron.* **10**, 97 (1967).

[53] A. S. Epstein, *Trans. AIME* **239**, 370 (1967).

[54] J. E. Nordman and H. Kvinlaug, *J. Appl. Phys.* **39**, 7 (1968).

resistance mechanism as being associated with an increase in ambipolar mobility has appeared.[55]

Ovshinsky created a stir when reports appeared in the news media of his paper[56] concerned with switching phenomena in glass films. The mechanism controlling these devices is not clear, but the devices are similar to those of Pearson.[17–17b] In many ways, some forms of these glass devices resemble single-crystal semiconductors. They also correspond to the evaporated polycrystalline silicon devices reported by Braunstein *et al.*[49] Negative-resistance switching devices which remain in the addressed state (called nonvolatile memory), and resemble double-injection-type devices, are being studied by Richardson[57] for evaporated GaAs.

A more detailed report on filaments in GaAs appeared in 1968.[6] An application of these GaAs devices to a light-emitting diode display[58] and a more detailed analysis of the basic device[59] have been reported.

2. Simplified Lampert Theory

A model based on the simplified double-injection theory of Lampert[16] will be briefly developed here as a starting point for the filamentary model. The prebreakdown region and minimum voltage V_M are based on Lampert's "semiconductor" model, which is reduced in the postbreakdown region to his "insulator" model. The result is a "semi-insulator" model. A more detailed explanation of Lampert's recent work appears in Chapter 1 this volume.

The negative-resistance devices discussed in this section are the result of two-carrier injection into semi-insulators, in which the lifetime of the carrier of one sign is significantly longer than that of the carrier of the opposite sign. The most common example of a semi-insulator possessing these characteristics is a semiconductor with a deep-lying trap level. In this case, the concentration of thermal carriers of one sign is very different from the concentration of carriers of the opposite sign. In addition, a significant percentage of the thermal carriers are trapped, yielding high-resistivity material. Figure 5 shows the energy-band diagram of a semi-insulator with a deep trap level and two injecting contacts, similar to the double-injection devices to be studied in this section. In our notation, E_c and E_v represent, respectively, the conduction and valence band energy levels, E_f is

[55] V. V. Osipov and V. I. Stafeev, *Fiz. Tekhn. Poluprov.* **1**, 1795 (1967) [*English Transl.*: *Soviet Phys.–Semicond.* **1**, 1486 (1968)].

[56] S. R. Ovshinsky, *Phys. Rev. Letters* **21**, 1450 (1968).

[57] J. R. Richardson, *J. Vacuum Sci. Technol.* **5**, 169 (1968).

[58] A. M. Barnett, H. A. Jensen, and V. F. Meikleham, *Govt. Microelectron. Appl. Conf. Dig. Papers* **1**, 154 (1968).

[59] A. M. Barnett, H. A. Jensen, V. F. Meikleham, and H. C. Bowers, *in* "Gallium Arsenide" (Proc. 2nd Intern. Symposium, Dallas, Texas, 1968), p. 136. Inst. of Phys. and Phys. Soc., London, 1969.

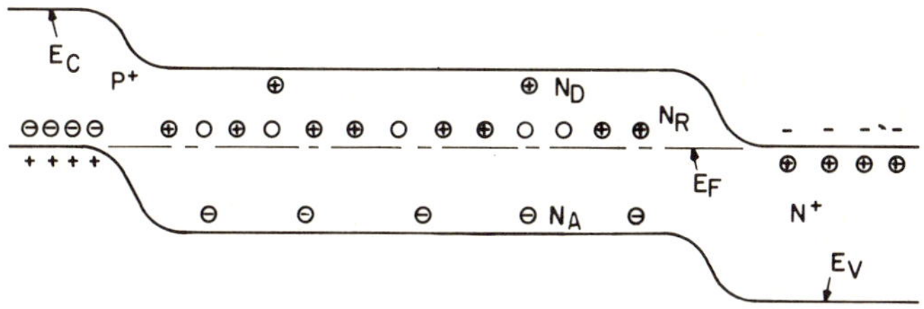

FIG. 5. Energy-band diagram for a double-injection device with a deep-level impurity.

the Fermi level, n_0 and p_0 are the thermal equilibrium concentrations of electrons and holes, respectively, N_D and N_A are the concentrations of shallow donors and acceptors, respectively, and N_R is the concentration of deep impurity-recombination centers.

The simplified theory to be reviewed here considers the space charge of the injected carriers and ignores the space-charge effects of the recombination centers. Considering the model of Fig. 5, the recombination center will be considered to be a deep donor. Therefore, for this device, $n_0 > p_0$, which is the consequence of $N_R > N_A > N_D$. This deep-donor model is the case for oxygen-doped GaAs. Lampert[2] considers the more common deep-acceptor model.

Ignoring the voltage drop across the junction, the low current is directly proportional to the voltage. The semi-insulator is acting as a resistor, according to the relationship

$$J = en_0\mu_n(V/L), \tag{1}$$

where J is the current density, e is the electronic charge of a single carrier, μ_n is the electron mobility, V is the voltage applied to the semi-insulator, and L is the device length. As the hole (minority carrier) lifetime τ_p becomes equal to the transit time, the bulk current conduction is changed from an ohmic to a one-carrier space-charge-limited current mechanism. This transition occurs as V_Ω,

$$V_\Omega = \frac{L^2}{\mu_p\tau_p} = \frac{L^2\sigma_p{}^0 v_{tp} N_R{}^0}{\mu_p}, \tag{2}$$

where μ_p is the hole mobility, v_{tp} is the thermal velocity of holes, $N_R{}^0$ is the concentration of neutral recombination centers, and $\sigma_p{}^0$ is the capture cross section for holes of the neutral recombination center.

Beyond this transition, the current is controlled by one-carrier space-charge-limited current, as represented by the relation

$$J = \tfrac{9}{8}\varepsilon[\mu_p(V^2/L^3)], \tag{3}$$

where ε is the dielectric constant.

Equation (3) was developed for one-carrier space-charge-limited current for a solid insulator by Mott and Gurney.[60] The V^2 region continues until the majority-carrier transit time is equal to its lifetime. At this point, the conduction mechanism shifts from one-carrier space-charge-limited current to two-carrier space-charge-limited current, and is marked by the onset of negative resistance. This threshold voltage can be described by setting the majority-carrier transit time equal to its lifetime, τ_n:

$$V_{\mathrm{TH}} = L^2/\mu_n\tau_n, \tag{4a}$$

$$\tau_n = 1/\sigma_n{}^+ v_{tn} N_{\mathrm{R}}{}^+, \tag{4b}$$

$$V_{\mathrm{TH}} = L^2\sigma_n{}^+ v_{tn} N_{\mathrm{R}}{}^+/\mu_n, \tag{4c}$$

where $N_{\mathrm{R}}{}^+$ is the concentration of ionized recombination centers, v_{tn} is the thermal velocity of electrons, and $\sigma_n{}^+$ is the capture cross section of electrons for an ionized recombination center.

The work of Ashley and Milnes[28] included the space-charge effects associated with recombination centers, and results in a larger value of V_{TH} and a modification of the prebreakdown square-law current density. These space-charge effects are strongly a function of the impurity levels and defect states in the material used. Equations describing these effects tend to change the magnitude of V_{TH} and J, but have little effect on the formation of the filament, and will not be included here.

Majority-carrier transit occurs because the majority-carrier traps are filled. Two effects are noticed: (1) The lifetime of the majority carriers increases, and (2) the space charge is partially relaxed by the presence of a similar number of carriers of both signs. These effects allow the voltage to decrease as the current increases. The minimum voltage V_{M} has been associated with the ratio of lifetimes and has been described by Lampert as

$$V_{\mathrm{M}} \approx (\sigma_p{}^0/\sigma_n{}^+)V_{\mathrm{TH}}. \tag{5}$$

The model presented to this point conforms to the Lampert model for an insulator with traps, which he calls[16] "double-injection into a semiconductor." The model describes the voltage–current characteristic of Fig. 2 through the low-current $I \propto V$ and $I \propto V^2$ regions to the threshold voltage V_{TH}, and through the negative-resistance region to the minimum voltage V_{M}.

[60] N. F. Mott and R. W. Gurney, "Electronic Processes in Ionic Crystals," p. 172. Oxford Univ. Press, London and New York, 1940.

For the high-current, postbreakdown region, beyond V_M, Lampert's model for an insulator will be considered. This model neglects the effect of thermal carriers and considers the traps to be filled (or the trapping effect to be saturated).

The author recognizes that this combination of two different physical models can be critical. The primary justification depends on experimental results for semi-insulating silicon.[3] From these results, it was noted that the hole traps were filled when the injected hole concentration was equal to 3×10^{11} carriers/cm^3, which is an order of magnitude greater than the thermal hole concentration. It was analytically determined, and experimentally confirmed (see Section 10 and Table I), that the carrier concentration in the postbreakdown region (at the filament center) is always greater than or equal to the traps-filled concentration. Therefore, the trapping effect is saturated for additional carriers. Finally, the experimental model contains both ionized and neutral trap centers, while the Lampert "semiconductor model" considers only initially ionized traps.

Perhaps the greatest value of this "semi-insulator model" lies in the tractability of the mathematics and the agreement with experiment. (This "semi-insulator model" is extended to current filaments in Section 5, and the "semiconductor model" has been adapted to a computer solution predicting current filaments in Section 6.)

Therefore, the description of the current density in the postbreakdown region will be identical to Lampert's treatment of double injection into an insulator. The following assumptions and approximations are made[60a]:

1. The thermal carrier concentrations are negligible.
2. The injected electrons and holes approximately neutralize each other, $n \approx p$. In effect, a plasma of mobile electrons and holes is injected into the insulator.
3. The space charge is caused by the difference in injected carrier concentration (primarily near the contact) and is not affected by trapped charges.
4. The current is also limited by recombination through the traps.
5. An average carrier concentration occurs at the middle of the insulator, due to the approximate neutrality of the injected plasma. (Experiments have shown that this assumption leads to an error of less than 1 %.)
6. The lifetimes of the two carriers are equal and dependent on the carrier concentration.
7. The transit time for both injected carriers is much less than its lifetime.

Lampert uses these approximations and assumptions to calculate the space charge, and, therefore, the voltage by integrating the difference in concentration of the two carriers.[60a] This yields the following expressions for the

[60a]These assumptions are justified in the work of Lampert,[16] pp. 1787–1789.

average carrier concentration $\bar{n}$ and the current density J:

$$\bar{n} = 8\varepsilon\tau \frac{\mu_n \mu_p}{\mu_n + \mu_p} \frac{V^2}{L^4}, \tag{6}$$

$$J = e\bar{n}(\mu_n + \mu_p)\frac{V}{L} = 8\varepsilon\tau\mu_n\mu_p \frac{V^3}{L^5}, \tag{7}$$

where τ is the average lifetime.

At these high injection currents, a bimolecular recombination mechanism, which concludes that the lifetime of carriers of one sign is dependent on the concentration of the carriers of the opposite sign, is postulated.[61] Therefore, in this region, the lifetime may be expected to be inversely proportional to the carrier concentration multiplied by a recombination constant C, which includes the recombination capture cross sections and the thermal velocities. Thus, $\tau = 1/nC$.

Inserting the lifetime relationship into Eqs. (6) and (7) results in

$$\bar{n} = \left[\frac{8\varepsilon\mu_n\mu_p}{eC(\mu_n + \mu_p)}\right]^{1/2} \frac{V}{L^2} \tag{8}$$

$$J = \left[\frac{8e\varepsilon\mu_n\mu_p(\mu_n + \mu_p)}{C}\right]^{1/2} \frac{V^2}{L^3}. \tag{9}$$

Equation (9) predicts a square-law voltage–current characteristic for double injection with a uniform current distribution.

This simplified model predicts two break points in the characteristic curve. At very low currents, the mechanism is ohmic until a point is reached, V_Ω, where space-charge-limited current sets in and the current is proportional to the square of the voltage. The current increases to a threshold voltage V_{TH}, when the majority-carrier lifetime becomes equal to its transit time. Beyond this threshold, negative resistance occurs until a minimum voltage V_{M} is reached. The current then increases as a function of V^2, as determined by two-carrier space-charge-limited current. This high-current square-law region for uniform current distribution occurs at higher current levels than the $I \propto V^{1.5-2.0}$ region characteristic of filamentary conduction.

3. The Nature of the Postbreakdown Region

A large number of the experimental results showing current-controlled negative resistance due to double-injection breakdown have deviated from the Lampert theory in the postbreakdown region. At low postbreakdown currents, a region with current increasing at nearly constant voltage has been observed.

[61] R. H. Parmenter and W. Ruppel, *J. Appl. Phys.* **30**, 1548 (1959).

Two possible explanations of this phenomenon have been developed. One model suggests the formation of current filaments, similar to the observations of Melngailis and Milnes,[18] for impurity impact ionization in compensated germanium at 4.2°K. Ridley[4] has suggested the possibility of current filaments in the negative-resistance region, based on thermodynamic considerations for the general case of current-controlled negative resistance. It can be postulated that the filament of the negative-resistance region can be sustained in the succeeding postbreakdown positive-resistance region.

Another model, which attributes negative resistance to the internal reabsorption of recombination radiation, has been proposed by Dumke.[24] This mechanism utilizes one-carrier space-charge-limited current throughout the breakdown region. The thermodynamics and reabsorption of recombination radiation models will be discussed in this section.

a. The Possibility of Filaments

Ridley[4] suggests that in the case of current-controlled differential negative resistance, the electric current can be multivalued. He discusses the transition from the standpoint of irreversible thermodynamics, invoking the principle of minimum entropy production to identify the steady state. He then studies electrical stability. (In the same paper, he predicts the domain formation associated with voltage-controlled differential negative resistance and observed in Gunn-effect devices.)

Beginning with the second law of thermodynamics, Ridley[4] develops the following equation:

$$dS/dt = -\operatorname{div} J_s + \dot{s}. \tag{10}$$

Equation (10) states that the rate of change of entropy is equal to the entropy production rate minus the rate of entropy flow, where S is the total entropy, J_s is the entropy flow, and $\dot{s}$ is the entropy production rate. Ridley expects the entropy production rate to be as small as possible since he assumes that the steady state will be as near the equilibrium state as the various constants will allow. This argument is the basis of the principle of the production of least entropy, which is used to define the stability of current filaments.

The principle leading to stable filament formation can also be defined as the point where joule heating is at a minimum. Therefore, the product of the current and the electric field will be a minimum.

Due to the difficulties involved in establishing a unique working point in the negative-resistance range, Ridley concludes that the necessary condition for describing stable dc filament formation leads to a system that is ac unstable. Therefore, he concludes that stable filament formation is difficult to obtain in practice.

This discussion considers filaments throughout the dynamic negative-resistance region. To date, there have been no reports of stable current filaments in this region. Reports of observations of current filaments beyond the dynamic negative-resistance region can be attributed to the fact that the filamentary currents, described by Ridley[4] through the negative-resistance region, remain stable once the device is biased into the postbreakdown positive-resistance region. Therefore, Ridley's calculations are consistent with observations of current filaments beyond the negative-resistance region in double-injection devices.

b. Reabsorption of Recombination Radiation

Dumke[24] noted that the recombination-limited current model of Lampert predicts postbreakdown currents that are several orders of magnitude greater than those usually observed. In addition, a two-carrier space-charge-limited current model predicts that the current would be proportional to the square of the voltage in the postbreakdown region, but observations show that the current increases at constant voltage in this region. Finally, in many cases, the Lampert theory predicts voltages considerably greater (up to two orders of magnitude) than those observed with *P–I–N* diodes.

For these reasons, Dumke presents another mechanism to account for the characteristics of *P–I–N* diodes with negative resistance, and he has developed a simple mathematical theory based on the suggested mechanism. He has applied this theory to the GaAs *P–I–N* diodes that were studied by Weiser and Levitt.[23] A review of this mechanism, for *p*-type material, is presented below.

Dumke refers to the prebreakdown current modification of the Lampert model given by Ashley and Milnes[28]:

$$J = \frac{9}{8} \frac{\tau_n}{\tau_{\mathrm{rel}}} \frac{\varepsilon \mu_n}{L} V^2 = aV^2, \tag{11}$$

where τ_{rel} is the dielectric relaxation time for free holes. He then considers the effect of photons produced by electrons recombining on the *p*-side, in addition to those recombining through the deep traps.

Assuming that the light is absorbed uniformly and that the current produced by photoconductivity may be added to the space-charge-limited current of Eq. (11), one obtains

$$J = aV^2 + \tau_{pa} \phi \gamma J \mu_n (V/L^2), \tag{12}$$

where τ_{pa} is the lifetime of an excess bound hole, ϕ is the fraction of photons that are reabsorbed, and γ is the internal quantum efficiency. Now, on the assumption that $\tau_{pa} = \tau_n$, one can solve for the current,

$$J = \frac{aV^2}{1 - \phi\gamma(\tau_n/\tau_t)}. \tag{13}$$

In this equation, $\tau_t = L^2/\mu_n V$, which is the electron transit time. When $\phi\gamma\tau_n/\tau_t \approx 1$, electrical breakdown occurs.

This condition for breakdown may be maintained at progressively lower voltages if the quantum efficiency is an increasing function of current. The negative resistance, therefore, reflects the increase in quantum efficiency.

Appropriate numbers were applied to the experimental devices of Weiser and Levitt. A voltage–current characteristic resembling their experimental devices was predicted. An increase in light output was observed through the negative-resistance region, as predicted by the increase in quantum efficiency. In addition, this mode, which considers reabsorption of photons, accounts for the decreasing threshold voltage observed in most double-injection studies with the application of external radiation.[14]

c. Comparison of the Models

The two models described above offer a plausible explanation for the nature of the postbreakdown region for double-injection diodes. Unfortunately, it appears that they are mutually exclusive. The existence of current filaments with high current densities and, therefore, bright localized areas of light emission, seems to exclude the reabsorption of recombination radiation as a mechanism, because the internally generated light would cause the filament to spread and uniformly fill the sample. Therefore, it is postulated that if filaments exist in a series of experimental *P–I–N* devices, the controlling mechanism of these devices is double injection, and not the reabsorption of recombination radiation. Additional evidence for the filamentary model would include the observation of the high-current power-law region.

It is important to include the observation of reduced threshold voltages and light sensitivity in the development of a complete space-charge-limited model.

It will be the purpose of the following sections to develop a model that is consistent with the reported experimental results, supporting the space-charge-limited filament model for a general class of double-injection devices. Experimental results of negative-resistance devices which do not appear to be filamentary will also be discussed.

II. Theory of Current Filament Formation

In this section, the conditions leading to the negative-resistance region will be discussed, and a simple model, which correlates satisfactorily with the experimental observations of current filaments in the high-current power-law region, will be developed.

Finally, a new analytical solution to the basic double-injection equations, which successfully predicts current filaments through the negative-resistance region, will be presented.[5]

4. The Breakdown Condition

Both the Lampert[2] theory and the Ashley–Milnes[28] modification assume that breakdown occurs when the majority-carrier transit time is equal to its lifetime. The prebreakdown square-law region is attributed to the space-charge-limited current due to injected minority carriers (modified by the space charge of injected majority carriers). This condition for breakdown is equivalent to assuming that the injected majority carriers, at the far boundary from their injecting contact, are sufficient in number to effectively neutralize the charged recombination centers, or that the traps are filled. Further, it is postulated that breakdown occurs at some device inhomogeneity.

The variation in threshold voltage for seemingly identical devices is often as great as $\pm 20\%$ Variations of this magnitude have been reported for silicon devices[3] and for GaAs arrays.[59] Therefore, the basic breakdown condition must be sensitive to variations of $\pm 20\%$ throughout a group of devices, and at a specific point in a single device. Both the Lampert and Ashley–Milnes breakdown condition can be written as

$$V_{TH} = KL^2 N_{ex}, \tag{14}$$

where L is the device length and N_{ex} is the number of excess shallow impurities $N_A - N_D$. The constant K includes various constant parameters, such as dielectric constant, the charge of an electron, capture cross sections, various thermal velocities, etc. Also implicit in this expression is the relationship that the ratio of the un-ionized deep impurities (recombination centers) divided by the total number of deep impurities in equilibrium is approximately equal to unity, which is part of the Ashley–Milnes modification. Also, it is implicitly assumed in Eq. (14) that the concentration of thermal carriers is much less than the excess shallow impurities, which, in turn, is less than the total concentration of deep impurities, $n < N_{ex} < N_R$. Typical examples of these values are a thermal carrier concentration of 10^{10} carriers/cm^3, a shallow donor concentration N_D of 10^{13}/cm^3, a shallow acceptor concentration of 10^{14}/cm^3, and a deep-donor concentration of 10^{15}/cm^3.

Therefore, Eq. (14) predicts that length, or excess shallow impurity variations, could cause an inhomogeneity that would explain the variations in threshold voltage and the existence of a specific point of breakdown. The devices, typically, range from 2 to 20 milli-inches in length, and groups of devices have been made at particular lengths. It is probable that length

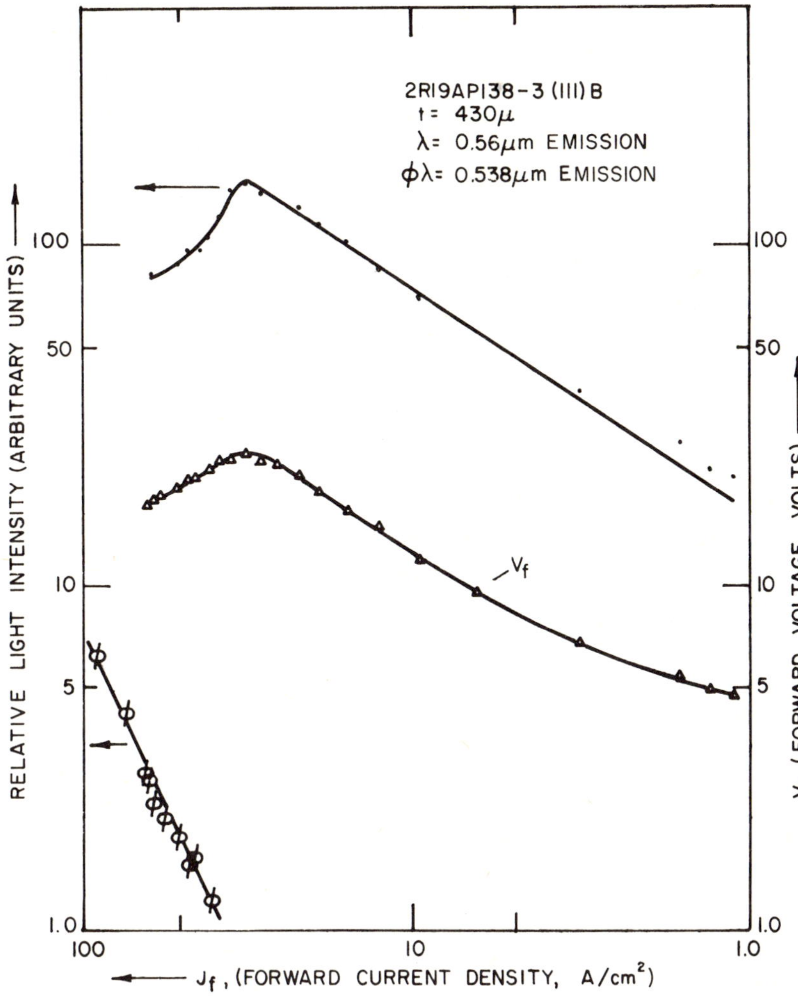

FIG. 6. Variation of voltage, light emission due to recombination centers (0.56 μm), and band-to-band recombination (0.538 μm) with current. (After Epstein.[53])

variations are a minor factor, since threshold variations have been randomly observed in large arrays (over 100 devices) on a monolithic substrate.[59]

Therefore, it is possible that a device under a uniform field will break down at the point where the excess shallow impurity concentration, near the contact opposite to the majority-carrier injecting contact, is at a minimum. This type of breakdown will occur when the majority-carrier concentration, at the contact opposite to the majority-carrier injecting contact, is sufficient to effectively neutralize the charge recombination centers.

It is also interesting to note that, whereas Lampert and Ashley–Milnes considered the injection of these carriers to be due to the uniform application of an external electric field, there are several other methods by which carriers can be injected. For example, carriers can be injected into a semi-insulator by the application of external light. Other methods include the direct injection of carriers by leakage current path, or the recombination of leakage currents giving rise to recombination radiation which, in turn, has been reabsorbed. This increase in the majority-carrier concentration, which lowers the threshold voltage and is a possible explanation for the photosensitivity of double-injection devices, has been reported by many. Techniques for lowering the threshold voltage will be discussed in Section 8c.

Additional evidence for a carrier-concentration model of breakdown comes from the observation of Melngailis and Rediker[12] that the breakdown voltage seems to depend on the density of injected carriers and not on the current density or electric field strength.

Perhaps the most direct evidence of the validity of the "traps-filled" model of breakdown are the studies carried out by Epstein[53] on GaP at liquid-nitrogen temperatures. He has correlated the radiation due to the deep traps (0.56 μm), the appearance of band-to-band radiation (0.538 μm), and the onset of negative resistance for samples doped with sulfur and silicon. He finds, for devices longer than 290 μm, that the onset of negative resistance occurs simultaneously with the saturation of the radiation due to trapping (0.56 μm) and the onset of band-to-band radiation (0.538 μm). These effects are shown in Fig. 6. This study is a clear example of the two dominant recombination mechanisms, before and after double-injection breakdown.

5. Simplified Filament Equation

In this section, a general form of the filament equation will be developed and will be solved by the method of separation of variables. Although this simplified solution is not a precise one, it offers insight into the physics of the problem and yields reasonable agreement with the experimental results described in Sections 7 and 9. A computer solution of a more general form of the filament equation will be given in Section 6.

The simplified Lampert theory developed in Section 1 considered only longitudinal drift currents. A more general solution considers the effect of diffusion currents, as shown in the following equation (this model will be developed for carriers of one sign, recognizing that analogous relationships can be developed for carriers of the opposite sign):

$$\mathbf{J}_n = en\mu_n\mathbf{E} + eD_n\nabla n. \tag{15}$$

Applying the continuity equation, we obtain

$$\nabla \cdot \mathbf{J}_n = eR, \tag{16}$$

where R is the recombination rate.

Expanding Eq. (16) and assuming angular symmetry, we can develop the following equation:

$$\frac{\partial}{\partial z} J_{nz} + \frac{1}{r} \frac{\partial}{\partial r}(rJ_{nr}) = eR. \tag{17}$$

This equation is expanded again, ignoring the longitudinal diffusion currents[61a] and considering the electric field in the radial direction to be negligible:

$$e\mu_n \frac{\partial}{\partial z}(nE_z) + D_n \frac{e}{r} \frac{\partial}{\partial r}\left(r \frac{\partial n}{\partial r}\right) = eR. \tag{18}$$

Rearranging variables, the following generalized form can be written:

$$D_n \frac{\partial}{\partial r}\left(r \frac{\partial n}{\partial r}\right) = rR - r\mu_n \frac{\partial}{\partial z}(nE_z). \tag{19}$$

By separation of variables, the two sides of Eq. (19) can be solved separately. The longitudinal drift equation solved by Lampert is

$$\mu_n \frac{\partial}{\partial z}(nE_z) = R_1. \tag{20}$$

The radial diffusion equation, or the simplified filament equation, is

$$\frac{1}{r} D_n \frac{\partial}{\partial r}\left(r \frac{\partial n}{\partial r}\right) = R_2. \tag{21}$$

Considering the average electron concentration to be equal to the average hole concentration as described in Section 2, Eq. (21) can be rewritten

$$\frac{1}{r} D_a \frac{\partial}{\partial r}\left(r \frac{\partial n}{\partial r}\right) = \frac{n}{\tau}, \tag{22a}$$

$$\frac{\partial^2 n}{\partial r^2} + \frac{1}{r} \frac{\partial n}{\partial r} = \frac{n}{D_a \tau}, \tag{22b}$$

where D_a is the ambipolar diffusion constant,[62] n is the concentration of carriers of both signs, and τ is the lifetime of both carriers, which is assumed to be the same for both.

[61a] This assumption is consistent with the Lampert development. Baron and Mayer, in Chapter 4 of this book, treat the effects of longitudinal diffusion currents more extensively. It should be pointed out that the diffusion lengths of the carriers, in the devices studied in this work, are much shorter than the cases studied by Baron and Mayer.

[62] W. Van Roosbroeck, *Phys. Rev.* **91**, 282 (1953).

The solution of Eq. (22) for constant τ is a Hankel function of zero order with an imaginary argument, $H_0^{(1)}$. This solution approaches infinity as the radius r approaches zero. Therefore, the concept of a minimum radius was introduced, so that $n = n_0$ at $r = r_0$. The following equation describes the radial variation of the carrier concentration:

$$n = kn_0 H_0^{(1)}(ir/D_a\tau), \tag{23}$$

where k is the reciprocal of $H_0^{(1)}$ at the minimum radius r_0.

The concept of r_0 is mathematically incorrect. It was used in the original publication on space-charge-limited current filaments[3] in order to demonstrate the radial diffusion model and to correlate the experimental results. A more precise treatment of this mathematical problem necessitates a computer solution. It is often difficult to emphasize the physics when developing a computer solution. For these reasons, this simplified solution is included.

An attempt was made to define a closed form of this relationship by further mathematical simplification, and eliminate the concept of r_0. An examination

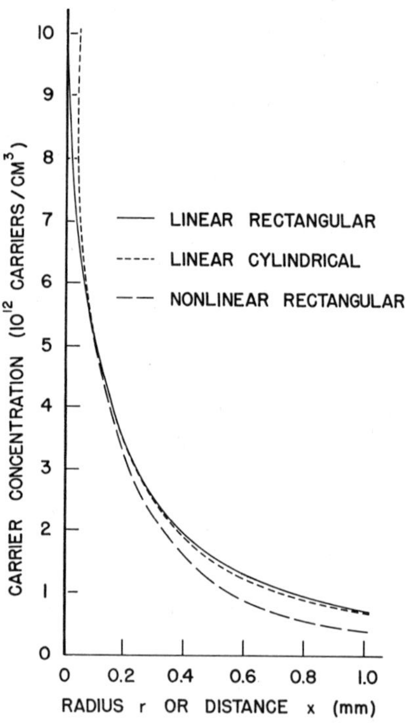

FIG. 7. Variation of carrier concentration with radius for three solutions of the diffusion equation: linear rectangular, linear cylindrical, and nonlinear rectangular. (After Barnett and Milnes.[3])

of Eq. (22) shows that the solution approaches infinity as r goes to zero, due to the $1/r$ variation of the second term. The physical boundary conditions demand that the solution must be finite throughout: therefore, the value of $\partial n/\partial r$ must approach zero as r approaches zero. The same $1/r$ variation eliminates this term at large r values.

It is postulated that the solution to Eq. (22b) without the second term is a good approximation to the exact two-dimensional cylindrical solution, Eq. (23). This abbreviated solution of Eq. (22b) is the one-dimensional case, or the rectangular solution.

The results of these two linear solutions are plotted in Fig. 7, with n_0 set equal to 10^{13} carriers/cm^3 and r_0 chosen as 0.035 mm.

The variation of carrier concentration is seen to be quite similar for the linear rectangular and linear cylindrical solutions, except near $r = 0$. This one-dimensional approximation yields an exact value for n at $r = 0$: it also gives a reasonable impression of the shape of the two-dimensional solution and permits an exact closed form for integration. The total error introduced at $r = 0$ will be shown to be small. Therefore, the nonlinear case is solved in rectangular coordinates only.

The nonlinear case is developed when bimolecular recombination is considered, and is discussed in Section 2 and by Parmenter and Ruppel.[61] It is assumed that the lifetime of carriers of one sign is inversely dependent on the concentration of carriers of the opposite sign, i.e., $\bar{\tau} = 1/\bar{n}C$, where C is a recombination constant. This condition is inserted into Eq. (22), which is reduced to a more general form:

$$D_a \nabla^2 n = n/\tau = n^2 C. \tag{24}$$

In rectangular coordinates, Eq. (24) becomes

$$d^2n/dx^2 = n^2(C/D_a). \tag{25}$$

The solution of Eq. (25) is

$$dn/dx = -[(2n^3C/3D_a) + A]^{1/2}, \tag{26}$$

where A is a constant of integration. Applying the boundary conditions $dn/dx = 0$ and $n = 0$ at $x = \infty$, and $n = n_0$ at $x = 0$, one can solve for A. Integrating again, applying the boundary conditions, and substituting τ for x, we get

$$n = \frac{n_0}{[1 + (n_0C/6D_a)^{1/2}r]^2}. \tag{27}$$

Equation (27), the nonlinear rectangular case, is plotted in Fig. 7. The carrier concentration at the center of the filament, n_0, is set equal to the conditions of Lampert,[16] Eq. (8), for average carrier concentration $\bar{n}$.

The use of the nonlinear rectangular case permits the calculation of the radial current distribution and total current, in a closed form, for the simplified case. The major disadvantage of this technique is the loss of detail around the filament center. Considering a real case, experimental studies in semi-insulating silicon (Section 7) have shown the contribution of the central core, radius <0.05 mm, to be less than 5% of the total current. The use of this simplified form seems justified by the ease with which the underlying physics can be observed and analyzed. A more rigorous mathematical solution of Eq. (19), with computer techniques, appears in the next section.

Substituting Eq. (8) into Eq. (27) for the variation of carrier concentration with radius, and then substituting this relationship for n into the current density equation, Eq. (7), the radial variation of current density becomes

$$J(r) = \frac{[8e\varepsilon\mu_n\mu_p(\mu_n + \mu_p)/C]^{1/2}(V^2/L^3)}{\{1 + [2\varepsilon\mu_n\mu_p C/e(\mu_n + \mu_p)]^{1/4}(V/3D_a)^{1/2}(r/L)\}^2}. \tag{28}$$

For a given voltage and device length, Eq. (28) may be written as

$$J(r) = B/(1 + ar)^2, \tag{29}$$

where

$$B = [8e\varepsilon\mu_n\mu_p(\mu_n + \mu_p)/C]^{1/2}(V^2/L^3),$$

and

$$a = [2\varepsilon\mu_n\mu_p C/e(\mu_n + \mu_p)]^{1/4}(V/3D_a)^{1/2}(1/L).$$

Integrating Eq. (29) yields an expression for the total current,

$$\begin{aligned} I &= \int_0^{2\pi}\int_0^R J(r) r\, dr\, d\theta \\ &= (2\pi B/a^2)[\ln(1 + aR) + 1/(1 + aR) - 1], \end{aligned} \tag{30}$$

where R is the device radius, or its equivalent radius, if the device is not circular in cross-sectional area.

The variation of I with voltage over any range of interest is closely approximated by a power law from Eq. (30) by simplifying the equation for values of $aR \ll 1$ or $aR \gg 1$, or by substituting typical values and plotting the results. The power law tends toward $I \propto V^{1.5}$ for aR values greater than 1.2, and $I \propto V^{1.9}$ for aR values less than 0.8. A power law with $I \propto V^x$, where x is greater than 1.9, has been observed in both silicon and gallium arsenide for multiple filament formation.

The equations that have been derived here characterize two-carrier space-charge-limited filaments in the region where current is dependent on the power of the voltage. The equations appear to provide satisfactory agreement with experimental studies in semi-insulating silicon. A model which extends the equation to include the region where the current rises at a constant voltage is discussed in Section 10.

6. Computed Solution of Filaments

In this section, Eq. (19), the one-carrier, longitudinal drift and radial diffusion equation, will be extended to include the case for the two-carrier solution. This equation will be manipulated to a form that is amenable to solution by computer techniques, in a manner similar to that of Muller[63] and Muller and Guckel[64] for avalanching *P–I–N* diodes. Preliminary results which predict the form of the negative-resistance region and the current density distribution within the filament will be presented.[5]

Equation (19) is reproduced as

$$D_n \frac{\partial}{\partial r}\left(r \frac{\partial n}{\partial r}\right) = rR - r\mu_n \frac{\partial}{\partial z}(nE_z), \tag{31}$$

and the similar equation that applies for holes is

$$D_p \frac{\partial}{\partial r}\left(r \frac{\partial p}{\partial r}\right) = rR + r\mu_p \frac{\partial}{\partial z}(pE_z). \tag{32}$$

The following relations are substituted:

$$n = J_{nz}/e\mu_n E_z, \tag{33a}$$

$$p = J_{pz}/e\mu_p E_z, \tag{33b}$$

assuming that the radial variation of E is negligible. Also, Eq. (31) is multiplied by μ_n and Eq. (32) by μ_p, and the ambipolar diffusion constant D_a is substituted for D_n and D_p. Now, upon adding, one obtains

$$\frac{1}{eE_z} D_a \frac{\partial}{\partial r}\left[r \frac{\partial}{\partial r}(J_{pz} + J_{nz})\right] = (\mu_n + \mu_p)rR + r\mu_p{}^2 \frac{\partial}{\partial z}(pE_z) - r\mu_n{}^2 \frac{\partial}{\partial z}(nE_z). \tag{34}$$

It is now noted that the right-hand sides of Eqs. (31) and (32) are the one-dimensional expressions treated by Lampert[2]; therefore, a one-dimensional recombination rate, R_{z1}, can be defined. The addition of the subscript 1 denotes the one-dimensional solution

$$R_{z1} = \mu_n \frac{\partial}{\partial z}(n_1 E_{z1}) = -\mu_p \frac{\partial}{\partial z}(p_1 E_{z1}). \tag{35}$$

[63] M. W. Muller, *Appl. Phys. Letters* **12**, 218 (1968).

[64] M. W. Muller and H. Guckel, *IEEE Trans. Electron Devices* **ED-15**, 560 (1968).

Letting R_{z1} approximate R in Eq. (34) and assuming that E_z is of the same form as E_{z1}, but modified by the actual voltage V_A, we have

$$E_z = E_{z1} V_A / V_{z1}, \tag{36}$$

where V_{z1} is the one-dimensional voltage. Substituting into Eq. (34), the following is the basic form of the computed solution:

$$\frac{D_a}{e}\frac{\partial}{\partial r}\left(r\frac{\partial J_z}{\partial r}\right) = (\mu_p + \mu_n) r R_{z1} E_{z1}\left(1 - \frac{V_A}{V_{z1}}\right)\frac{V_A}{V_{z1}}. \tag{37}$$

Equation (37) indicates that the radial distribution may be a function of z; therefore, we obtain an equation for the average profile by integrating along the device length L:

$$\frac{\partial}{\partial r}\left(r\frac{\partial J_z}{\partial r}\right) = \frac{e}{D_a L}(\mu_n + \mu_p) r\left(1 - \frac{V_A}{V_{z1}}\right)\frac{V_A}{V_{z1}}\int_0^L R_{z1} E_{z1}\, dz. \tag{38}$$

At this point, the radial distribution of current has been described as a function of the one-dimensional voltage and current modified by the actual voltage V_A. Thus, the solution depends on the particular model used for the one-dimensional solution. Lampert's "semiconductor model" was used.

Note that this model reduces to the "semi-insulator model" when the thermal carrier concentration and trap occupancy make a negligible con-

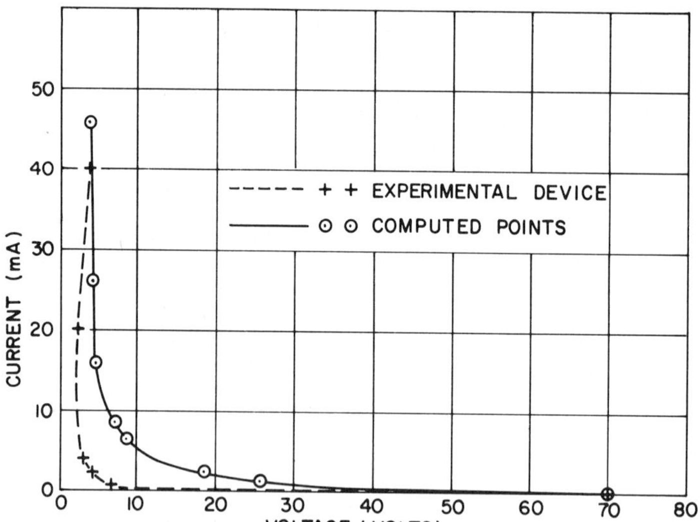

FIG. 8. Computed voltage–current characteristics for a low-current filament compared to an experimental GaAs device with similar characteristics.

tribution to the total space charge. The computed solution verified the validity of the "semi-insulator model" throughout the region of interest.

The voltage–current characteristic was calculated for a GaAs device, with $V_{TH} = 70$ V, $V_M = 2.1$ V, μ_n and μ_p equal to 10,000 and 500 cm^2/V-sec, respectively, $\langle v_{tn}\sigma_n{}^+\rangle$ and $\langle v_{tp}\sigma_p{}^0\rangle$ equal to 1.4×10^{-5} and 7.5×10^{-9}, cm^3/sec, respectively, $N_R = 10^{15}$/cm^3, $N_A = 10^{14}$/cm^3, device length $L = 10^{-2}$ cm, and the ambipolar diffusion constant $D_a = 11$ cm^2/sec.

The current density at the center of the filament and a value for V_A are given to the computer. The computer then determines the correct value of V_A by minimizing the value of the current density as the value of the radius approaches infinity.

The resultant calculated voltage–current characteristic at low currents is shown in Fig. 8: also shown are the characteristics of an actual device. The average of the radial current density distribution over the device length was also computed, and is shown in Fig. 9.

These computed three-dimensional solutions show a voltage–current characteristic and radial current distribution similar to the experimental observations that are discussed in Section 8. It would be premature to make any quantitative comparison between theory and experiment at this time. Several qualitative observations can be made about this computed solution

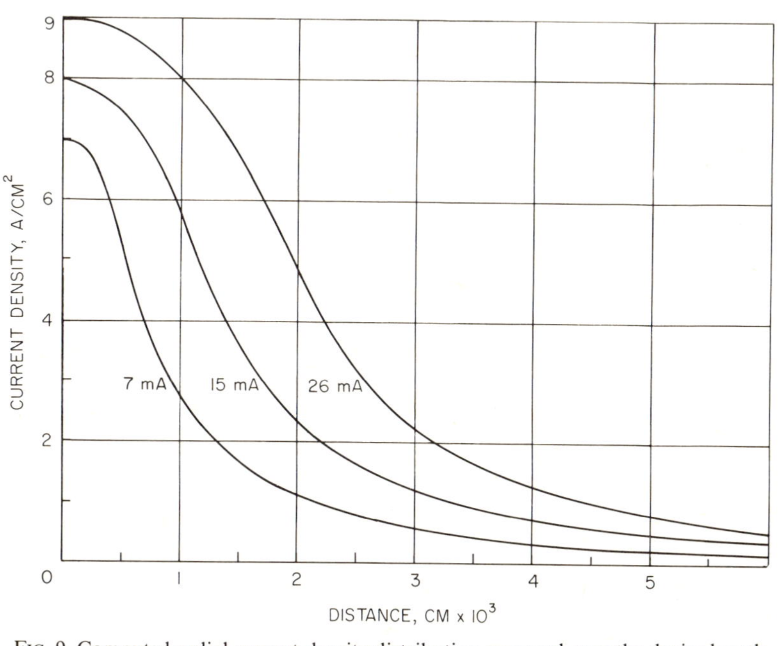

FIG. 9. Computed radial current-density distribution averaged over the device length.

and its comparison to the experiment described in Section 8. The computed solution successfully predicts filaments throughout the negative-resistance regime and gives a reasonable approximation to the voltage–current characteristic. This prediction of filaments through the negative-resistance region is similar to the thermodynamic model of Ridley[4] and the model for avalanching silicon p^+–i–n^+ junctions of Muller[63] and Muller and Guckel.[64] The observation of filaments in the positive-resistance region and in the higher-current power-law region suggests the need for additional investigation and extension of the model, the experiment, and the mathematical approximations

III. Experimental Observation of Current Filaments

Current filaments have been directly observed in Si, GaAs, ZnTe, $GaAs_xP_{1-x}$, and polycrystalline Si. These filaments have been observed by collecting the recombination radiation at one of the injecting contacts. Attempts have been made to correlate the recombination radiation to the current distribution. These experimental results are discussed in this part.

7. Current Filaments in Silicon Double-Injection Devices

The first direct observation of current filaments was reported for devices made from semi-insulating silicon at 77°K.[3] The silicon was doped with a single-level deep impurity, indium, at a concentration of approximately $6 \times 10^{16}\,\mathrm{cm}^{-3}$. This single deep acceptor level occurs 0.16 eV above the valence band. The material also contained an excess shallow donor concentration, $N_D - N_A$, of 4×10^{13} atoms/cm^3. The thermal hole concentration at 77°K was 2×10^{10} holes/cm^3, corresponding to a resistivity of approximately 10^4 ohm-cm.

a. Experimental Details

The P–I–N devices constructed for this experiment were stepped as shown in Fig. 10. The semi-insulator region was thicker near the wire contacts than elsewhere in the sample. The purpose of the step geometry is to force the filament to occur on a planar region of the sample removed from the contact wires. The experiments were conducted using 12 devices having semi-insulator lengths (at the thin region) between 140 and 220 μm. The other dimensions were approximately 1 by 2 mm. The n-type junction was provided by a 10-μm phosphorus diffusion; the p-type junction was provided by 2-μm boron diffusion after the step was formed. It was planned to collect the recombination radiation through the p^+ junction.

The device under study was suspended in liquid nitrogen with the p^+ junction facing the unsilvered section of a Dewar. Radiation emerging from

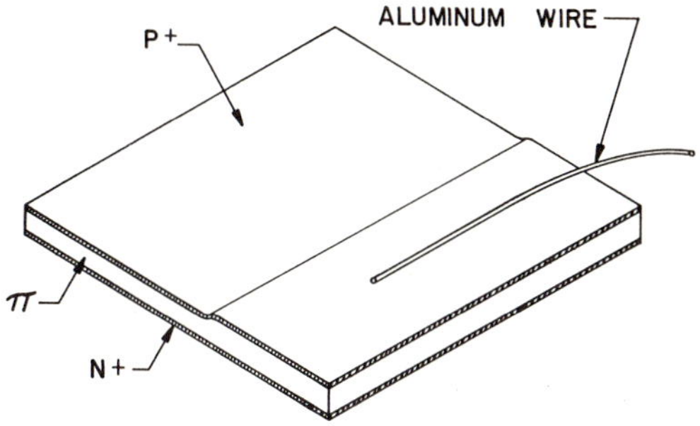

FIG. 10. Stepped geometry P–I–N device. (After Barnett and Milnes.[3])

this window was focused on the photoconductive surface of an S-1 image-converter tube. The sensitivity of the S-1 surface of this tube drops off rapidly in the wavelength range of 1.06–1.2 μm, which are the radiation peaks for indium-doped silicon.[65] Therefore, the image was derived from the 1.06-μm radiation peak. A simple lens was placed after the phosphor of the image-converter tube to magnify the photographs. The film employed was Tri-X,[65a] developed with Diafine,[65a] with exposure times ranging from minutes to hours. Figure 11 shows the experimental apparatus.

The optical system was aligned by illuminating the device with a light pipe inserted in the Dewar. The resultant background illumination system was

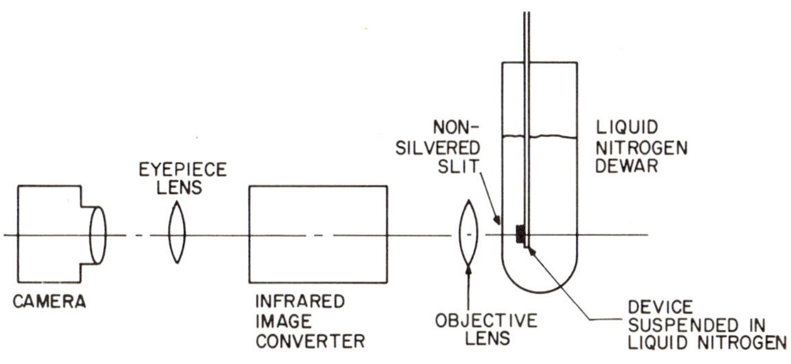

FIG. 11. Experimental apparatus for recombination radiation photographs of devices suspended in liquid nitrogen. (After Barnett and Milnes.[3])

[65] C. Benoit à la Guillaume and O. Parodi, *in* "Solid State Physics in Electronics and Telecommunications" (M. Desirant and J. L. Michiels, eds.), Vol. 1, Pt. 1, p. 294. Academic Press, New York, 1960.

[65a] Registered trademarks, respectively, of Eastman Kodak Co., Rochester, New York, and Baumann Photochemical Corp., Chicago, Illinois.

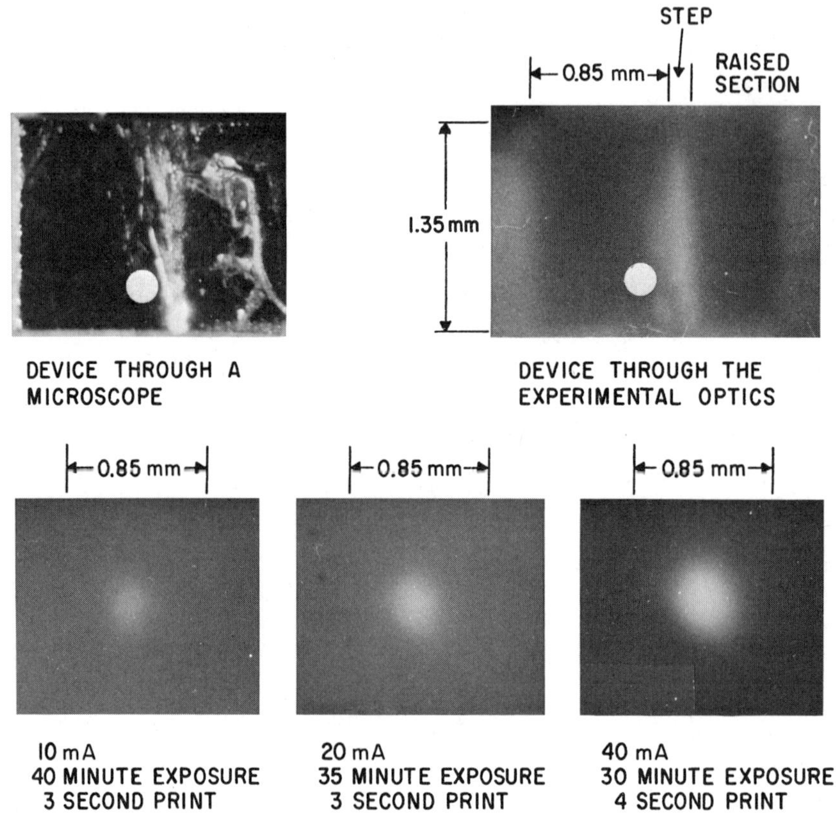

FIG. 12. Photographs of device No. 25-1 and 10-, 20-, and 40-mA filaments. The white dots superimposed on the device photographs indicate the filament location. (After Barnett and Milnes.[3])

used to obtain photographs of the device, which were then used to locate the position of the filament.

Filamentary breakdown was observed in all silicon devices that were from 140 to 220 μm long. Figure 12 shows photographs of these filaments at 10 mA (0.5 V), 20 mA (0.7 V), and 40 mA (1.1 V). These single filaments are observed in the high-current power-law region. The complete current–voltage characteristic of device No. 25-1 is discussed in Section 10. Included in Fig. 12 is a photograph of this device taken through a microscope and a photograph taken through the image-converter optical system (with the device removed from the Dewar). The vertical line of light represents reflections from the edge of the step; a white circle has been added to indicate the approximate nucleation point of the filament on the photograph of the device.

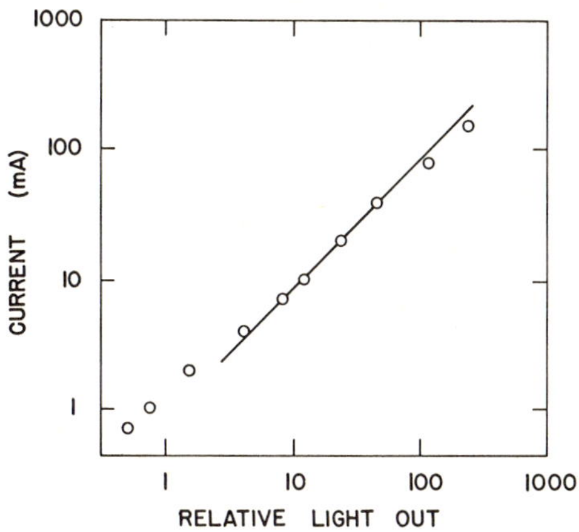

FIG. 13. Light output versus current for device No. 25-1 in the filamentary conduction region. (After Barnett and Milnes.[3])

b. Current Distribution

An attempt was made to convert the qualitative information obtained from the photographs of the filaments to a quantitative relationship between the current density and the light intensity.

The total current was found to be proportional to the light output, as measured by an S-1 surface photomultiplier, for the silicon diode being studied, as shown in Fig. 13. This result agrees reasonably well with the observations of Benoit à la Guillaume and Parodi,[65] who found a linear relationship between the 1.06-μm recombination radiation of indium-doped silicon and the current density. The experimental technique involves the conversion of infrared radiation to visible light, and the subsequent recording of this light in terms of film density.

Film density may be directly proportional to light intensity for very low exposure conditions.[66] This relationship was studied by increasing the exposure time of various diodes in the low-film-density region, and by taking low-level photographs of a calibrated light source. It was shown[3] that, at very low levels, the assumption that the film density varies linearly with light intensity is a reasonable one. Therefore, it was concluded that, within a suitable range of light intensity and exposure time, the relationship between the current density J and the film density D may be represented by

$$J = mD + \text{const}. \tag{39}$$

[66] W. J. Moore, "Physical Chemistry," 2nd ed., p. 611. Prentice-Hall, Englewood Cliffs, New Jersey, 1965.

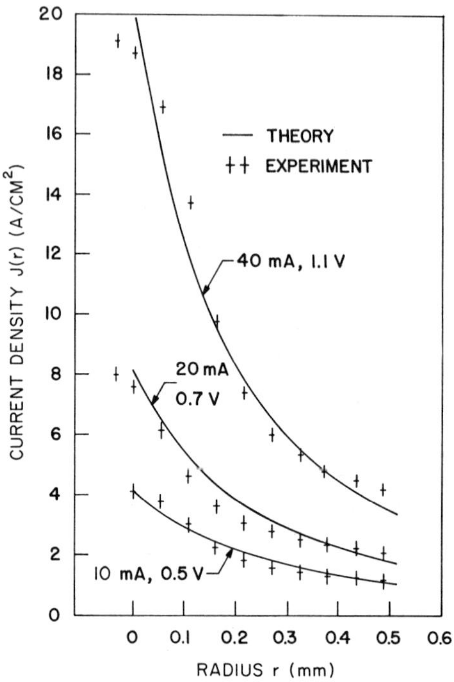

FIG. 14. Radial current distribution compared to the simplified filament equation for device No. 25-1 at 10, 20, and 40 mA. (After Barnett and Milnes.[3])

Equation (39) is only valid for very low film densities. For photographs that illustrate filament growth in silicon, longer exposures were made in order to obtain contrasts that are suitable for reproduction. The slope m and the constant were determined from the 10-mA filament data and were then used (with adjustments for exposure times) for the 20- and 40-mA filaments (as well as the 80- and 160-mA data presented in Section 11). The predictions of the simplified filament equation developed in Section 5 are compared with this experiment in Fig. 14. The 10-mA filament curve has been fitted at two points, and the theoretical curves for the 20- and 40-mA filaments are derived without further fitting. These higher-current curves are positioned to compare favorably with the one-dimensional equation, which is inaccurate as r approaches zero. Therefore, some values are plotted at $r < 0$ to show experimental detail in this region.

Note that the experimental curves of current density distribution show a decrease in slope as the radius approaches zero, corresponding to the computed solution shown in Fig. 9, but deviating from the closed-form, one-dimensional equation developed in Section 5. If the radius of this central core is considered to be 0.05 mm, one can calculate that it contributes less than 5% of the total current, which justifies for this example the approximation made in Section 5 which ignores the two-dimensional effects at $r < 0.035$ mm.

The agreement between theory and experiment, particularly with respect to the magnitude of the current density at the center and the general shape of the curves, is quite satisfactory under these conditions.

8. Current Filaments in Gallium Arsenide Double-Injection Devices

There have been many observations of current-controlled negative resistance in GaAs *P–I–N* devices. In this section, the evidence on current filaments is presented, and the results of these filament studies are correlated with other reports of GaAs double-injection breakdown.

a. Experimental Details

The first studies of current filaments in GaAs were reported for step-geometry devices,[6] similar to the geometries used in studies on silicon devices.[3] These devices were made using oxygen-doped GaAs, in which the oxygen level has been described as a single-level deep donor.[67] The room-temperature resistivity was 10^8 ohm-cm. The n^+ junction was grown by liquid-phase epitaxy techniques. For this series of devices, the n^+ layer was 0.003 in. thick, the semi-insulator in the thin region was 0.005 in. thick, and a 1-μm p^+ junction was formed by zinc diffusion. The voltage–current characteristic curve for a typical device is shown in Fig. 15 (where the junction voltage has been subtracted). At low voltages, the voltage is directly proportional to the current, which is typical for these gallium arsenide devices with a clean surface. The prebreakdown relationship $I \propto V^2$ has been observed for these gallium arsenide devices, and a higher-power-law prebreakdown region has also been observed. This characteristic, which is discussed in greater detail in Section 8c, is presently attributed to surface leakage. The postbreakdown characteristic is quite similar to the theoretical curve of Fig. 2.

The current filament was observed by photographing it through the p^+ junction using Polaroid infrared film. Figure 16 shows photographs of the device when illuminated, and the illuminated device with the filament "on" showing the filament position and the filaments at 4, 6, and 8 mA. These photographs were taken with a $\frac{1}{8}$-sec exposure; the emitted light is approximately 8800 Å. The recombination mechanism is, apparently, a transition from the conduction band to the shallow acceptor. The apparent diameter

[67] J. Blanc and L. R. Weisberg, *Nature* **192**, 156 (1961).

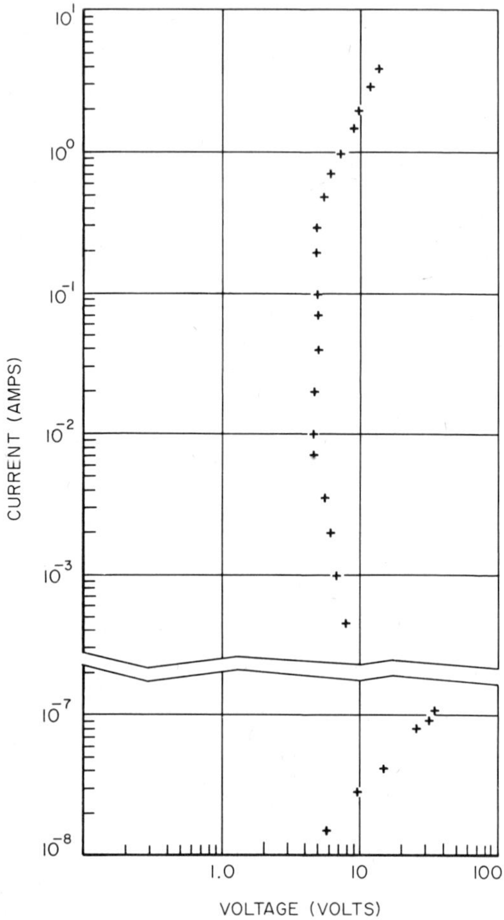

FIG. 15. Voltage–current characteristics for device No. SLES-1, with the junction voltage subtracted. (After Barnett and Jensen.[6])

of this filament is 0.05 mm, which is smaller than that observed in silicon, and is expected, due to the shorter carrier lifetime.

These filaments were observed in the low-current region, where the current rises approximately at constant voltage. As the current is increased, multiple filaments are observed. These will be discussed in Section 11.

These devices, with step-type geometries, have proven to be unsatisfactory for the study of current filaments in semi-insulating gallium arsenide because the threshold voltage is often dominated by high leakage currents along the edges of the device. As a consequence, filaments often seem to form near the physical boundaries of the device. Therefore, a guard-ring geometry was

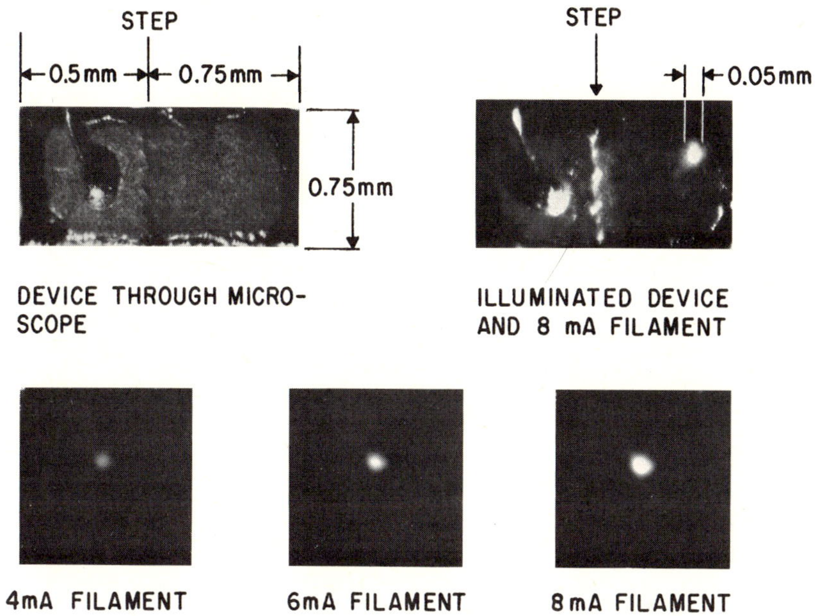

FIG. 16. Photographs of device No. SLES-1 and filaments at 4, 6, and 8 mA, at the same magnification. (After Barnett and Jensen.[6])

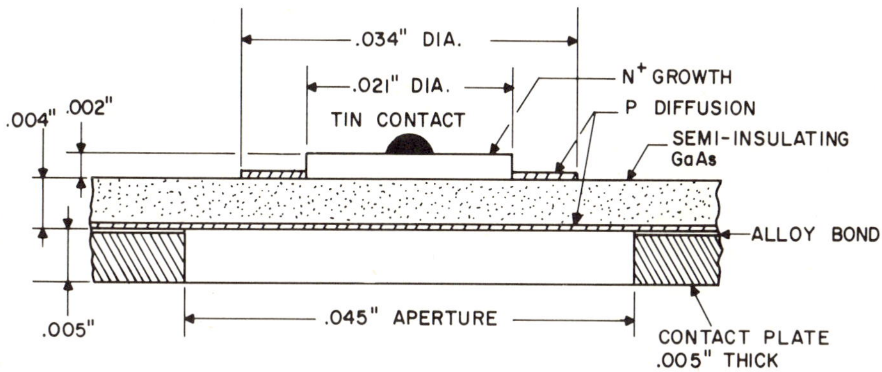

FIG. 17. Drawing of the *P–I–N* device with a guard-ring to minimize leakage current. (After Barnett *et al.*[59])

developed to study the filamentary device. The guard-ring type devices were fabricated on a 10^7 ohm-cm, oxygen-doped GaAs substrate. The n^+ junction was formed by liquid-phase epitaxial growth. This injecting contact was 0.002 in. thick and 0.021 in. in diameter. A *p*-type guard-ring was placed around the n^+ junction to minimize surface leakage current. The semi-

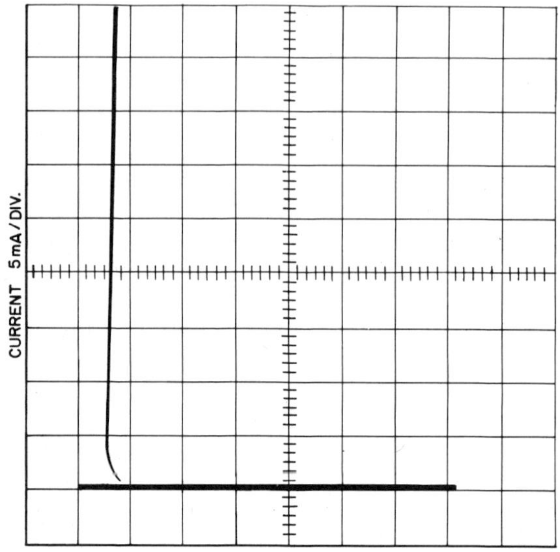

FIG. 18. Voltage–current characteristics of guard-ring device LES-J2. (After Barnett *et al.*[59])

insulator substrate was 0.004 in. thick, and a 1-μm p^+ junction was diffused into the opposite face. The p^+ junction was alloyed to a gold-plated molybdenum contact with a 0.045-in.-diameter aperture. Figure 17 is a schematic drawing of the device.

It is interesting to note that Marsh and Viswanathan,[68] in their study of space-charge-limited hole currents, developed a complex geometry to eliminate similar surface leakage effects.

Figure 18 shows the voltage–current characteristic of the experimental device. Photographs of filaments taken through the p^+ junction at currents of 6, 10, and 15 mA are shown in Fig. 19.

b. Current Distribution

The large amount of light emitted by gallium arsenide devices allows for the direct measurement of the recombination radiation. A simple lens is used to focus the device on a silicon photocell (30:1 magnification), which is covered by an opaque material except for a small aperture. The size of the aperture determines the resolution of the experiment. It was found that resolutions greater than 10 μm did not yield more information about the filament, but simply increased the noise level. This apparent smearing is possibly caused by the varying depth of the formation of the recombination

[68] O. J. Marsh and C. R. Viswanathan, *J. Appl. Phys.* **38**, 3135 (1967).

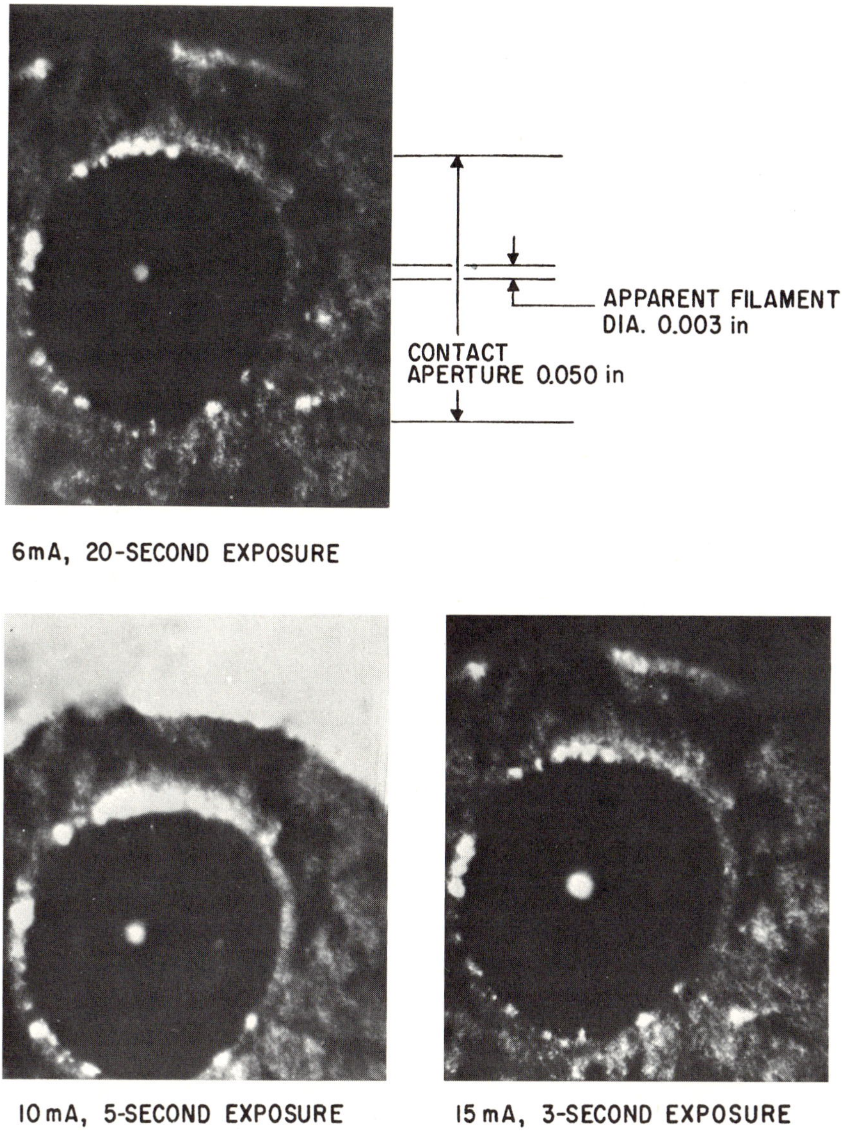

FIG. 19. Infrared film photographs of current filaments of device LES-J2. (After Barnett *et al.*[59])

radiation. The signal into the photocell is chopped and the photocell output is fed into a lock-in-amplifier. The photocell is scanned automatically across the plane of a device image. The experiment was calibrated by varying the

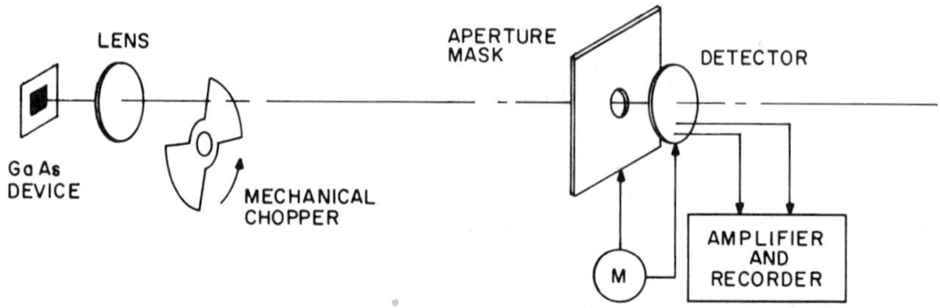

FIG. 20. Experimental apparatus for the direct measurement of current distribution.

position of the diode relative to the detector. The system was found to be linear over a reasonable range. The experimental setup is shown in Fig. 20.

It is also necessary to correlate the current-density values with the intensity values for the recombination radiation. The recombination radiation for a GaAs *p–n* junction diode is reasonably linear if we ignore the effects of heating and nonradiative recombinations. Measurements of filamentary GaAs diodes have shown that the intensity of the light output is proportional to the square of the total current density. A reasonable model for light output

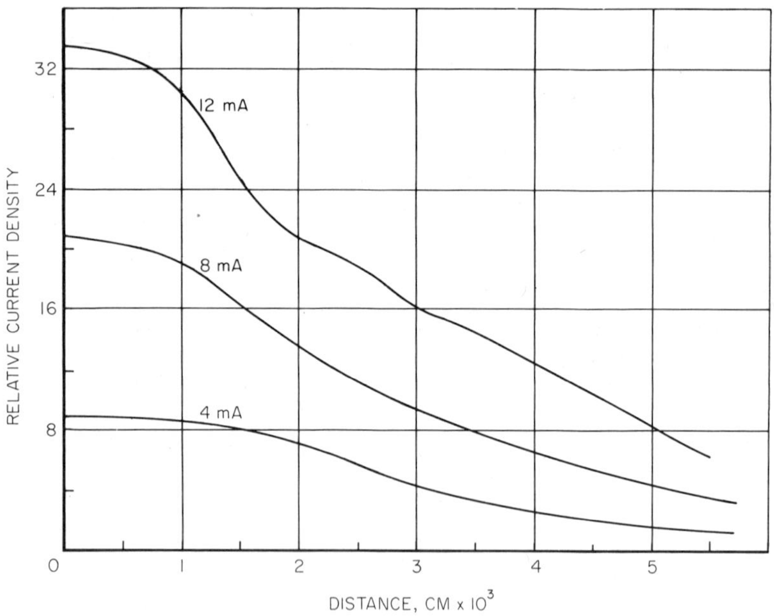

FIG. 21. Relative current density versus distance for device LES-J2.

of this nature is to assume that recombination is occurring in an area dominated by free injected holes p and free injected electrons n. One would then expect the level of recombination to be the product of p and n, as opposed to the product of the free-carrier concentration and the number of traps. The simplified Lampert theory shows that, in the postbreakdown region, one could resonably expect the number of injected free holes to be equal to the number of injected free electrons, and, thus, the total current would be proportional to the total number of carriers. Therefore, one could postulate that the light intensity would be proportional to the square of the current density. This assumption is supported by noting a square-law dependence of total light on total current in the region of interest. The assumption also corresponds to studies made in GaP.[53]

The square root of the directly measured values of optical output was translated to relative current density. Figure 21 shows the adjusted current-density values versus the radial distance for the present experiment. Note that, at 12 mA, a second filament is beginning to emerge. These experimental data are of the same form as the computed curves of Fig. 9. It would be presumptuous, at this early date, to attempt any closer correlation between theory and experiment; therefore, the current density has been left in arbitrary units. The computed curves of Fig. 9 show more detail toward the center of the filament than could be determined with the present experiment. The practicality of this experimental technique has been demonstrated; it is still too early to place much weight on the quantitative results.

The observation of current filaments in gallium arsenide is similar to the case in silicon.

c. Difficulties and Variations in GaAs Double-Injection Experiments

Clean, reproducible fabrication techniques are not as common for GaAs as for Si. Therefore, there is a large variation in the interpretation of experimental results for *P–I–N* devices made from GaAs and other compound semiconductors. The basic device is usually fabricated by placing an electron-injecting contact on one face of the semi-insulator and a hole-injecting contact on the opposite face. Experience has shown that surface impurities can cause the formation of another layer, at the interface between the substrate, and either a liquid- or vapor-phase epitaxially grown junction. The characteristics of these sublayers often dominate the device. In addition, zinc diffusion, which is commonly used to form the p^+ junction, causes a band of dislocations to form at the interface, where the zinc diffusion front changes from interstitial to substitutional. This band of dislocations and other imperfections has been identified through stain-etching techniques, and can dominate the characteristics of these devices. At the General Electric Company's Electronics Laboratory, Jensen has fabricated negative-resistance structures by diffusing zinc into n-type GaAs.[69] He has noted the formation of a high-resistivity sublayer during the processing. Weiser[31] also noted this high-resistivity sublayer effect after the original work of Dumke[24] and of Weiser and Levitt.[23] Meyerhofer *et al.*[41] have described a switching diode that was first obtained by diffusing zinc at 900°C into 0.1 ohm-cm, n-type GaAs layers. This device, with negative resistance, operates in a fashion similar to the double-injection diodes discussed in this section. It will be discussed again in Section 13a.

It has also been discovered that surface leakage current can affect the magnitude of the breakdown voltage. The guard-ring device LES-J2 yields a

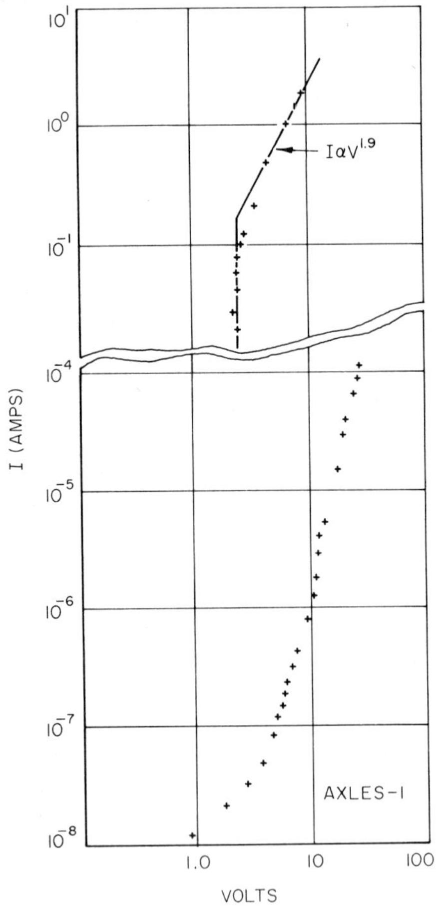

FIG. 22. Voltage–current characteristic for GaAs device with high surface leakage current.

[69] H. A. Jensen, Private communication, 1966.

threshold voltage of 70 V and a minimum voltage of 6 V. The same device with the guard-ring around the n^+ solution regrown mesa removed exhibited a threshold of 30 V. The minimum voltage remained unchanged. The size of the device has been reduced to a square 0.005 in. on a side, and the mesa structure was eliminated so that the injecting contacts had the same cross-sectional area as the entire device. These devices have very low (approximately 10 V) threshold voltages. Threshold currents of devices that have little surface leakage are usually in the 1-μA range. Devices of small cross-sectional area show leakage currents up to several hundred milliamperes. Figure 22 shows a device with a high surface leakage current that is quite similar, except for its small geometry (5 mils on a side), to the large-area step device of Fig. 15. (The low leakage current exhibited by this unpassivated stepped structure was fortuitous and is not generally reproducible.)

Threshold voltages in excess of 400 V have been observed for guard-ring devices similar to LES-J2. These high-voltage devices show currents below 1 μA at 400 V. It is probable that the guard-ring has improved the threshold characteristics of LES-J2, but that leakage currents are still present.

At this point, it is known that leakage current can reduce the threshold voltage, but the mechanism is not clear. The leakage path may permit majority carriers to directly fill traps near the minority-carrier injecting contact. A more plausible explanation could be that the surface leakage current causes recombination radiation at the p^+ junction, and it is the reabsorption of this radiation that permits the lowering of the threshold voltage. The sensitivity of these devices to light has been shown by many investigators, including Weiser and Levitt[23] and Holonyak.[14] Holonyak has shown this effect in semi-insulating GaAs, Co-doped silicon, and Cu-compensated germanium. The effect is shown clearly in Fig. 23, which is a reproduction of the Holonyak data showing light sensitivity of a Co-doped silicon *P–I–N* device. Figure 24 shows typical data, from Weiser and Levitt, showing relatively high prebreakdown leakage currents.

The model which considers that surface leakage causes recombination radiation that is reabsorbed would indicate that GaAs and other direct-energy-gap semiconductors are more sensitive to surface leakage effects than silicon or other indirect-gap semiconductors. Further, all negative-resistance devices would be sensitive to light, since it was postulated in Section 3 that the carrier concentration causes breakdown.

The results of many workers who have been studying GaAs *P–I–N* devices seem to be compatible with the double-injection filamentary theory. The reabsorption of recombination radiation may satisfactorily explain the low threshold voltages that are commonly observed in GaAs. At this time, there is little published evidence to contradict the two-carrier space-charge-limited current filament model for the postbreakdown region. The existence

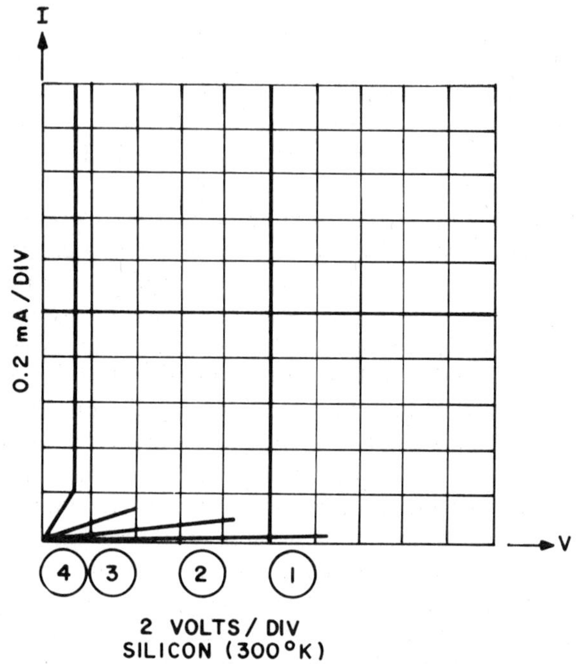

FIG. 23. Sensitivity of threshold voltage to light of cobalt-doped silicon device. Light is increased with increasing number. (After Holonyak.[14])

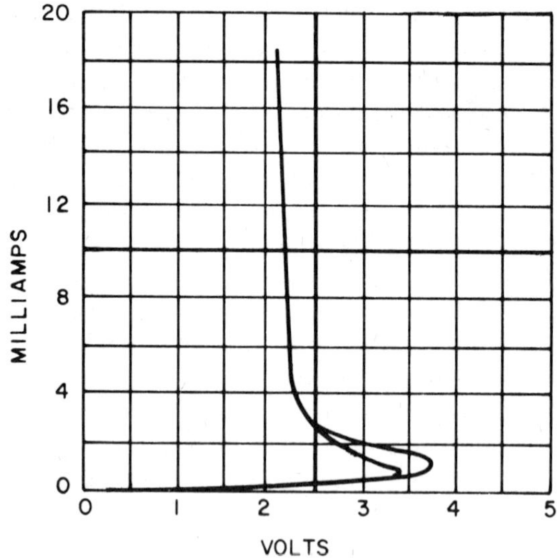

FIG. 24. High prebreakdown currents for GaAs *P–I–N* device. (After Weiser and Levitt.[23])

of the filament depends on both the cross-sectional and the longitudinal geometries, which are described in Section 12.

The nature of *P–I–N* breakdown in gallium arsenide has yet to be completely resolved, and Haisty[70] and Selway and Nicolle[71] have both noted that the breakdown observed in chromium-doped GaAs is different from that reported for oxygen-doped material. They have suggested the possibility that the Dumke theory more nearly fits their observed experimental results.

9. Additional Evidence of Current Filaments

Additional evidence of current filaments has been directly observed in at least four different reports. In this section, these observations and experiments will be reviewed. While all of these authors reported spots of light through a transparent injecting contact, only two of these reports attributed the spot of light to the possibility of current filaments. Ing *et al.*[25] reported the emission of a small spot of 8800-Å radiation through the zinc-diffused *p* contact in gallium arsenide *P–I–N* devices. The n^+ junction was an alloyed tin sphere. They reported that they were unable to determine whether the geometrical configuration of the diode had any effect on the location and size of the region that emitted light.

Crowder *et al.*[38] reported evidence of current filaments in their analysis of ZnTe avalanche injection devices with negative resistance. These devices exhibited a threshold voltage of 15–20 V at liquid-nitrogen temperatures and a minimum voltage of 8–15 V. Discussing the efficient injection electroluminescence at 77°K, they said that the light emission observed under the negative electrode comes from a small spot, so that local current densities may be orders of magnitude higher than predicted by uniform current densities. Following the publication of the work on silicon described in Section 7, current filaments were directly observed in evaporated silicon films by Braunstein *et al.*[49] These workers attributed the negative resistance to double injection, based on their observation of current filaments.

Gerhard and Jensen[47] noted negative resistance and visible light from $GaAs_{1-x}P_x$ devices that were fabricated with alloy n^+-injecting contacts and p^+-diffused junctions, similar to earlier GaAs devices. They comment that the voltage threshold for switching was not satisfactorily correlated with any existing theories. However, in the high conducting state, they observed that the light emission appears to result from a filamentary conduction current.

Current filaments have been observed frequently in several different materials. More observations of these filaments would probably have been

[70] R. W. Haisty, Private communication, 1968.

[71] P. R. Selway and W. M. Nicolle, *J. Appl. Phys.* **40**, 4087 (1969).

reported, except for the fact that it is not common to observe the light output in complex device structures through a transparent injecting contact. It is often a more efficient technique to remove this light output from a junction edge, thereby eliminating the possibility of observing filaments.

IV. Some Boundary Conditions

Some of the boundary conditions for filament formation are discussed in this section. The model referred to in Section 5 is expanded to provide a complete description of a single filament. The observation of multiple filaments and the device length limitations are also reported.

10. Complete Description of a Single Filament

Equation (30) is a simplified form that describes a single filament quantitatively in the high-current power-law region. It is now necessary to develop a model that extends the current–voltage characteristic predicted by these equations into the region where the current increases at constant voltage. It is postulated that if the minimum voltage and minimum current, V_M and I_M, can be found, then this point can be joined to the power-law equation by assuming that the current increases at constant voltage until it intersects the high-current power-law region. This model is offered as an interim solution to the problem. The complete solution should result from an extension of the computed solutions described in Section 6.

Lampert has described the minimum voltage V_M after breakdown as being approximately related to the product of the threshold voltage and the capture cross-section ratio, as shown in the following equation for the case where $\tau_n \gg \tau_p$:

$$V_M \approx (\sigma_n{}^0/\sigma_p{}^-)V_{TH}. \tag{40}$$

Experimental results, however, do not support this expression.[3,14] It is not clear, at this time, whether the discrepancies are caused by a fault in the theoretical model or in the experimental vehicles. Contact resistance and irregular diffusion layers can cause increases in the values of the observed minimum voltages. Various problems associated with the experimental observations of threshold voltage were discussed in Section 8. The study carried out on semi-insulating silicon did allow some interesting observations to be made on the value of the minimum current.

The current-density expression for the filament, Eq. (28), derived from Eqs. (27) and (7) with $\bar{n} = n$, can be written in the form

$$J(r) = \frac{en_0(\mu_n + \mu_p)(V/L)}{[1 + (n_0 C/6D_a)^{1/2} r]^2}, \tag{41}$$

where n_0 is the carrier concentration at the center of the filament. Equation

(41) has been successfully used to account for variations in current density of filaments in the $I \propto V^{1.5-1.9}$ region. In the minimum-voltage region, the voltage serves to establish the carrier transit time and excess space charge. It is, additionally, postulated that the carrier concentration n_0 at the center of the filament decreases with decreasing current (since it is no longer controlled by the voltage as in the power-law region) until it reaches a minimum value. Beyond this minimum value of injected carrier concentration, the filament extinguishes.

It has been postulated that this minimum carrier concentration, n_0', is the free majority-carrier concentration that exists when majority-carrier traps have been filled or saturated, which is the same situation as the condition for breakdown. This number was determined for semi-insulating silicon by computing the hole concentration at V_{TH}. A form of the Ashley–Milnes equations was arranged by Schibli for computer use.[72] The capture cross sections for indium were taken from the work of Wagener and Milnes.[32] The hole concentration at threshold for device No. 16-5 was 3×10^{11} cm^{-3} at a point that was approximately 90% of the device length from the p^+ junction. The assumptions involved in this calculation render this value as an approximation, probably valid within a factor of three or, possibly, less.

Equation (41) was integrated for devices of radius R using the above value for the minimum n_0'; the resultant values of the minimum voltage were compared to the observed values for a number of devices. Since most of the devices were not circular, R was taken as $(A/\pi)^{1/2}$, where A was the actual device cross-sectional area. The experimental and theoretical values of minimum current are shown in Table I; note that they are in closer agreement than might have been expected. The functional agreement over the range of devices used appears to favor the minimum-carrier-concentration concept presented above.

At this point, minimum-current experiments have not been carried out for GaAs. Devices of similar length and area show variations in minimum current of several orders of magnitude. These variations are presently attributed to the imperfect junctions formed during the epitaxial growth and diffusion processes. In many cases, heat treatment, similar to that used for semiconductor laser diodes,[72a] has lowered the minimum current by as much as an order of magnitude. Although GaAs devices are excellent experimental vehicles for studying the low-current region, processing variations often dominate the semi-insulator characteristics.

The prebreakdown work of Lampert[2] and Ashley–Milnes[28] can be combined with this postbreakdown model. Figure 25 shows the complete charac-

[72] E. Schibli, Private communication, 1965.

[72a] R. O. Carlson, *J. Appl. Phys.* **38**, 661 (1967).

TABLE I

MINIMUM CURRENT I_M, THEORETICAL AND EXPERIMENTAL

Device number	Length (10^{-2} cm)	Radius, R (10^{-2} cm)	V_M (V)	I_M experiment (mA)	I_M theory (mA)
17.3	1.4	6.7	0.15	0.43	0.39
17.9	1.4	2.7	0.18	0.14	0.11
18.3	1.7	5.2	0.44	0.48	0.50
18.7	1.7	9.8	0.52	1.89	1.54
16.5	2.2	4.4	0.94	0.72	0.71
16.8	2.2	1.75	1.6	0.52	0.58
25.1	1.65	4.4	0.25	0.26	0.25

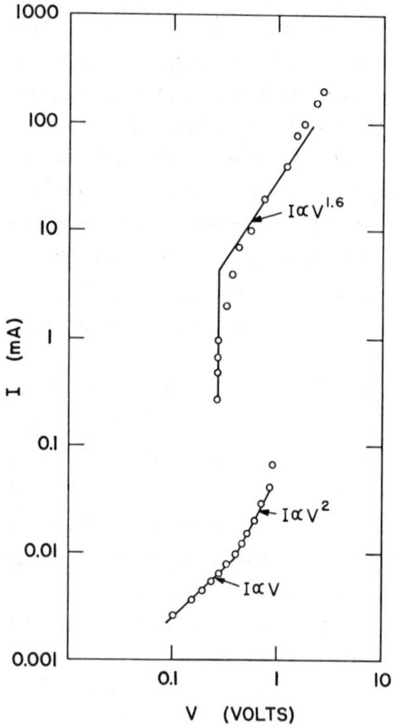

FIG. 25. Characteristic I–V curve for device No. 25-1. The solid lines represent the prebreakdown characteristic of Lampert–Ashley–Milnes and the postbreakdown characteristic of the complete single-filament model. (After Barnett and Milnes.[3])

teristic curve for device No. 25-1. At very low currents, the device exhibits ohmic conduction, followed by minority-carrier space-charge-limited current. Beyond the threshold voltage, negative resistance sets in and the current increases from a minimum value, predicted by this model, at nearly constant voltage, to a power-law region, $I \propto V^{1.6}$. Beyond approximately 50 mA, the experimental results depart from the theoretical line because of the onset of a second filament. The emergence of the second filament is discussed in the following section.

11. Multiple Filament Formation

Figure 26, which is a curve-tracer plot for device No. 25-1, accents the breakdown of the second filament, and also shows that the retrace path maintains the formation of both filaments. Recombination radiation photographs of multiple filaments are shown in Fig. 27.

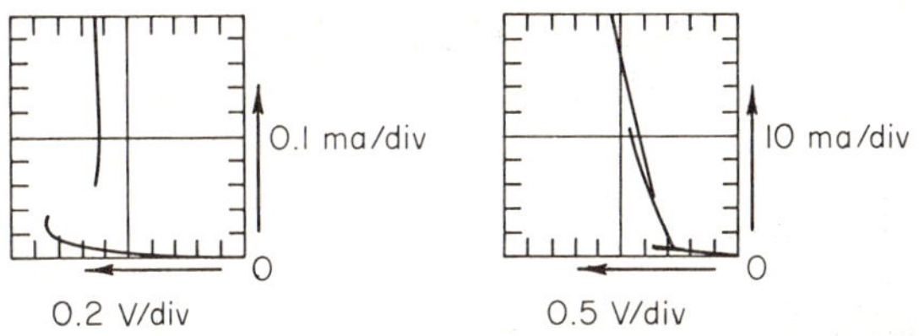

Fig. 26. Curve-tracer plot of device No. 25-1.

The formation of secondary filaments is always observed when the device dimensions are sufficiently large. Secondary filaments may form (1) when the voltage across the primary filament is as large as the threshold voltage, (2) when the carrier concentration due to the radial diffusion of carriers from the primary filament is sufficient to create the second filament, (3) from thermal considerations, or (4) in direct-gap materials, from the reabsorption of recombination radiation. The voltage across device No. 25-1 was used to predict the current distribution of the primary filament at both 80 and 160 mA. The current concentration at the center of the second filament was determined experimentally. A comparison between the theoretical and experimental values is shown in Fig. 28. The solid line joining the two filaments has been determined by the addition of the equation for single filaments. At 160 mA, interaction between the filaments is quite large, which is consistent with Ridley's[4] model of minimum joule heating.

Multiple filaments have also been observed in GaAs. Photographs of these filaments for device LES-J2 are shown in Fig. 29. These secondary filaments, which occur at slightly lower currents in GaAs, may be caused by greater

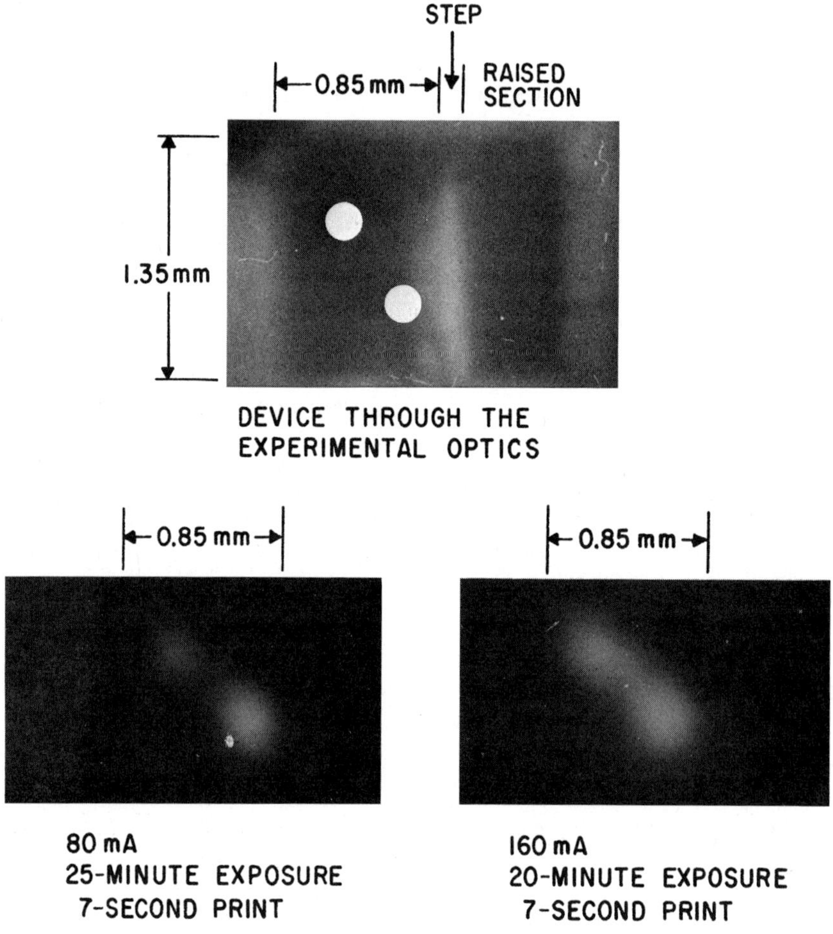

FIG. 27. Recombination radiation photographs of multiple filaments of silicon device No. 25-1. (After Barnett and Milnes.[3])

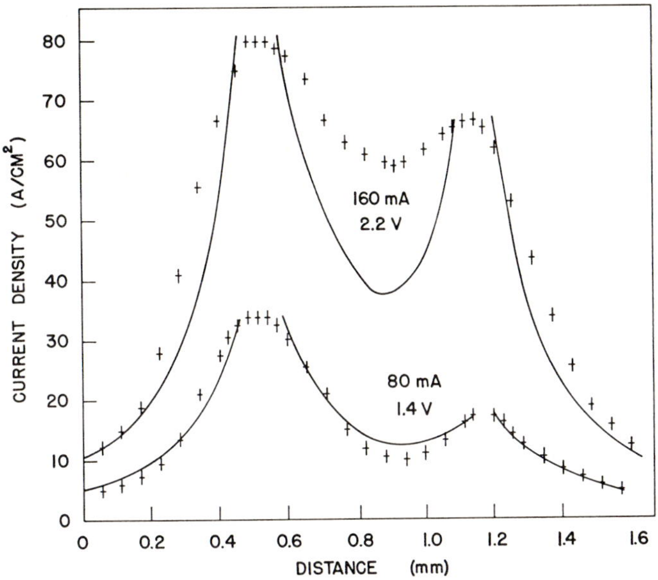

FIG. 28. Current density versus distance for the multiple filaments of silicon device No. 25-1. (After Barnett and Milnes.[3])

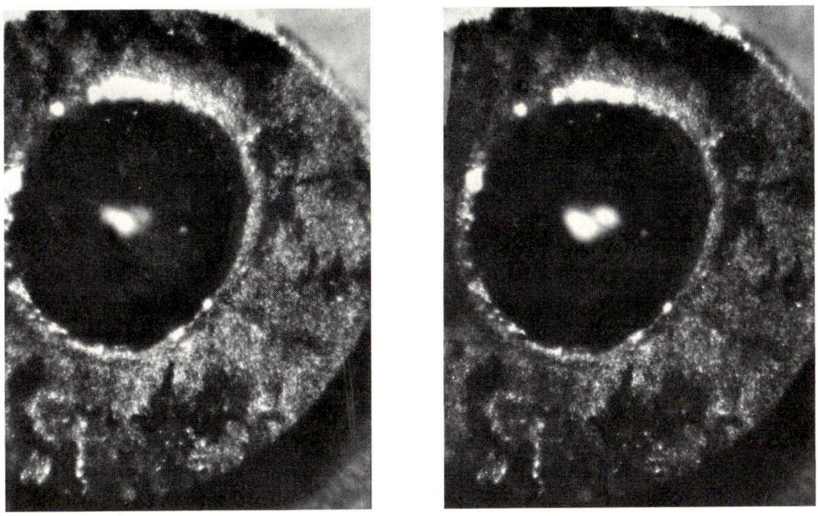

20 mA
2-Second Exposure

30 mA
2-Second Exposure

FIG. 29. Multiple-filament photographs from GaAs device No. LES-J2.

local heating for these room-temperature devices or by the internal re-absorption of the recombination radiation of the primary filament.

The onset of a second filament may be accompanied by a small discontinuity in the current–voltage characteristic and/or by a slope change in the characteristic curve.

12. Device Length Limitations

As the semi-insulator is decreased in length (thickness), a minimum length is reached for the establishment of filamentary conduction. There is also a maximum device length, which has been associated, in semi-insulating silicon, with an electric-field-dependent breakdown phenomenon. Both the minimum and maximum length limitations are discussed in this section.

a. Minimum Length

For all of the devices in this study, the junction voltage has been subtracted from the total device voltage to yield the voltage drop across the semi-insulator region. There is a minimum length of the semi-insulator material, approximately 0.002–0.004 in. depending on the material involved, beyond which it is difficult to separate the characteristics of the semi-insulator region from those of the junction. For a limited range of these nonnegative-resistance devices, the nucleation of filaments is still observed.[72b] As device length is still further decreased, the characteristic curve becomes identical to that of a *p–n* junction diode, and the current distribution in the forward direction is uniform throughout. Therefore, the device length, just before current distribution becomes uniform, is the minimum length.

b. Maximum Length

A maximum device length for filamentary conduction has been observed. Due to limitations in the experimental techniques, this length has not been exactly determined for semi-insulating gallium arsenide, but appears to be of the order of 0.4 mm. This maximum length depends largely on the material parameters, such as deep-impurity concentrations and the level of shallow acceptor compensation. In the semi-insulating silicon studied, this maximum length was observed to be 0.22 mm.

The postbreakdown characteristic curve for long, nonfilamentary devices is similar to the $I \propto V^2$ region originally predicted by Lampert. Lampert's model assumes that the current density is uniformly distributed throughout the device. Recombination radiation photographs taken in this apparently nonfilamentary region have indicated a uniform current density. This region is similar to the high-current $I \propto V^2$ region that Mayer *et al.*[7] have reported

[72b] A practical application for devices that exhibit no negative resistance, but still exhibit filamentary conduction, will be discussed in Section 13c.

for their long silicon devices. The length of their devices is 0.2–0.8 mm, which is consistent with this study.

It has been possible to experimentally correlate maximum length with a critical electric field. As the device length is increased, the electric field strength of the postbreakdown region also increases. There is evidence that shows that there is a critical electric field value that provides an upper limit on filament formation. If the filament would normally be formed at an electric

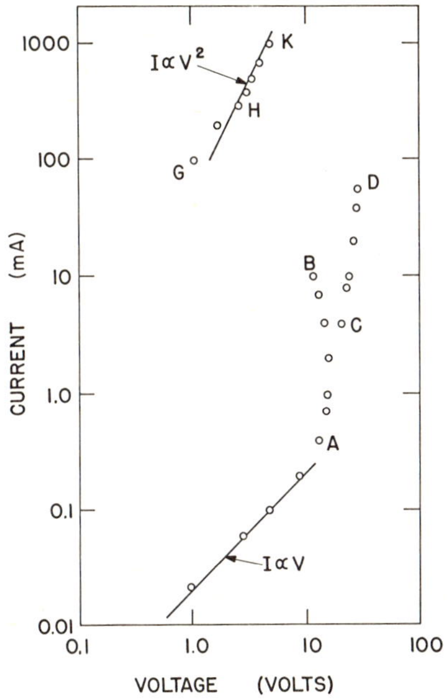

FIG. 30. Voltage–current characteristic for a long, nonfilamentary silicon device.

field higher than this critical field, the device will be forced into a nonfilamentary conduction mode.

A typical characteristic curve for longer devices is shown in Fig. 30. The ohmic region extends to point *A*, and is followed by the section *A*–*B*. As the voltage is increased beyond point *B*, the device characteristic switches back to that observed at point *C*. From point *C*, increasing the voltage generates the curve that passes through point *D*. Beyond point *D*, double-injection breakdown causes the characteristic curve to switch to point *G*. It was necessary to use pulse techniques to trace the characteristic curve from *G* through *K*, since the device showed considerable heating effects in this region.

The theoretical high-current $I \propto V^2$ region has been calculated using the original Lampert equation. With the same lifetime conditions that were used in the filament study, the straight line in the high-current region can be drawn. The lower voltage at point G has been attributed to heating effects,

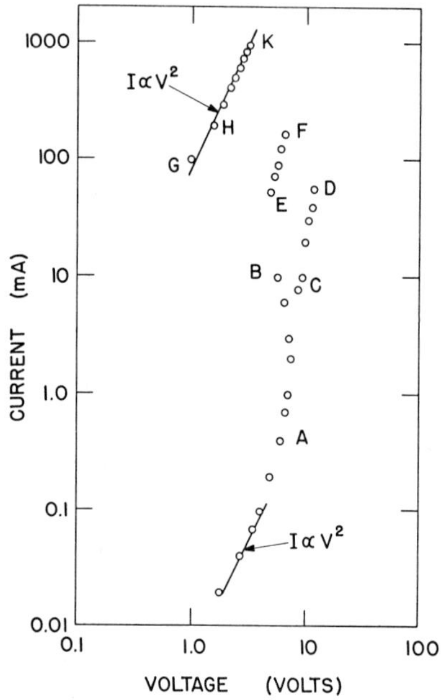

FIG. 31. Voltage–current characteristic for a device with a partially filamentary postbreakdown region.

and, for that reason, additional points at higher currents were plotted using pulse techniques.

An attempt was made to apply a high electric field across an already established filament. Figure 31 shows the characteristic curve of this device. The prebreakdown region is traced along the path A–B–C–D, similar to that of Fig. 30. At point D, breakdown occurs up to a point along the path E–F. The path E–F represents the nearly constant voltage filamentary region (as shown by recombination radiation photographs). The path G–H–K is the nonfilamentary conduction region predicted by Lampert's theory.

The electric field across the filament, which terminates at point F of Fig. 31, is the same as the electric field that appears in the discontinuity of Figs. 30 and 31 from point A to B. The effect of this electric field was studied

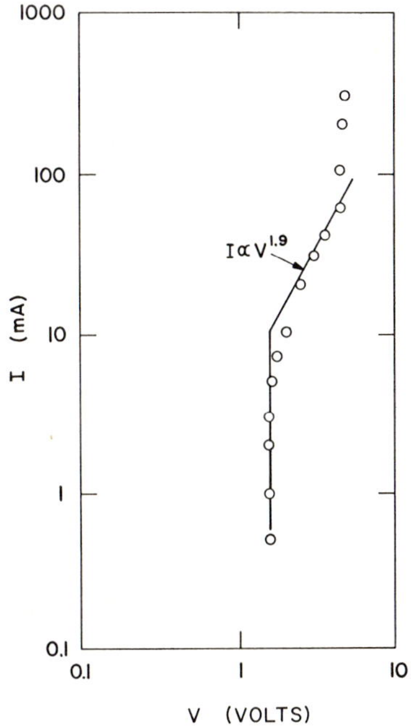

FIG. 32. Voltage–current characteristic for a contained filament showing the effect of a high electric field. (After Barnett and Milnes.[3])

using a device of very small cross-sectional area, so that the filament completely filled the device when the critical field was reached. The prebreakdown characteristic of this device was dominated by leakage currents, and is not shown. The postbreakdown characteristic of the device is shown in Fig. 32. If the current is increased beyond the critical electric field strength, the voltage remains nearly constant. Current increase at nearly constant voltage is a characteristic of impurity impact ionization for material with a low level of impurity compensation.[73]

c. Impurity Impact Ionization

Impurity impact ionization has been invoked to explain the significance of the critical electric field strength. It would be convenient if one could calculate, from the energy level of the impurity and the velocity of the carriers, the theoretical value of this electric field. Unfortunately, the experimental evidence available at this time does not permit this. Bok *et al.*[74] and

[73] W. Kaiser and G. H. Wheatley, *Phys. Rev. Letters* **3**, 334 (1959).

[74] J. Bok, J. C. Sohm, and A. Zylberstejn, *Proc. Intern. Conf. Semicond. Phys., Prague, 1960*, p. 138. Czech. Acad. Sci., Prague and Academic Press, New York, 1961.

Sohm[75] have shown that the electric field strength is dependent on the compensation, or the ratio of impurities of one type to impurities of the other type. For compensated material, the critical electric field strength can be quite low (values as low as 100 V/cm have been observed) for low levels of compensation, i.e., $N_D/N_A = 5 \times 10^3$. The electric field strength for similar material with a high level of compensation, $N_D/N_A = 10$, can be quite high (1400 V/cm). In addition, for highly compensated material, a negative resistance has been observed. This is the impurity impact ionization mechanism that Melngailis and Milnes[18] showed to be filamentary for compensated germanium at 4.2°K.

When the level of compensation is low, one can expect the current to increase at nearly constant voltage beyond the critical electric field strength. Using the semi-insulating silicon used for the filament study, $p^+-\pi-p^+$ devices have been constructed in order to separate this effect from possible double-injection breakdown effects. A typical characteristic curve for a device of this type is shown in Fig. 33. The curve departs from linearity between 2 and 3 V, which corresponds to an electric field strength of approxi-

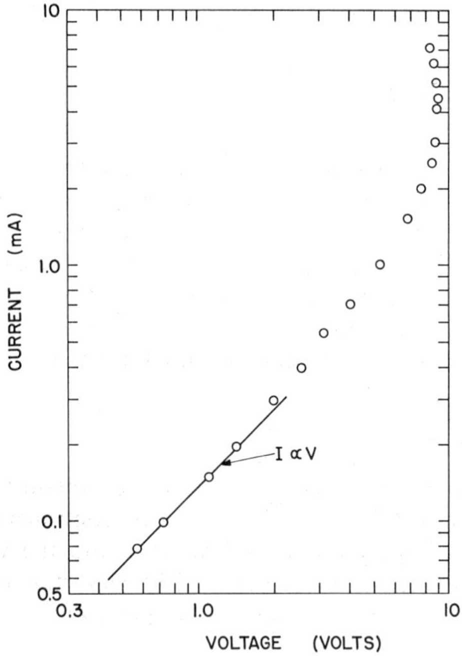

FIG. 33. Voltage–current characteristic for a $p^+-\pi-p^+$ device.

[75] J. C. Sohm, *J. Phys. Chem. Solids* **18**, 181 (1961).

mately 200 V/cm. The characteristic curve then breaks into a current rise, at nearly constant voltage, at a field strength of approximately 800 V/cm.

Studies using devices that vary in length by a factor of four show that the voltage in the region where the current rises at nearly constant voltage is directly proportional to the device length. Therefore, there seems to be an electric field effect. This effect cannot be the trap-filled-limited effect associated with single injection, which is transit-time dependent and, therefore, would depend on the square of the device length.

The nature of this material breakdown phenomenon has not been fully determined. It is, unfortunately, necessary to invoke related experimental results, instead of related analytical results, to explain the limiting electric field phenomenon. Nevertheless, experimental evidence points to the onset of an impurity impact ionization mechanism as a limiting condition for filamentary conduction. A similar effect has been observed in semi-insulating GaAs.[76] The physical mechanism that causes this phenomenon to interact with the filament and inhibit its formation has not been fully investigated.

V. Concluding Remarks

This section describes some device applications of filamentary double-injection devices. In addition, some of the problems and suggested techniques for their solution will be discussed.

13. Electronic Devices

In this section, devices whose basic characteristics are similar to those described for filamentary devices will be discussed. These devices are grouped in the class of filamentary devices, but do not include the basic two-terminal electronic device, although the interaction of two-terminal devices with external effects, such as multiple terminals and photons, will be discussed.

a. Light-Activated Switches

Several authors have shown that threshold voltage can be lowered when the devices are irradiated. Figure 23 shows the data of Holonyak[14] for a cobalt-doped silicon device under increasing amounts of radiation. As the light intensity is increased, the threshold voltage decreases from 13 to 4 V. The minimum voltage seems to remain constant at a little more than 1 V.

Marsh *et al.*[77] have been able to turn off a double-injection device by using light of a critical wavelength. It has been postulated that this is the wavelength that is necessary to change the trap occupancy. This effect was

[76] B. F. Stein and J. F. Seitchik, *Bull. Am. Phys. Soc.* **12**, 406 (1967).

[77] O. J. Marsh, R. Baron, and J. W. Mayer, *Appl. Phys. Letters* **7**, 120 (1965).

observed in nonfilamentary devices, but has not been reported for filamentary devices.

b. Negative-Resistance Light Emitters

Reasonable quantities of emitted light have been reported in the high-conductance state by various workers. It is postulated that an individual device could be biased in the bistable mode and triggered into the high-conductance mode, where the light output is observed, or turned back into the low-conductance mode, where no light is emitted and negligible current is observed. This device could be used as a light emitter with inherent memory.

c. Filamentary Illuminator

When the device length is decreased, a region is encountered where current filaments are still observed, but without negative resistance. A glance at the simplified filament equation, Eq. (28), shows that the current density at the filament center is inversely proportional to the cube of the device length. Therefore, it is postulated that very high current density and, therefore, efficient light-emitting devices can be fabricated. The advantages of such a device over a normal *p–n*-junction illuminator are the small spot size and the relatively high reverse breakdown voltage typical of a *P–I–N* structure. The characteristic curve for a GaAs device of this type is shown in Fig. 34. At present, efficiencies as high as 1% have been observed in these devices.

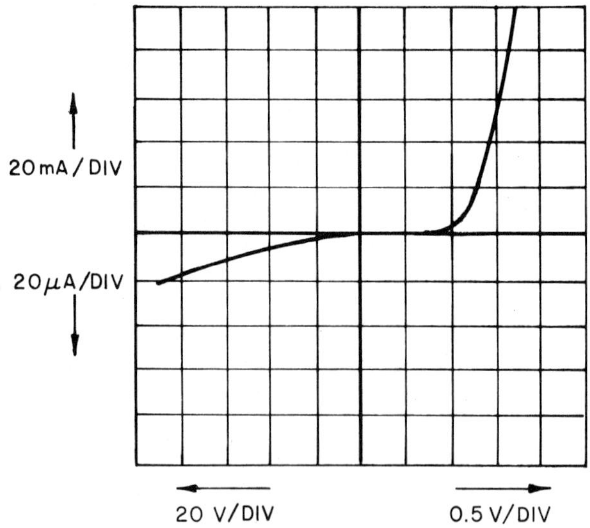

FIG. 34. Voltage–current characteristic for a filamentary illuminator.

d. Three-Terminal Switch

Several authors have reported a device in which the threshold voltage can be lowered by the application of a third terminal. Ing and Jensen[21] and Weiser and Levitt[23] have built GaAs light-emitting diodes on the same substrate as the negative-resistance device. They used the light-emitting diode to lower the threshold voltage. Shohno[37] controlled threshold voltage by applying a gate connection at a point between the *n* and *p* contacts in the *I* region of a silicon *P–I–N* diode. He was able successfully to lower the threshold voltage from 9.8 to 5.6 V.

More recently a three-terminal GaAs device has been developed[78] which has shown average turn-on speeds below 1 μsec and turn-off times less than 15 μsec. Threshold voltages for this device commonly exceed 200 V and gate turn-on currents are typically in the 1 μA to 1 mA range.

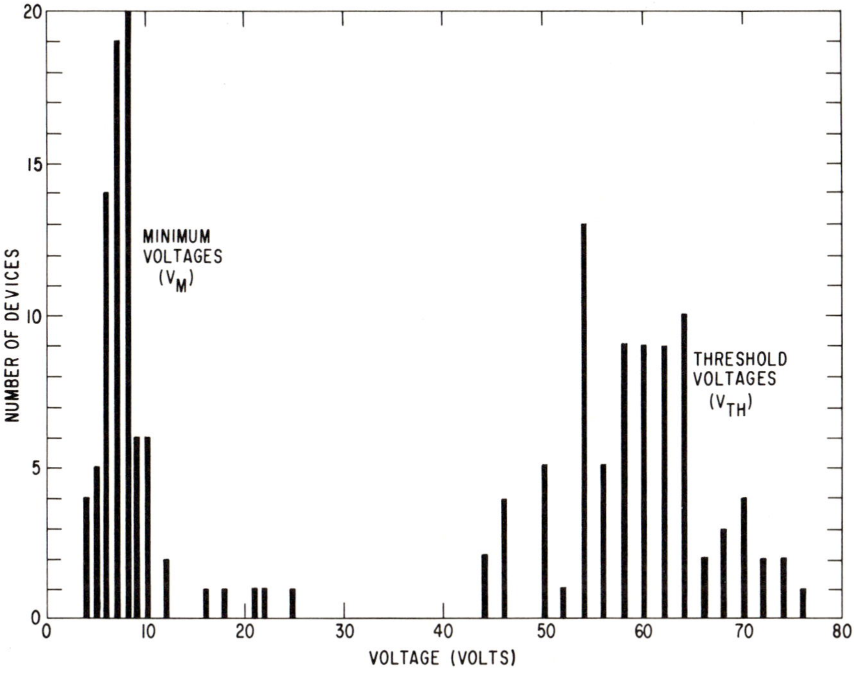

FIG. 35. Distribution of threshold and minimum voltages for LES-IC P49B. (After Barnett and Glusick.[79a])

[78] A. M. Barnett and H. A. Jensen, unpublished.

[79] A. M. Barnett, R. E. Glusick, H. A. Jensen, V. F. Meikleham, and D. C. Osborn, *IEEE Intern. Solid-State Circuits Conf., Digest Tech. Papers* **12**, 56, IEEE Cat. No. 69C9-ISSCC (1969).

[79a] A. M. Barnett and R. E. Glusick, *IEEE J. Solid-State Circuits* **SC-4**, 346 (December 1969).

e. Light-Emitting Switch-Integrated Circuit

The basic negative-resistance light-emitter has been used to make large arrays of matrix-addressed light-emitting elements with inherent memory.[58,59,79,79a] Arrays as large as $\frac{1}{2}$ in. on a side with up to 100 elements on 0.050-in. centers have been successfully developed. It was necessary to achieve a high level of uniformity of the threshold voltage and of the minimum voltage in the course of developing these arrays in order to permit matrix (row and column or half-select) address. Figure 35 shows the distribution of threshold and minimum voltage for the 81 devices of LES-IC P49B. This array can be operated with a 12-V steady-state bias and 32-V pulses on the rows and columns. Seventy-seven of the 81 devices can be successfully operated under these matrix-address conditions. The remaining four devices return to the low-conductance state after the turn-on pulse voltage is removed. It is believed that the uniformity achieved in the characteristics for these large arrays, as shown in Fig. 35, can be achieved routinely in devices of this nature, utilizing conventional processing techniques.

f. Madistor

Melngailis and Rediker[13] were able to show the effects of large magnetic fields on InSb *P–I–N* diodes. They were able to change both the threshold and minimum voltages by applying fields of 20–80 G. They have proposed a device, called the Madistor, which would utilize this effect. The characteristic curve for this device under the influence of various magnetic fields is shown in Fig. 36. They have proposed a structure that involves mounting the semiconductor device in the air gap of a small ferromagnetic toroid.

They have also proposed several devices that utilize the switching characteristic of this filament from one point to another. Included in these devices is a novel stepping switch with a centrally located n^+–p junction surrounded by 10 branches with p^+–p junctions.

g. Infrared Detector

Zuleeg[80] has suggested a device with intrinsic photoresponse in silicon by proper introduction of deep-lying donor and acceptor impurities. He would use a change in current in the prebreakdown region at a fixed voltage to measure the photoexcitation. He is presently fabricating devices with various levels of deep impurities in order to evaluate the existence of significant intrinsic photoresponse beyond the intrinsic threshold of 1 μm. Zuleeg postulates that it would be possible with such devices to detect near- and far-infrared spectra with specific doping.

[80] R. Zuleeg, Private communication, 1967.

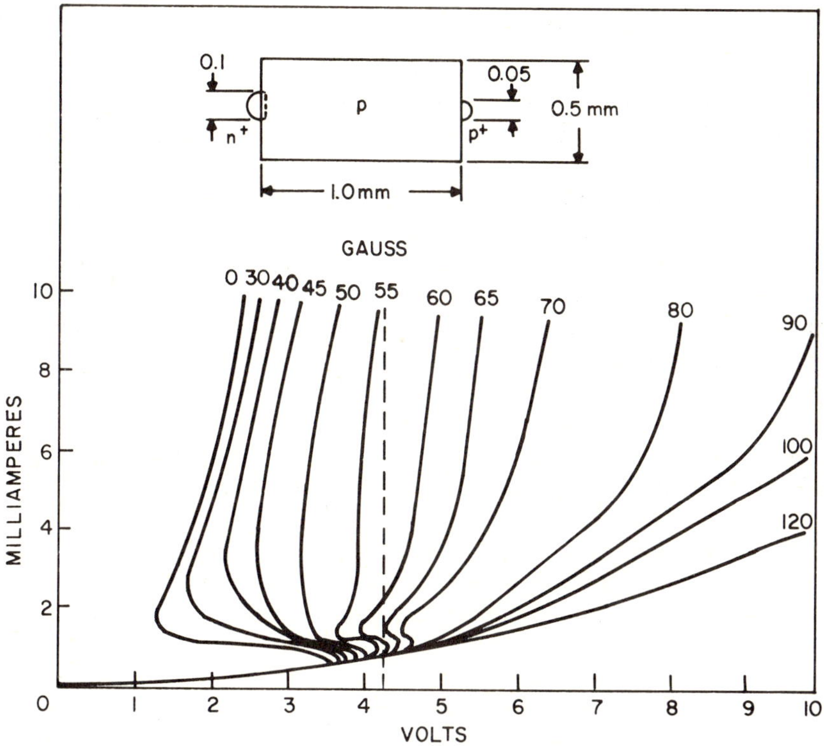

FIG. 36. Voltage–current characteristic for InSb Madistor at 77°K as a function of magnetic field. (After Melngailis and Rediker.[13])

h. Oscillating Devices

Holonyak and Bevacqua[20] first reported oscillations in the prebreakdown, postbreakdown, and secondary breakdown regions of *P–I–N* devices made from cobalt-compensated silicon. Studies of the prebreakdown oscillations have continued.[42,43,81,82] If a suitable model can be developed for these observed prebreakdown oscillations, useful devices might be developed. The author believes that a model for these oscillations might be based on the incomplete nucleation of a current filament. The space-charge effects due to unstable or extinguishing filaments could successfully explain the observations.

[81] J. S. Moore, C. M. Penchina, N. Holonyak, Jr., M. D. Sirkis, and T. Yamada, *J. Appl. Phys.* **37**, 2009 (1966).

[82] J. S. Moore, N. Holonyak, Jr., M. D. Sirkis and M. M. Blouke, *Appl. Phys. Letters* **10**, 58 (1968).

14. Accomplishments and Problems

Current filaments have been directly observed in single-crystal silicon, gallium arsenide, zinc telluride, gallium arsenide–phosphide, and polycrystalline silicon. There is also evidence of filamentary conduction in other materials.

A simplification of Lampert's theory called the "semi-insulator model" has successfully predicted the general voltage–current characteristic indigenous to filaments. This model has also provided reasonable agreement with experiment for the voltage–current characteristic and the current distribution of single filaments in semi-insulating silicon. A computer solution of the basic equations predicts current filaments through the negative-resistance region.

Although satisfactory agreement between theory and experiment has been achieved over a range of experimental conditions and several of the boundary conditions for filament formation have been described, several problems remain to be solved.

The basic device has not been completely analytically described at this time. The sensitivity of the device to light of specific wavelengths should, together with recombination radiation, prove to be valuable experimental tools for its study.

An experimental apparatus similar to that shown in Fig. 20 would enable one to measure the dynamic formation of a filament. The computed solutions can be extended to include these transient effects.

A better understanding of the boundary conditions is necessary. Specifically, the mechanism that causes a critical electric field to inhibit filament formation must be analyzed. Optical transient studies of extinguishing filaments may be useful.

In conclusion, it seems that space-charge-limited filaments provide a fertile area for the study of current-transport phenomena in semiconductors. There seems to be an abundance of novel effects, as well as problems with tractable solutions, connected with current-filament formation.

Acknowledgments

Many people have directly or indirectly contributed to this work. I would like to thank Dr. I. A. Lesk of Motorola, Inc. and Prof. C. T. Sah of the University of Illinois for timely advice and encouragement. Prof. A. G. Milnes of Carnegie-Mellon University provided guidance throughout the original study; his direction proved to be invaluable. My association with H. A. Jensen of the General Electric Company has enhanced my understanding of the capabilities and limitations of related experiments. I would also like to thank the Technical Information Unit of the General Electric Company's Electronics Laboratory for editorial assistance.

CHAPTER 4

Double Injection in Semiconductors

R. Baron and J. W. Mayer

I. Introduction

1. General Discussion

In insulators and lightly doped semiconductors, current flow in many cases can be dominated by injected carriers of one or both types. This class of problems leads to a number of phenomena, including power-law currents, abrupt transitions in current, negative resistance, and oscillations. In addition, because injection effects are sensitive to recombination and trapping centers, they can be used to study the properties of such centers.

These problems fall into two broad categories, single-carrier and two-carrier (double) injection phenomena. Single-carrier injection occurs when current flow is due to injected carriers of one type and the behavior is not dependent on the presence of carriers of opposite type. In this case, the injected carriers establish a space charge which reduces the field at the injecting contact and leads to the well-known single-carrier space-charge-limited currents. Double injection covers the cases when injected carriers of both types play a role in determining the behavior, even if the current is "carried" by only one type of carrier.

In double injection, the space charge is often reduced to a small fraction of the free-carrier density. The mechanism by which this is accomplished may be attributed to a rearrangement of either mobile or trapped carriers. When this relaxation occurs (in a majority of double-injection cases), the current can greatly exceed that of the single-carrier case. Double injection depends on the presence of both carriers because recombination kinetics are the controlling factor in determining the behavior. The recombination may either be bimolecular (i.e., band-to-band hole–electron recombination) or may occur through one or more sets of localized recombination centers. The latter case, with which we shall be concerned in this chapter, exhibits a greater variety of behavior than the bimolecular case because the changing occupation on the recombination centers with changes in injection level can affect the current–voltage characteristics in various ways.

Recombination-center occupancy affects behavior in two ways: (1) it affects the lifetimes, and thus the recombination kinetics, directly; and (2) it affects the amount of charge which resides on the recombination centers and enters through Poisson's equation. This last mechanism is important only when the density of deep levels is large compared with thermal carrier densities. The discussion herein will follow this natural division. The first part (Sections 3–9) will thus be concerned with the simpler case, where the number of deep centers is small compared with thermal densities. Because this case is simpler, it is here that the effects of carrier diffusion have been investigated both theoretically and experimentally. We will then treat the case (Sections

10–16) where the number of deep centers is large; this often leads to negative resistance and a large number of distinct behavior regimes. To date, the theoretical treatment is by no means exhaustive, and only a few regimes have been verified experimentally, notably the work of Milnes and co-workers at Carnegie-Mellon University. In the last portion (Sections 17–21), we discuss nonsteady-state phenomena which provide both insight into injection phenomena and a useful method of measuring lifetime.

Double injection is not a new concept. As early as 1952, Hall[1] had pointed out that injection and recombination of electrons and holes occurred in the intrinsic region of a thin *p–i–n* structure. This case of injection into a thin *i*-region was later studied in great detail by Herlet and Spenke.[2] Their results were extended to still higher injection levels by Kleinman.[3] Starting in 1959, the theory of double injection in long structures was developed by Lampert and Rose.[4–7] Lampert first considered double injection into an insulator,[4] where, in analogy to single-carrier space-charge-limited currents, space-charge effects are important (although for different reasons). Lampert and Rose[5] then considered double injection into a semiconductor. The influence of deep levels on double injection in insulators and the resulting negative resistance was then treated.[6] The work of Lampert and Rose developed the basic concepts of double injection and provided a framework for the more detailed investigations that followed. An excellent phenomenological review of these basic concepts was given by Lampert.[7] Rose[8] has presented a "comparative anatomy" of high-level double-injection models.

The early experimental work on double-injection phenomena dealt mainly with negative resistance. The paper by Holonyak[9] reviews the experimental situation as of 1962. Since that time, Ashley and Milnes[10,10a] and Brown and Jordan[11] have studied the prebreakdown region (before the onset of negative

[1] R. N. Hall, *Proc. IRE* **40**, 1512 (1952).

[2] A. Herlet and E. Spenke, *Z. Angew. Phys.* 7, 99, 149 (1955); A. Herlet, *Z. Physik* **141**, 335 (1955).

[3] D. A. Kleinman, *Bell System Tech. J.* **35**, 685 (1956).

[4] M. A. Lampert, *RCA Rev.* **20**, 682 (1959).

[5] M. A. Lampert and A. Rose, *Phys. Rev.* **121**, 26 (1961).

[6] M. A. Lampert, *Phys. Rev.* **125**, 126 (1962).

[7] M. A. Lampert, *Proc. IRE* **50**, 1781 (1962).

[8] A. Rose, *J. Appl. Phys.* **35**, 2664 (1964).

[9] N. Holonyak, *Proc. IRE*, **50**, 2421 (1962).

[10] K. L. Ashley and A. G. Milnes, *J. Appl. Phys.* **35**, 369 (1964).

[10a] K. L. Ashley, Investigation of the effects of space charge on the conduction mechanism of double injection in semi-insulators, Ph.D. Thesis, Dept. of Eng., Carnegie Inst. of Technol., Pittsburgh, Pennsylvania, 1963.

[11] J. M. Brown and A. G. Jordan, *J. Appl. Phys.* **37**, 337 (1966).

resistance), and Holonyak and co-workers[12–16a] have investigated the oscillations that occur under certain conditions in prebreakdown regions.

This review is primarily concerned with the development of double injection that occurred after 1963. The details of the framework were being filled in during this period, and the first quantitative experimental verification of the long structure regimes was obtained.

2. General Theoretical Approach

The steady-state equations which define the problem of current flow in semiconductors are well known. They are the current flow equations

$$J_n = e\mu_n\left[(n_0 + \delta n)\mathscr{E} + \beta\frac{dn}{dx}\right], \tag{1a}$$

$$J_p = e\mu_p\left[(p_0 + \delta p)\mathscr{E} - \beta\frac{dp}{dx}\right], \tag{1b}$$

$$J = J_n + J_p, \tag{1c}$$

the particle conservation (continuity) equations

$$\frac{1}{e}\frac{dJ_n}{dx} = r, \tag{2a}$$

$$-\frac{1}{e}\frac{dJ_p}{dx} = r, \tag{2b}$$

Poisson's equation[16b]

$$\frac{\varepsilon}{e}\frac{d\mathscr{E}}{dx} = \delta p - \delta n + \delta p_{\rm R}, \tag{3}$$

and

$$\frac{d\psi}{dx} = -\mathscr{E}. \tag{4}$$

[12] N. Holonyak, S. W. Ing, Jr., R. C. Thomas, and S. F. Bevacqua, *Phys. Rev. Letters* **8**, 426 (1962).

[13] N. Holonyak and S. F. Bevacqua, *Appl. Phys. Letters* **2**, 71 (1963).

[14] J. S. Moore, C. M. Penchina, N. Holonyak, M. D. Sirkis, and T. Yamada, *J. Appl. Phys.* **37**, 2009 (1966).

[15] J. S. Moore, N. Holonyak, M. D. Sirkis, and M. M. Blouke, *Appl. Phys. Letters* **10**, 58 (1967).

[16] J. S. Moore, N. Holonyak, and M. D. Sirkis, *Solid-State Electron.* **10**, 823 (1967).

[16a] M. M. Blouke, N. Holonyak, B. G. Streetman, and H. R. Zwicker, *Solid-State Electron.* (to be published).

[16b] Note that $\delta n_{\rm R}$ is not included in Eq. (3) since, by definition, $\delta n_{\rm R} \equiv -\delta p_{\rm R}$. This comes from the fact that $p_{\rm R} + n_{\rm R} \equiv N_{\rm R}$, and thus, either $n_{\rm R}$ or $p_{\rm R}$ can be used to reflect the state of charge on the recombination centers.

The symbols are defined in a list at the end of the chapter (note that $\beta = kT/e$ here).

These equations are completely general and apply to all problems of current flow in semiconductors and insulators. The appropriate definition of the recombination rate density and the appropriate boundary conditions must be added to them (the choice of boundary conditions will be discussed in Section 4c). Unless otherwise stated, only the case of recombination through a single set of localized recombination centers will be discussed. The Shockley–Read equations for the steady-state recombination rate density may be written as

$$\begin{aligned} r &= [\delta n\, p_{\mathrm{R},0} + \delta p_{\mathrm{R}}(\delta n + n_0 + n_1)]c_n \\ &= [\delta p\, n_{\mathrm{R},0} - \delta p_{\mathrm{R}}(\delta p + p_0 + p_1)]c_p. \end{aligned} \tag{5}$$

The case of bimolecular recombination has been discussed by Parmenter and Ruppel.[17]

This last choice—confining the discussion to cases involving recombination through a single set of recombination centers—illustrates an important point. The above equations, because of their generality, are intractable to simple analysis as they stand. In order to simplify these equations to the point where an analytic solution can be obtained, it is necessary to choose a physical model which permits a sufficient number of restrictive assumptions. When this is done, the resulting behavior represents a limiting case, which, in some cases, may be closely realized by an appropriate choice of experimental parameters. In addition, the simple, analytic nature of the solution allows an insight into the physical nature of the behavior that would otherwise be difficult to obtain. This insight, in turn, allows an understanding of the behavior even when the conditions for the limiting case are not exactly satisfied.

When more exact solutions to the problem are needed (i.e., a solution to a model involving fewer restrictive assumptions), it is necessary to resort to the computer. Such computer solutions are often necessary in the case of double injection because it may be difficult to satisfy the conditions under which the simple solutions mentioned above hold accurately. Thus, it becomes important to consider more general solutions in many cases.

As mentioned earlier, the role of the deep centers is twofold. On the one hand, the occupation of these centers determines the lifetimes and recombination rate. This dependence is explicitly given in Eq. (5) through the presence of δp_{R} (the change from thermal equilibrium of the hole density on the recombination center). On the other hand, δp_{R} also appears in Poisson's equation.

[17] R. H. Parmenter and W. Ruppel, *J. Appl. Phys.* **30**, 1548 (1959).

As the injection level becomes large, δp_R increases from zero to its high-level value given by

$$\delta p_R = \frac{c_p n_{R,0} - c_n p_{R,0}}{c_p + c_n}.$$

This clearly can produce large changes in the lifetimes regardless of the magnitudes of N_R. However, Poisson's equation cannot be significantly affected by these changes in δp_R unless N_R is large compared with the thermal densities of free carriers, n_0 and p_0.

II. Small Density of Deep Centers—Theory

3. General Discussion

We will first consider the cases which arise when N_R is small, so that δp_R can be neglected in Poisson's equation. Changes in occupation of the recombination centers will affect only the lifetime. For these cases, it is convenient to take the sum of $1/\mu_n$ times Eq. (2a) and $1/\mu_p$ times Eq. (2b), which results in[17a]

$$\frac{b+1}{\mu_n} r = \underbrace{(n_0 - p_0)\frac{d\mathscr{E}}{dx}}_{A} - \underbrace{\frac{\varepsilon}{e}\frac{d}{dx}\left[\mathscr{E}\left(\frac{d\mathscr{E}}{dx}\right)\right]}_{B} + \underbrace{\beta\frac{d^2}{dx^2}(n + p)}_{C}, \tag{6}$$

where Poisson's equation has been used to eliminate $(\delta n - \delta p)$ from term B. Note that the current equation (1c), which becomes

$$J = e\mu_p\left[(bn_0 + p_0)\mathscr{E} + (b\,\delta n + \delta p)\mathscr{E} + \beta\frac{d}{dx}(bn - p)\right], \tag{7}$$

is a first integral of the difference of Eqs. (2a) and (2b). Thus, Eqs. (6) and (7) are sufficient, along with Eqs. (3) and (4), to solve the problem.

When the various regimes are being traced out, it should be recalled that at sufficiently low injection levels, the carriers present in thermal equilibrium, n_0 and p_0, will dominate, and the I–V characteristic will be ohmic if junction drops are negligible.

4. High-Level Regimes

We next consider the limit of high injection, so that δn and δp are much greater than any other carrier density, i.e., p_0, n_0, p_1, n_1, N_R, and the space

[17a] Any other approach will, of course, lead to the same result. For example, using Eq. (2a) instead of the above linear combination, one finds that in the high-level approximation, J_n is a constant to first order. Taking second-order terms, one is led to Eq. (6).

charge. Poisson's equation gives us $\delta p \approx \delta n$, and this equality can be used everywhere except where $(\delta p - \delta n)$ appears, as in term B of Eq. (6). In this limit, Eq. (5) leads to

$$r \approx \frac{\delta n}{\tau} \approx \frac{n}{\tau}, \qquad \tau = \frac{c_p + c_n}{c_p c_n N_R},$$

where τ is now the common high-level lifetime. Note that because $r \equiv \delta n/\tau_n \equiv \delta p/\tau_p$, $\delta p = \delta n$ naturally leads to $\tau = \tau_n = \tau_p$. The utility of the choice of this particular linear combination of the continuity equations [(2a) and (2b)] is now seen; the mobile-carrier densities have now been eliminated from all but term C in Eq. (6). Of the three right-hand terms in Eq. (6), A is dominant for semiconductors, B is dominant for insulators, and C is dominant at very high injection levels. This last case, when C is dominant, is sometimes known as the "short structure" case. Various aspects of it have been treated by Herlet and Spenke,[2] Kleinman,[3] and Hall.[1] This division of regimes, along with the treatment of the semiconductor and insulator regimes given below, follows the work of Lampert[4,6,7] and Lampert and Rose.[5]

In all cases, term C becomes dominant in the immediate vicinity of the injecting contacts, where n is very large and $\mathscr{E}$ is very small compared with the average values. In the semiconductor and insulator cases, the region of the sample dominated by term C extends only a few ambipolar diffusion lengths from the injecting contacts. Thus, the effects of term C near the contacts may be neglected if $L_a/L \ll 1$. Assuming that this is the case, the distinction between the semiconductor regime and the insulator regime corresponds to whether $(n_0 - p_0)$ is much larger or much smaller than $(\delta n - \delta p)$. This last term, as remarked above, is the space charge, and $(n_0 - p_0)$ is the fixed charge in the crystal. Note that in Eq. (6), terms A and B together are just $d[(n - p)\mathscr{E}]/dx$, where n and p are the *total* electron and hole densities. Since this term is proportional to the recombination rate if the contribution of the diffusion term is negligible, we must have $(n - p) \neq 0$ to provide the divergence in the current needed to support the recombination traffic. The split into terms A and B arises because the two physical mechanisms capable of supporting a difference in n and p are fixed charge in the crystal and space charge (i.e., Poisson's equation must be satisfied).

Note that a semiconductor will eventually enter the insulator regime as the injection level (and thus the space charge) increases, making term B dominate. A further increase in injection level eventually leads to term C dominating as the diffusion-dominated regions near the contacts spread into the bulk of the sample. Figure 1 indicates the progression from regime to regime of the current–voltage characteristic for a semiconductor. Each regime will be discussed below. It should be mentioned that one or more of these regimes may be missing, depending on the actual parameters, and that the upper

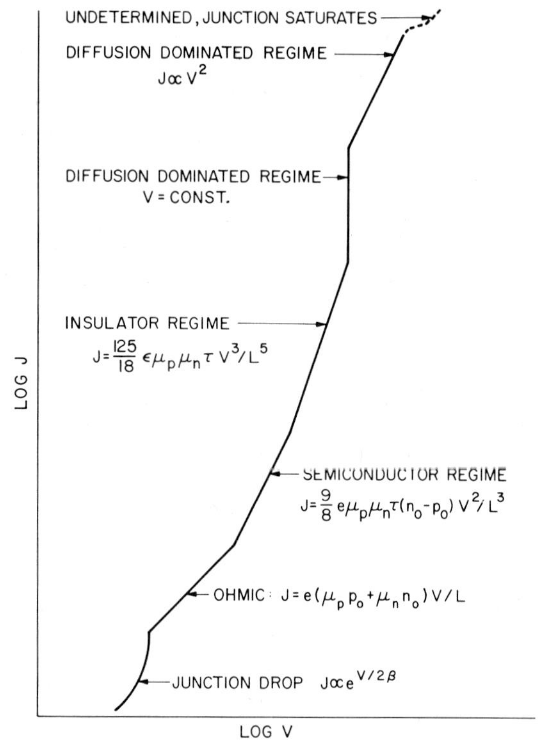

FIG. 1. Schematic representation of the possible regimes of high-level double-injection behavior in a structure with small density of deep centers, starting with the semiconductor regime. The manner in which the low-level regimes merge into the high-level regimes is also shown schematically (the "junction drop" regime shows up distinctly only for lower temperatures or for wide-band-gap materials, i.e., when E_g/kT is large enough). Depending on material parameters, some of these regimes may not appear.

reaches of the figure may not be obtainable because of contact limitations, power-dissipation limitations, or other practical considerations.

a. Semiconductor Regime

Let us first consider the limiting case of a semiconductor, where we retain only term A on the right side of Eq. (6). Equation (6) reduces to

$$\frac{(b+1)n}{\mu_n \tau} = (n_0 - p_0)\frac{d\mathscr{E}}{dx}. \tag{8}$$

The appropriate form of the current equation (setting $n = p$ and neglecting diffusion) is

$$J = e\mu_p(b+1)n\mathscr{E}. \tag{9}$$

Only one boundary condition is needed, since Eqs. (8) and (9) yield a first-order differential equation. The simplest such condition is to set $\mathscr{E} = 0$ at the contact where minority carriers are injected.[17b] These two equations can be solved easily to give

$$J = \frac{9}{8} e\mu_n\mu_p(n_0 - p_0)\tau \frac{V^2}{L^3} \tag{10}$$

and

$$\mathscr{E} = \pm\frac{3}{2}\frac{V}{L}\left(\frac{|x|}{L}\right)^{1/2}. \tag{11}$$

In Eq. (11), $x = 0$ is taken at the minority-carrier injecting contact (i.e., the hole-injecting contact if $n_0 > p_0$) and the sign is chosen to make the sign of Eq. (8) come out right.[17c]

The space charge needed to support this field distribution is given by [from Eq. (8)]

$$\begin{aligned}\frac{\varepsilon}{e}\frac{d\mathscr{E}}{dx} &= \frac{\varepsilon}{e}\frac{(b+1)}{\mu_n\tau(n_0 - p_0)} n \\ &= (b+1)\frac{\Theta_{\Omega,n}}{\tau} n,\end{aligned}$$

where $\Theta_{\Omega,n} = \varepsilon/e\mu_n(n_0 - p_0)$ is similar to a low-level dielectric relaxation time (the "Ohmic" dielectric relaxation of Lampert). Thus, the initial assumption that the space charge is a small fraction of the injected charge (necessary for high-level approximation) is valid as long as $(b + 1)\Theta_{\Omega,n}/\tau \ll 1$. The assumption that the space charge is a small fraction of $n_0 - p_0$ (necessary for term A to be dominant) is valid as long as

$$(b+1)\frac{\Theta_{\Omega,n}}{\tau}\frac{n}{(n_0 - p_0)} \ll 1.$$

Note that increasing the injection level increases n, thus leading to a breakdown of this condition.

[17b] Setting $\mathscr{E} = 0$ is reasonable, since the very high densities force a very small value of $\mathscr{E}$ in the diffusion region near the contacts. The approach to $\mathscr{E} \approx 0$ at the other contact is accomplished by the *neglected terms* and, as will be seen in the more exact treatment given later, occurs very quickly.

[17c] The field distribution and the dependence of the current on V^2/L^3 are identical to those found for single-carrier space-charge-limited current. This is coincidental, and arises because Eq. (8) has the same functional form as Poisson's equation, i.e., $n \propto d\mathscr{E}/dx$.

b. Insulator Regime

When term B becomes dominant, Eq. (6) becomes

$$\frac{(b+1)}{\mu_n}\frac{n}{\tau} = -\frac{\varepsilon}{e}\frac{d}{dx}\left(\mathscr{E}\frac{d\mathscr{E}}{dx}\right). \tag{12}$$

This can occur for an insulator, where $(n_0 - p_0)$ is zero or very small, or for a semiconductor at high enough injection levels, as shown previously. The high-level limit for r has been used again, and the current equation again reduces to Eq. (9), $J = e\mu_p(b+1)n\mathscr{E}$. Solution of these two equations becomes easy if, as suggested by Lampert and Rose,[5] we define a new variable y' by

$$\mathscr{E}\,dy' = dx. \tag{13}$$

In terms of this variable, Eqs. (9) and (12) become

$$\frac{J}{\varepsilon\mu_p\mu_n\tau} = K = -\mathscr{E}\frac{d}{dx}\left(\mathscr{E}\frac{d\mathscr{E}}{dx}\right) = -\frac{d^2\mathscr{E}}{dy'^2}. \tag{14}$$

We still need the boundary conditions. In the spirit of the previous subsection, we set $\mathscr{E} = 0$ at both contacts, since we have a second-order equation. Anticipating the symmetry of the solution, we define $x = y' = 0$ at the center of the sample and $x = \pm L/2$, $y' = \pm y_L'$ at the ends. Equation (14) immediately integrates to

$$\mathscr{E} = \frac{J}{2\varepsilon\mu_p\mu_n\tau}(y_L'^2 - y'^2), \tag{15}$$

and the connection to x is obtained by integrating Eq. (13) to give[17d]

$$x = \frac{J}{2\varepsilon\mu_p\mu_n\tau}y'\left(y_L'^2 - \frac{y'^2}{3}\right). \tag{16}$$

Integrating $\mathscr{E}\,dx$ and connecting y_L' to L by Eq. (16), we obtain

$$J = \frac{125}{18}\varepsilon\mu_n\mu_p\tau\frac{V^3}{L^5}. \tag{17}$$

Unlike the results for the semiconductor regime, the explicit expressions for $\mathscr{E}(x)$, $n(x)$, etc. become very involved. However, the main features of the solution can be extracted. The field distribution is symmetrical; thus, the space-charge distribution is antisymmetrical, with equal positive and negative

[17d] This is a cubic equation for y which can be solved explicitly to give $y = 2y_L\cos[\frac{1}{3} \times \cos^{-1}(3x/Ky_L{}^3) + \pi/3]$, allowing y to be eliminated from Eq. (15). However, this is not necessary to obtain the J–V characteristic and most of the features of the solution.

charges on either side of the center. This is in contrast to the semiconductor regime, where the space charge given by the solution was all of one sign, with the equal and opposite charge located in a very narrow region near the majority-carrier injecting contact, where the simple solution breaks down.

Although space charge is important, its role is very different from that in the single-carrier space-charge-limited regime. In the latter, the buildup of space charge reduces the field at the injecting contact to zero, thus limiting the current. In the present case, the space charge (by creating a difference in n and p) controls the divergence of the current and thus the recombination kinetics.

c. *Diffusion Dominated Regime and Boundary Conditions*

As mentioned above, at very high injection levels, the diffusion-dominated regions near the contacts grow into and dominate the entire sample. When this occurs, Eq. (6) becomes

$$\frac{b+1}{\mu_n}\frac{n}{\tau} = 2\beta\frac{d^2n}{dx^2}. \tag{18}$$

This must be coupled with the current equation, retaining now the diffusion term, i.e.,

$$J = e\mu_p[(b+1)n\mathscr{E} + \beta(b-1)\,dn/dx]. \tag{19}$$

The solution to Eq. (18) is

$$n = n'\cosh[(x-x')/L_a], \tag{20a}$$

where $L_a = [2\beta\mu_n\tau/(b+1)]^{1/2} \equiv$ the ambipolar diffusion length, and n' and x' are constants to be determined from the boundary conditions. This may be expressed in a more convenient form for computation as

$$n = \frac{n(0)\sinh[(L-x)/L_a] + n(L)\sinh(x/L_a)}{\sinh(L/L_a)}, \tag{20b}$$

since $n(0)$ and $n(L)$ are directly the densities at the boundaries.

At this point, it is necessary to discuss the boundary conditions in greater detail, because the simple approximations used in the previous section are no longer valid. In fact, it is necessary to assume a specific physical model for the contacts in order to obtain sufficient information. We shall consider the case where both injecting contacts are forward-biased junctions, so that the structure under consideration is a *p–i–n*.

Let us first consider the *p–i* junction. The basic boundary condition for this or any other type of contact is the continuity of the current across the contact. For the *p–i* junction it is convenient to consider the electron current

J_n. In the p region, the field is very small because of the heavy doping, so that J_n is a pure diffusive current given by

$$J_n = e\mu_n\beta[n_p(0)/L_n], \tag{21}$$

where $n_p(0)$ is the electron density in the p region at the edge of the space-charge layer (see Fig. 2). In the i region,

$$\begin{aligned} J_n &= e\mu_n\left[n\mathscr{E} + \beta\frac{dn}{dx}\right] \\ &= e\mu_n\left[\frac{J}{e(b+1)\mu_p} + \frac{2\beta}{b+1}\frac{dn}{dx}\right], \end{aligned} \tag{22}$$

eliminating $n\mathscr{E}$ by using Eq. (19).

Equating Eqs. (21) and (22),[17e] and using the Boltzmann condition $n_p(0)p_p - n(0)^2$, the boundary condition for the p–i junction becomes[18]

$$\frac{J_n}{e\mu_n} = \frac{\beta n(0)^2}{L_n p_p} = \frac{J}{e\mu_p(b+1)} + \frac{2\beta}{b+1}\frac{dn}{dx}. \tag{23}$$

Consider first the case where the boundaries are sufficiently far apart that the solution of Eq. (18) is $n = n(0)\exp(-x/L_a)$; $L \gg L_a$. Equation (23) then reduces to

$$\frac{\beta}{L_n}\frac{n(0)^2}{p_p} = \frac{J}{e\mu_p(b+1)} - \frac{2\beta}{(b+1)L_a}n(0). \tag{24}$$

The asymptotic solutions of Eq. (24) are

$$n(0) = \frac{JL_a}{2e\mu_p\beta} \qquad \text{for} \quad n(0) \ll \frac{2L_n}{(b+1)L_a}p_p \tag{25}$$

and

$$n(0) = \left[\frac{JL_n p_p}{e\mu_p(b+1)\beta}\right]^{1/2} \qquad \text{for} \quad n(0) \gg \frac{2L_n}{(b+1)L_a}p_p. \tag{26}$$

It should be noted that the solution given in Eq. (26) is valid only for $2L_n/(b+1)L_a \ll 1$, since the assumptions made earlier on the nature of the space-charge layer depend on the assumption that $n(0) \ll p_p$. If this last is not satisfied [i.e., $n(0) \gtrsim p_p$], the junctions wash out and the relevant boundary conditions become those at the metal contacts—changing the entire nature of the problem. Similar relations hold for the n–i junction.

[17e]Assuming that no recombination occurs in the junction itself.

[18] R. Baron, *Phys. Rev.* **137**, A272 (1965).

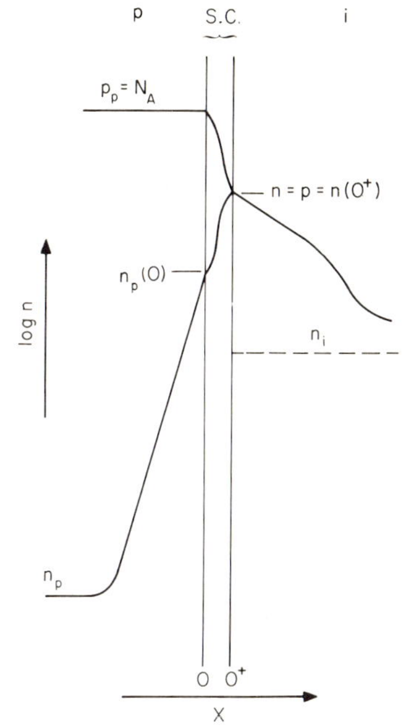

FIG. 2. A sketch of the electron and hole density distribution in the vicinity of a p–i junction under conditions of high forward bias. The space-charge region between 0 and 0^+ is expanded greatly in the x direction. (After Baron.[18])

It follows from Eq. (20) that the density everywhere in the i region depends on a term proportional to $n(0)$ added to a term proportional to $n(L)$ (given by the boundary conditions above). Thus, for relatively low injection levels, where Eq. (25), rather than Eq. (26), is valid, and both $n(0)$ and $n(L)$ are proportional to J, the carrier density in the i region is everywhere proportional to J. The field may then be found by substituting Eq. (20b) in Eq. (19) and using Eq. (25) for $n(0)$ [and the corresponding expression for $n(L)$]. It is then seen that $\mathscr{E}$ is independent of J, and the voltage drop across the i region (found by integrating $\mathscr{E}$) is a constant given by

$$V_1 = \beta \frac{b^{1/2}\pi}{b+1} \exp \frac{L}{2L_a} \qquad (L \gg L_a), \tag{27a}$$

or

$$V_1 = \beta \frac{2b}{(b+1)^2} \left(\frac{L}{L_a} \right)^2 \qquad (L \ll L_a). \tag{27b}$$

Of course, the voltage drop across the junctions must be added to the voltage drop in the bulk, particularly for $L \ll L_a$.

As shown by Mayer *et al.*,[19] the voltage drop across the junctions is given by

$$V_J = \frac{2b\beta}{b+1} \ln \frac{n(0)}{n_i} + \frac{2\beta}{b+1} \ln \frac{n(L)}{n_i}.$$

For this case, where Eq. (25) is valid at the p–i junction and the corresponding relation is valid at the n–i junctions, this becomes

$$\begin{aligned} V_J &= \frac{2b\beta}{b+1} \ln \frac{JL_a}{2e\mu_p n_i \beta} + \frac{2\beta}{b+1} \ln \frac{JL_a}{2e\mu_n n_i \beta} \\ &= 2\beta \ln \frac{JL_a}{2e\mu_p n_i \beta} - \frac{2\beta}{b+1} \ln b. \end{aligned}$$

Two comments should be made:

1. Equation (25) is valid when the left-hand side of Eq. (24) is negligible compared with either of the terms on the right-hand side. This is equivalent to assuming that $J_n \approx 0$ (i.e., $J_n \ll J$) at the junction. The p–i junction is then a perfect injector of holes (since the injection ratio is unity). This also means that all recombination takes place in the i region, and none in the diffusion wing in the p region. In order to obtain Eq. (27), the corresponding assumption for the n–i junction must also be valid.

2. Although the density distribution is determined by the diffusion term in the continuity equation, the current is field driven. The contribution of the diffusion term in the current is only significant very near the junctions, where it causes J_n to go to zero at the p–i junction and J_p to go to zero at the n–i junction.

For higher injection levels, where Eq. (26) is valid, it must be used for $n(0)$ in Eq. (20b) along with the corresponding expression for $n(L)$, and $n \propto J^{1/2}$ everywhere. Thus, when Eq. (20b) is substituted in Eq. (19) and the resulting equation is integrated, we obtain

$$J = AV^2, \tag{28}$$

with

$$\begin{aligned} A &= \frac{4e\mu_p(b+1)p_pL_p \exp(-L/L_a)}{\pi^2 \beta L_a{}^2} \\ &= \frac{2e(b+1)^2 p_p L_p \exp(-L/L_a)}{\pi^2 \beta^2 b\tau} \qquad \text{for} \qquad L \gg L_a \end{aligned}$$

[19] J. W. Mayer, O. J. Marsh, R. Baron, R. Kikuchi, and J. M. Richardson, *Phys. Rev.* **137**, A295 (1965).

and

$$A = \frac{e\mu_p(b+1)p_pL_p}{\beta L^2} \qquad \text{for} \qquad L \ll L_a$$

(where a symmetrical structure has been assumed, $p_p = n_n$, $L_p = L_n$).

Again, V in Eq. (28) is the voltage drop across the bulk, and the voltage drop V_J across the junctions must be added to give the applied voltage. In this case, where Eq. (26) is valid at the p–i junction and the corresponding relation is valid at the n–i junction, the voltage drop across the junctions is given by

$$\begin{aligned} V_J &= \frac{b\beta}{b+1}\ln\frac{JL_np_p}{e\mu_p(b+1)\beta n_i^2} + \frac{\beta}{b+1}\ln\frac{JL_pn_n}{e\mu_p(b+1)\beta n_i^2} \\ &= \beta\ln\frac{J}{e\mu_p(b+1)\beta} + \frac{b\beta}{b+1}\ln\frac{L_n}{n_p} + \frac{\beta}{b+1}\ln\frac{L_p}{p_n}. \end{aligned}$$

In contrast to the other case, the diffusion current is negligible even at the junctions, and the current is entirely field driven. Thus, $J_n = bJ/(b+1)$ at the p–i junction [the injection ratio is $b/(b+1)$] and most of the recombination takes place in the diffusion wings.

This $J \propto V^2$ dependence has no relation to the earlier $J \propto V^2$ behavior characteristic of the semiconductor regime. The $J \propto V^2$ behavior of the semiconductor regime is imposed by volume recombination; the behavior described in Eq. (28) is imposed by the boundary conditions. This is the regime considered by Kleinman.[3] The current at which the transition from a constant voltage to a $J \propto V^2$ behavior takes place is found by substituting Eqs. (27) into Eq. (28), giving

$$J_{tr} = \frac{4e\mu_np_pL_p\beta}{(b+1)L_a^2} \qquad \text{for} \qquad L \gg L_a, \tag{29}$$

$$J_{tr} = \frac{4eb\mu_np_pL_p\beta L^2}{(b+1)^3L_a^4} \qquad \text{for} \qquad L \ll L_a. \tag{30}$$

Note that for $L \gg L_a$, J_{tr} is independent of L; for $L \ll L_a$, however, $J_{tr} \propto (L/L_a)^2$. This makes it much easier to reach the $J \propto V^2$ regime for $L \ll L_a$.

In summary, the most general possible high-level current–voltage characteristic is sketched in Fig. 1. Depending on the parameters of an actual sample (L, τ, L_a, L_n, L_p, b), any of these regimes may be missing. For example:

1. If the junction saturation current is sufficiently high, the junction drop will be everywhere negligible.
2. If $|n_0 - p_0|$ is small enough, the semiconductor regime will not appear.

3. If L/L_a is small enough, the ohmic regime will be followed immediately by the $V = V_1$ regime.

4. If L/L_a is still smaller, only the junction drop followed by the $J \propto V^2$ regime described by Eq. (28) will appear.

5. If $2L_p \gtrsim (b + 1)L_a$, the $J \propto V^2$ regime described by Eq. (28) will not appear before the junctions saturate.

5. Effects of Diffusion and Thermal Generation

In practice, it is often difficult to satisfy the requirements necessary to observe the pure limiting cases described above. For the insulator and semiconductor regimes, with which we will be concerned in this section, the two most important corrections involve diffusion and thermal-generation effects. The simple theory for these two regimes, presented above, assumed that L was sufficiently greater than L_a that it was possible to neglect the diffusion-dominated regions near the junctions. This allowed the replacement of the exact boundary conditions by $\mathscr{E} = 0$ at one or both junctions. The simple theory also assumed that the injection level was high enough that the thermal generation implied by the presence of thermal carriers and its effect on the lifetimes could be neglected. It will be seen below that the semiconductor regime cannot be discussed properly without including the effect of both diffusion and thermal generation. For this reason, we shall first discuss the insulator regime, for which the effect of diffusion alone can be considered.

a. *Insulator Regime (Diffusion Only)*

The use of the junction boundary conditions discussed in Section 4c implies very high carrier concentrations and very low fields near the junctions. Thus, the portions of the *i* region closest to the junctions will be dominated by the diffusion term in the continuity equation, and n will decrease exponentially away from the junctions. In the center of the *i* region, the diffusion term becomes negligible and n closely approximates that of the simple theory for the insulator regime. Typical results for $n(x)$ and $\mathscr{E}(x)$ $(L/L_a \approx 28)$ are shown in Fig. 3. Nearly all of the voltage drop occurs across the field-dominated central portion. Thus, the primary effect of the diffusion wings is to effectively shorten the sample by the combined length of the diffusion wings. As a first approximation, we may define an effective length $L_{\text{eff}} = L - 2(QL_a)$, where QL_a is the length of one of the diffusion wings, and Q is about 2–4. This L_{eff} is then used in place of L in Eq. (17) for the insulator regime. Since $J \propto L^{-5}$, this correction can be very significant.

(1) *Assumptions and Equations.* We continue with many of the previous assumptions, i.e., (i) no trapping; (ii) recombination through one set of localized centers; (iii) high level, $n \approx p \gg n_0, p_0, N_R$; and (iv) insulator

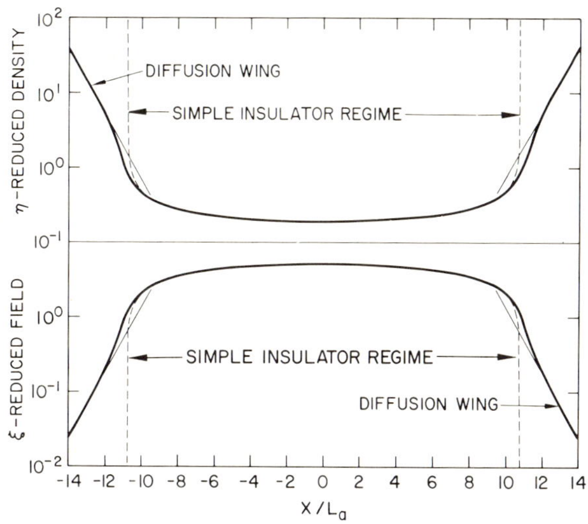

FIG. 3. Typical density and field distribution for a long p–i–n structure (insulator regime), showing the presence of diffusion wings extending into the sample from the junctions and the central high-field region.

regime, $n_0 - p_0 \approx 0$. This allows the continuity equation (6) to be written

$$\frac{(b+1)n}{\mu_n \tau} = -\frac{\varepsilon}{e}\frac{d}{dx}\left(\mathscr{E}\frac{d\mathscr{E}}{dx}\right) + 2\beta\frac{d^2 n}{dx^2}, \tag{31}$$

where we now retain the diffusion term. As shown by Baron,[18] the current equation may be simplified to

$$J = e\mu_p(b+1)n\mathscr{E} \tag{32}$$

for the purpose of eliminating $\mathscr{E}$ in Eq. (31). This neglect of the diffusion term in Eq. (32) is allowable because this term only becomes important at values of n and $\mathscr{E}$ where the diffusion term in Eq. (31) completely dominates. Thus, in the only region where Eq. (32) is not accurate for eliminating $\mathscr{E}$, the term in Eq. (31) involving $\mathscr{E}$ is negligible. In other words, the current is field driven not only in the central region of high field, but also over most of the region where the diffusion term in the continuity equation dominates. As seen in Section 4, the diffusion term in the current equation is only necessary at the junctions in order to determinate the boundary conditions.

This smallness of the diffusion term in the current away from the immediate vicinity of the junction is demonstrated below. Near the p–i junction, the density is given by

$$n = n(0)\exp(-x/L_a)$$

with $n(0) = JL_a/2e\mu_p\beta$ being the appropriate boundary condition as shown in Eq. (25). The field near the junction can then be determined from Eq. (19) and is given by[19a]

$$\mathscr{E} = \frac{\beta}{L_a(b+1)}\left[2\left(\exp\frac{x}{L_a}\right) + (b-1)\right]; \tag{33}$$

the ratio of the diffusion and field-driven contributions to the current density is given by

$$\frac{J_{\text{diff}}}{J_{\text{field}}} = \frac{(b-1)}{2[\exp(x/L_a)] + (b-1)}. \tag{34}$$

As we move away from the junction, the field quickly becomes proportional to $1/n$ and the contribution of the diffusion term to the current becomes negligible, thus justifying Eq. (32).

(2) *Results.* The equation obtained for n, when $\mathscr{E}$ is eliminated between Eqs. (31) and (32), is best expressed in terms of a reduced density $\eta = n/n^*$ and reduced length $y = x/L_a$ and is given by

$$\eta = -\frac{3}{\eta^4}\left(\frac{d\eta}{dy}\right)^2 + \left(\frac{1+\eta^3}{\eta^3}\right)\frac{d^2\eta}{dy^2}, \tag{35}$$

where

$$n^* = \left(\frac{\varepsilon}{2e\beta}\right)^{1/3}\left[\frac{J}{e\mu_p(b+1)}\right]^{2/3}.$$

This eliminates all constants from the equation, which allows all possible density distributions to be represented by a one-parameter family of curves in $\eta(y)$. A convenient choice of parameter is the minimum reduced density, η_0. Although this equation is exactly solvable,[18] the exact solutions are complicated expressions involving elliptic functions and are not of much use. Equation (35) may be reduced to quadrature. The resulting integral for $y(\eta)$ may be expanded for $\eta \gg 1$ and for $\eta \ll 1$. This allows the derivation of approximate solutions for $J(V)$ and $n(x)$ which give insight into the behavior of the solutions and the manner in which they deviate from the simple theory for the insulator regime. These approximations are valid in the limit of large L/L_a. For smaller values of L/L_a, it is necessary to supplement them with numerical results.

Figure 4 is a plot of the reduced density versus the reduced length, showing the change in behavior in going from the insulator regime to the diffusion-

[19a]Note that, just as in the simple diffusion-dominated regime (Section 4), the field is independent of the current.

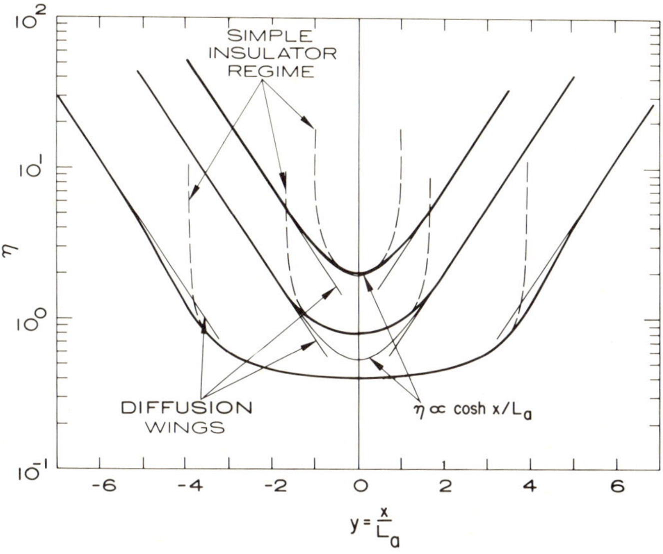

FIG. 4. The reduced density distribution for three values of η_0. The progression from field- and diffusion-dominated (insulator regime) at low injection levels to diffusion-dominated at high injection levels is shown. The dashed lines are the density distributions for the simple insulator regime. The light lines represent the diffusion-dominated solution. (After Baron.[18])

dominated regime. For values of $\eta < 1$, the density distribution is given by the simple theory for the insulator regime; for values of $\eta > 1$, the density distribution is given by the diffusion-dominated regime. The progression from $\eta_0 = 0.4$ to $\eta_0 = 2$ illustrates the case as the injection level is increased and the diffusion-dominated wings spread into the sample. It should be noted that a change in η of a factor of 10 in the wings corresponds to a change in the current by a factor of 10^3; thus, it may not be possible to see the entire progression in any one sample.

Figure 5 is a plot of the reduced density versus the actual length for various values of L/L_a (i.e., letting τ vary). A constant value of η at the boundary implies that the injection level was chosen to satisfy $J \propto L_a^{-3}$. The two top curves ($L/L_a = 6$ and 9.54) are completely in the diffusion-dominated regime. The bottom curve, with $L/L_a = 67.24$, satisfies the condition $L \gg L_a$ required for the insulator regime. However, even at this large ratio of L/L_a, the diffusion-dominated wings occupy 0.1 of the length of the sample.[19b] The very fast rise in n and consequent fast fall in $\mathscr{E}$ is the basis for the effective-

[19b]The exponential diffusion wing, appearing as the straight-line segment of the curve adjacent to the junction, extends from $x/L = 0.5$ to $x/L \approx 0.45$ for the $L/L_a = 67.24$ case. Since the curve is symmetrical, the diffusion wings occupy $x/L = 0.1$ of the sample.

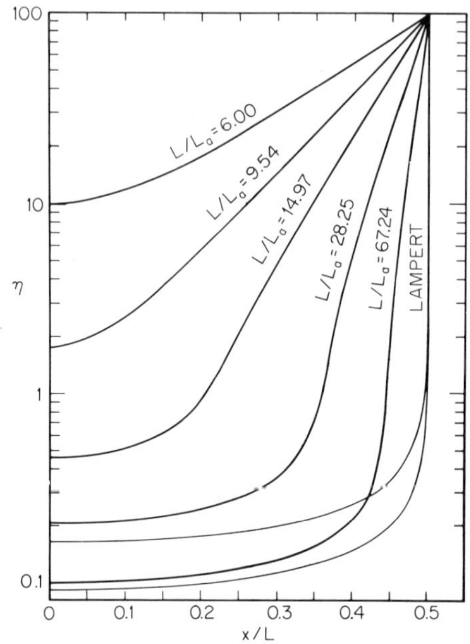

FIG. 5. The function η versus x/L (rather than x/L_a) plotted from the center of the sample. This shows the effect of varying L_a (or τ) while keeping L fixed. The Lampert and Rose curves for equal values of L/L_a are plotted for comparison for the $L/L_a = 67.24$ and 28.25 cases. (After Baron.[18])

length approximation mentioned above, since it means that the voltage drop in the diffusion-dominated region is negligible. For this value of $L/L_a = 67.24$ (and this injection level), $L_{\text{eff}} = 0.9L$; this leads to a not insignificant correction to the current of 1.7.

The approximate solutions referred to above lead to the following expression for the current:

$$J \approx \frac{125}{18} e\mu_n\mu_p\tau \frac{V^3}{[L - 2L_a(\frac{1}{3}\ln K'J)]^5}, \tag{36}$$

where

$$K' = \frac{L_a{}^3(b+1)^2}{4\varepsilon\beta\mu_n b^2}.$$

The effective-length approximation sets $L_{\text{eff}} = L - Q(2L_a)$, so $Q \approx \frac{1}{3}\ln K'J$. This is not a constant, but $\ln K'J$ varies slowly, so that an effective-length approximation is possible. This logarithmic growth of the diffusion wings follows from the fact that $\eta(0) = (K'J)^{1/3}$ at the boundary; since the diffusion wing ends at $\eta \approx 1$, the length of the diffusion wing L' is given by solving

$$1 \approx \eta(0)\exp(-L'/L_a),$$

or

$$L' \approx L_a \ln \eta(0) = L_a(\tfrac{1}{3}\ln K'J).$$

This variation in Q with J changes the functional dependence of J from $J \propto V^3$ to

$$J \propto V^{3/(1-\gamma)}; \qquad \gamma = 10L_a/3L. \tag{37}$$

It should be stated again that these approximations are valid only in the limit of large L/L_a. The resulting $J(V)$ characteristics are given in Fig. 6. The full curves are exact (numerical calculation), the straight lines are of the slope given in Eq. (37), and the points are a better approximation given in Eq. (60) of Baron.[18] In this figure, $y_2 = L/2L_a$. The strong upward trend of the curves for $y_2 = 4$ and 6 is the approach to the diffusion-dominated regime.

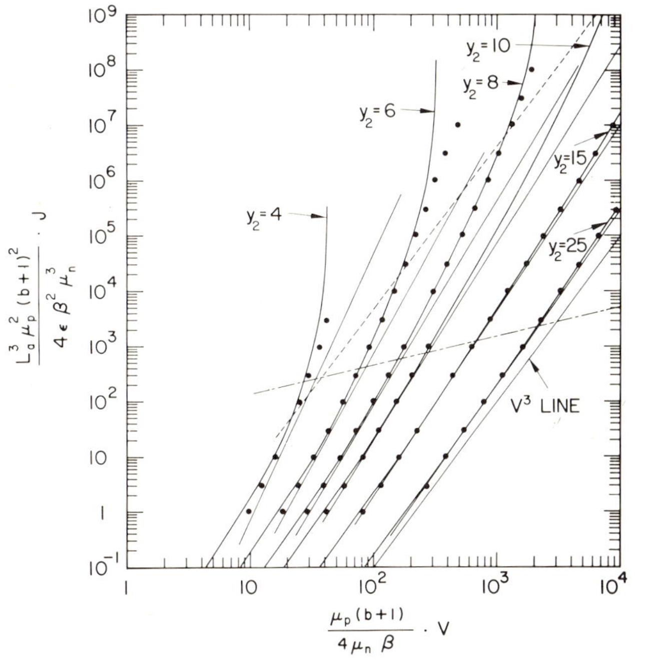

FIG. 6. The theoretical current–voltage characteristic for the high-level insulator regime when the effects of diffusion are included. The parameter $y_2 = L/2L_a$. The full curves are from numerical calculation, the straight lines are from Eq. (37), and the solid points are an approximation given in Eq. (60) of Baron. (After Baron.[18])

b. Insulator Regime (Effect of Thermal Generation)

As shown in the following section, the effect of thermal generation on the semiconductor regime results in a very extended transition region between the ohmic[19c] and the semiconductor regimes. In a similar fashion,[20] thermal generation produces a transition region in the I–V characteristics for an intrinsic insulator, as shown in Fig. 7. Unlike the semiconductor-regime case discussed in the next section, the ohmic line could appear anywhere on either Fig. 6 or 7 by an arbitrary choice of n_i. The constant $\mathscr{P}_i$ in Fig. 7 is related to n_i by $\mathscr{P}_i = e\mu_n \tau n_i/\varepsilon$, which is like τ/Θ_i (where Θ_i is the intrinsic dielectric relaxation time). Figure 7 is for a fixed choice of $\mathscr{P}_i = 69.333$ and $L/L_a = 30$, but for several values of τ_l/τ, where τ_l and τ are the low- and high-level lifetimes, respectively, and are discussed in the next section, with τ_l related to the initial occupation of the recombination centers.

The increased slope and current for $L/L_a = 30$ ($\nu_2 = 15$ in Fig. 6) is evident, as well as a transition region extending over more than six orders of magnitude

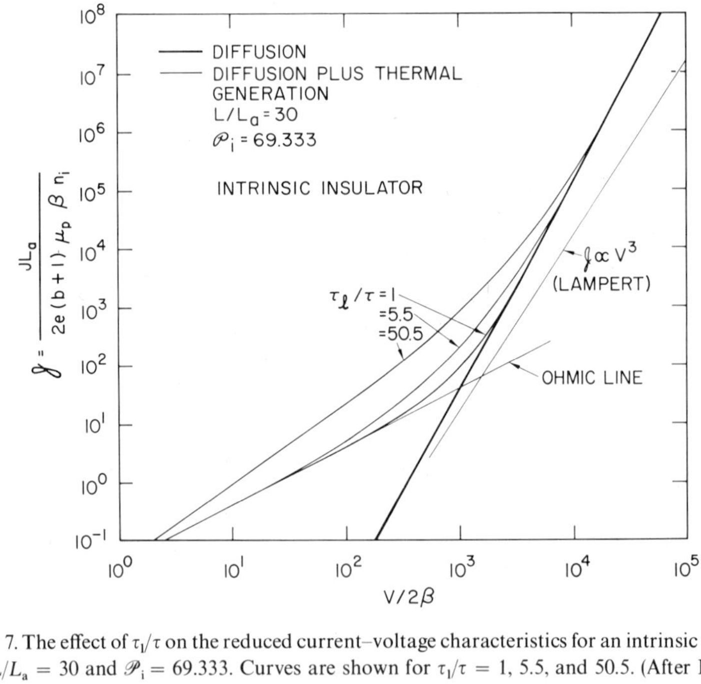

FIG. 7. The effect of τ_l/τ on the reduced current–voltage characteristics for an intrinsic insulator with $L/L_a = 30$ and $\mathscr{P}_i = 69.333$. Curves are shown for $\tau_l/\tau = 1$, 5.5, and 50.5. (After Baron.[20])

[19c] In general, the voltage drop across the junctions must be considered at low values of $V/2\beta$, particularly for low temperatures and materials with a large bandgap, where E_g/kT is large.

[20] R. Baron, *J. Appl. Phys.* **39**, 1435 (1968).

in the current for $\tau_1/\tau = 50.5$. The transition region also may have a portion extending over about two orders of magnitude in the current, in which the current is approximately proportional to V^3. It is in this portion of the transition region (where $J \propto V^3$) that the effective-length approximation has its greatest validity. This transition region may easily be mistaken for true high-level behavior, which actually is proportional to $V^{3/(1-\gamma)}$, as shown previously.

c. Semiconductor Regime (p–π–n)

As with the insulator regime, the junction boundary conditions (Section 4c) imply very high carrier concentrations and very low fields near the junctions. Again, this means that the regions nearest the junctions will be dominated by the diffusion term in the continuity equation, so that n will decrease exponentially away from the junctions. In the center of the π region, n is much smaller and $\mathscr{E}$ is much larger, so that the diffusion term becomes negligible and n and $\mathscr{E}$ closely approximate those of the simple theory for the semiconductor regime. Thus, we have, as for the insulator regime, a central region closely approximated by the simple theory, with diffusion wings on either side.[20] One difference is that the distribution is now asymmetrical, with a very abrupt transition to the diffusion wing adjacent to the p–π junction and a gentle transition to the diffusion wing adjacent to the n–π junction (see Fig. 8). An effective-length approximation is again indicated, although the more gentle transition to the diffusion wing on the n–π side makes it less clear-cut. In addition, since $J \propto L^{-3}$, the effect is less pronounced.

An important difference between the insulator and semiconductor regimes occurs when the effects of thermal generation are considered. Both the semiconductor-regime current and the ohmic current are proportional to p_0. Thus, the transition from ohmic to semiconductor–regime behavior occurs at a voltage which is independent of p_0. This means that the ohmic regime cannot be pushed to arbitrarily low values with respect to the semiconductor regime (as it could with respect to the insulator regime, where n_i could be made arbitrarily small without affecting the current in the insulator regime).

(1) *Assumptions and Equations.* We again assume (i) and (ii) noted previously. However, we now assume $p_0 \gg n_0$ (semiconductor regime) and relax the requirement of high level (since we now wish to consider the effects of thermal generation). We still assume that N_R is small. Under these conditions, Baron[20] shows that Eqs. (6) and (7) can be replaced by

$$\frac{(b+1)}{\mu_n} r = -p_0 \frac{d\mathscr{E}^\dagger}{dx} + 2\beta \frac{d^2(\delta n)}{dx^2} \tag{38}$$

and

$$J = e\mu_p[p_0 + (b+1)\,\delta n]\mathscr{E}^\dagger . \tag{39}$$

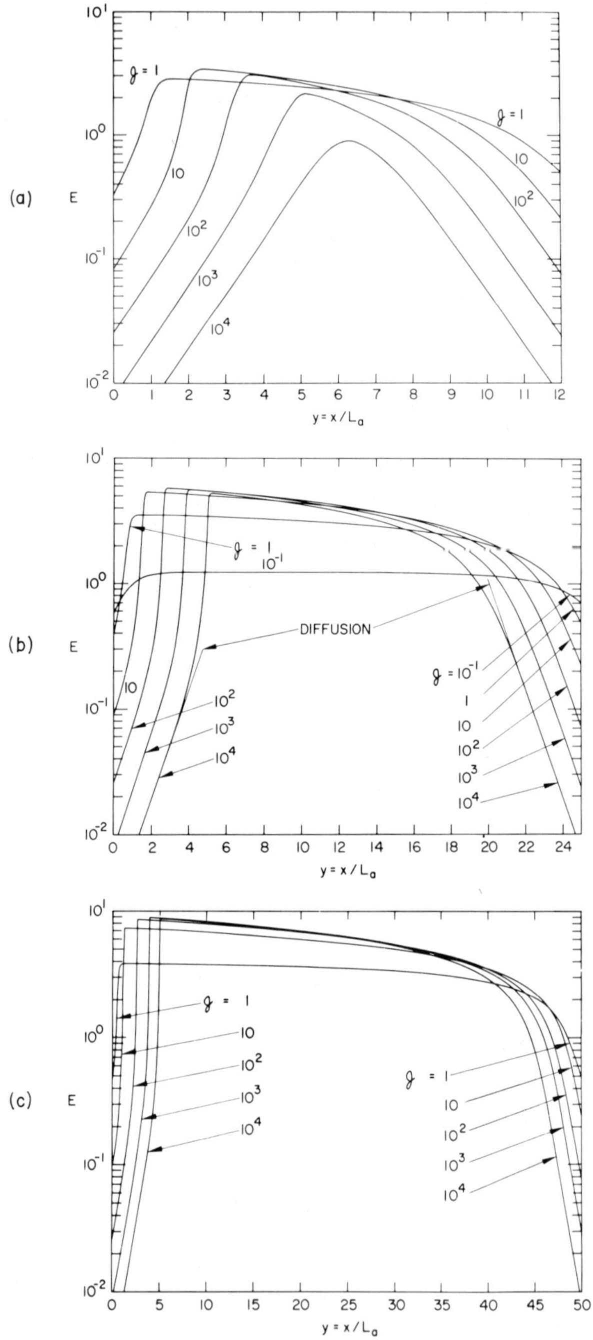

FIG. 8. The reduced field E versus the reduced distance y (semiconductor regime) for $\tau_1/\tau = 1$, with the reduced current $\mathscr{J}$ as a parameter. The curves for $L/L_a = 12$, 25, and 50 are shown in (a), (b), and (c), respectively. The minority-carrier injecting contact is on the right. Note that the diffusion regions almost meet in the center for the higher currents of the $L/L_a = 12$ case. (After Baron.[20])

The quasineutrality condition $[(\varepsilon/e)(d\mathscr{E}/dx)$ very small] and the assumed smallness of N_R (and thus δp_R) were used in Poisson's equation to set $\delta p = \delta n$. The new variable $\mathscr{E}^\dagger \equiv dV/dx$ is the measured field (V is the potential as measured by an infinite impedance probe). It is related to the electrostatic field $\mathscr{E} \equiv -d\psi/dx$ approximately by

$$\mathscr{E}^\dagger = \mathscr{E} + [(b-1)/(b+1)](\beta/n)(dn/dx), \tag{40}$$

as shown by Baron.[20] Except near the junctions, where $\mathscr{E} \to \beta/L_a$, $\mathscr{E}^\dagger$ and $\mathscr{E}$ are nearly equal and may be used interchangeably. Near the junctions, the $\mathscr{E}^\dagger$ term in Eq. (38) contributes negligibly, so that, once again, no error is made (see Baron[20]).

The high-level approximation for r can no longer be used. Using the above assumptions, the following may be obtained from Eq. (5):

$$r = \frac{\delta n(p_0 + \delta n)}{\tau[(\tau_1/\tau)p_0 + \delta n]}, \tag{41}$$

where τ_1, the low-level minority-carrier lifetime ($r \to \delta n/\tau_1$ as $\delta n \to 0$), is given by

$$\frac{\tau_1}{\tau} = \frac{n_1(c_n/c_p) + p_1 + p_0}{p_0} \tag{42}$$

for an acceptor-type center ($c_n/c_p \ll 1$) and by

$$\frac{\tau_1}{\tau} = \frac{(p_1 + p_0)(c_p/c_n) + n_1}{p_0} \tag{43}$$

for a donor-type center ($c_p/c_n \ll 1$). In solving the equations, these last two expressions are not needed except to relate the results to an experimental situation, since only the parameter (τ_1/τ) is needed.

(2) *Results.* Equations (38), (39), and (41) may be combined to give a single equation for $\mathscr{E}^\dagger(x)$. This last must be numerically integrated using the boundary conditions given in Section 4c. Because of the number of parameters involved, it is not possible to cover all cases. However, we have explored[20] the effects of varying L/L_a for $\tau_1/\tau = 1$ and the effect of varying τ_1/τ for $L/L_a = 25$.

The calculated field distributions for $\tau_1/\tau = 1$ and $L/L_a = 12$, 25, and 50 are shown in Fig. 8. In this figure, $\mathscr{J}$, E, and y are, respectively, reduced current, field, and distance variables defined by

$$\mathscr{J} = JL_a/[(b+1)e\mu_p 2\beta p_0], \qquad E = \mathscr{E}^\dagger L_a/2\beta\mathscr{J}^{1/2},$$

and $y = x/L_a$. The corresponding I–V characteristics are shown in Fig. 9. The curves for $L/L_a = 12$ clearly show the transition to the diffusion-

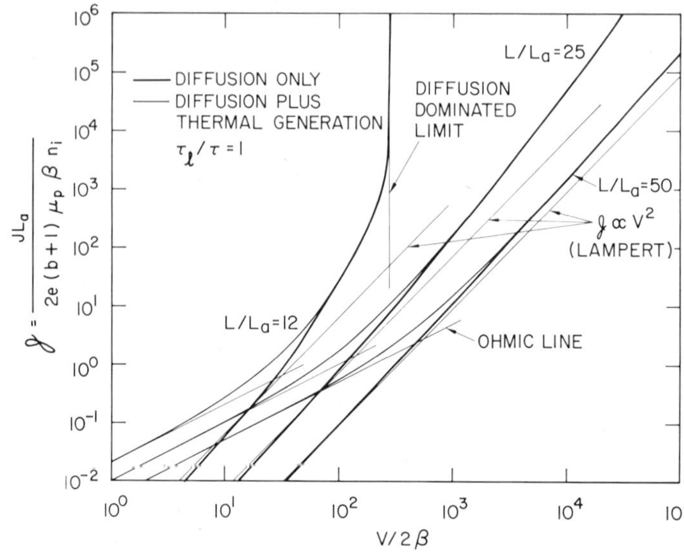

FIG. 9. Reduced current–voltage characteristics (semiconductor regime) for $L/L_a = 12, 25$, and 50. The ohmic curve, the Lampert curve, the curve for the case of adding only diffusion to the equations, and the curve for the case of adding both diffusion and thermal generation (for $\tau/\tau_1 = 1$) to the equations are shown for each value of L/L_a. Thermal generation provides the transition to ohmic behavior. In addition, the limiting curve for complete domination of the equations by diffusion is shown for $L/L_a = 12$. (After Baron.[20])

dominated limit—the vertically rising I–V characteristic (Fig. 9) and the spreading of the diffusion wings into the center of the sample with increasing current (Fig. 8). Even at $L/L_a = 50$, the current rises appreciably faster than $J \propto V^2$ and the movement of the diffusion wings into the center of the sample is evident. At a reduced current of $\mathscr{J} = 10^4$, the diffusion wings occupy about 20% of the length, and the current is 1.8 times that predicted by the simple theory for the semiconductor regime. Figure 9 also shows the results of including only the diffusion term in the high-level limit (i.e., neglecting thermal generation). The amount of error introduced is easily seen from the extent of the transition region. This is particularly evident in Fig. 10, where the I–V characteristics for $L/L_a = 25$ and several values of τ_1/τ are shown. The $\tau_1/\tau = 10^{-3}$ case exhibits an incipient negative resistance whose nature is identical to that of the negative resistance discussed by Lampert[6,7] for the case where thermal carriers are neglected. That is, the short low-level lifetime and consequent high recombination rate prevent the transit of injected carriers until the injection level near the contacts is high enough to alter the occupation of the recombination centers toward longer lifetime. The $\tau_1/\tau = 10$ case exhibits the opposite behavior. The long low-level lifetime and conse-

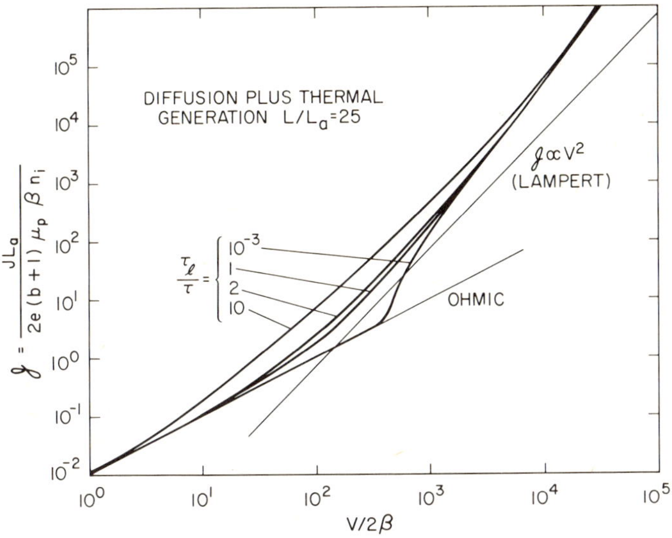

FIG. 10. The effect of τ_l/τ on the reduced current–voltage characteristics (semiconductor regime) for $L/L_a = 25$. Curves are shown for $\tau_l/\tau = 10^{-3}$, 1, 2, and 10, along with the Lampert and ohmic curves. (After Baron.[20])

quent small recombination rate allow freer transit of the carriers, which, in turn, means higher currents. These higher currents persist until the injection level becomes large enough to alter the occupation of the recombination centers toward the shorter high-level lifetime. Note the extreme range of the transition region for this case—over the current range $10^{-2} \lesssim \mathscr{J} \lesssim 10^6$. The high value of τ_l/τ allows injection across the sample at low values of $\mathscr{J}$; it is apparently harder to force the recombination-center occupation toward shorter lifetimes than toward longer lifetimes (as for the $\tau_l/\tau = 10^{-3}$ case), so that the higher low-level lifetime persists to very high currents.

The inclusion of thermal generation leads to a large transition region between the ohmic and high-level regions. The nature of this transition region is such that it greatly influences the interpretation of experimental results. When the effects of diffusion alone are taken into account, the results of the simple semiconductor-regime treatment of high-level double injection in a p^+–π–n^+ structure become significantly affected for L/L_a values as large as 50. The findings are similar to those found in Section 5a for the effect of adding diffusion to the insulator regime; i.e., the I–V characteristic is shifted to higher currents and increases faster than the simple prediction of $J \propto V^2$, going nearly as some power law $J \propto V^{2+s}$ (where s increases monotonically with L_a/L). The $\mathscr{E}^\dagger(x)$ curve becomes modified by diffusion-dominated regions near the contacts, where $\mathscr{E}^\dagger \propto \exp(\pm x/L_a)$, which merge into the central region, where the field approximates the behavior of the simple semiconductor regime $\mathscr{E}^{\dagger} \propto (x - x_0)^{1/2}$, as shown in Fig. 8. This central region occupies a greater proportion of the total length for larger values of L/L_a, leading to increasing validity of the simple theory for the semiconductor regime. The logarithmic growth of the length of diffusion-dominated regions with current is responsible for the effects mentioned above. For sufficiently short samples, the diffusion-dominated regions meet in the center, and the behavior approaches that of a pure diffusion-dominated, "short" structure: the current rises vertically at a constant voltage determined by L/L_a, and the field behaves as $\mathscr{E}^{\dagger} \propto [\cosh(x - x_0)]^{-1}$.

The major result of including thermal generation is the existence of the extremely large transition region between the ohmic and high-level regimes. This region can extend over very large ranges of current, by as much as a factor of 10^8 in one case. The characteristics of this transition region depend on the ratio between the low- and high-level lifetimes, τ_l/τ. Depending on what type of charge states the recombination center may assume (acceptor or donor type) and where it lies in the bandgap, τ_l/τ and the transition region may depend explicitly on one or more of the following properties: (1) the energy range within which the level lies; (2) the exact energy of the level; and (3) the ratio of the capture probabilities.

The most important property of the transition region for $\tau_l/\tau \gtrsim 1$ is that over ranges of current of about two orders of magnitude, the transition-region I–V characteristic has an almost exact $I \propto V^2$ dependence, but at a current level greater than that predicted by the simple theory. The effective-length approximation is most valid within this range, since $Q(I)$ [in $L_{\text{eff}} = L - Q(2L_a)$] passes through a broad minimum, and thus is nearly constant in this range.[20] This is in contrast to the high-level regime, where $Q \propto \ln I$. An experimenter might easily mistake this region of approximate $I \propto V^2$ dependence for true high-level behavior, especially as it may often be difficult or impossible to extend experimental I–V curves to currents more than three or four orders of magnitude greater than the transition current between the high-level and ohmic curves (which corresponds to only two to three orders of magnitude greater than the current at which the approximate $I \propto V^2$ region starts). Even if the measurements are extended above the transition region by an order of magnitude in the current, the possibility that this same error may be made is still high, because the start of the semiconductor-regime high-level behavior with $I \propto V^{2+\text{const}}$ could be mistaken for the transition to the insulator regime.

6. Finite Cross Section

In the above discussion, we have considered only a one-dimensional analysis (planar current flow). The one-dimensional analysis is adequate for samples of sufficiently large cross section. When the transverse dimensions of the sample are comparable to a diffusion length, as they may often be in practice, it is necessary to consider the effect of surface recombination. This problem has been treated for the semiconductor regime by Hirota *et al.*[21] for the steady-state case and by Baron *et al.*[22] for the transient case.

The essential result is that the form of the solution is identical to that resulting from the one-dimensional analysis, except that the lifetime is replaced by an effective lifetime. Thus, the steady-state current density (average) is given by

$$J = \frac{9}{8} e\mu_p \mu_n (n_0 - p_0)\tau_{\text{eff}} \frac{V^2}{L^3} \tag{44}$$

and the transient response to a small voltage step is given approximately by

$$J = J_{\text{f}} \left[1 - \frac{V_{\text{f}} - V_{\text{i}}}{V_{\text{f}}} \exp -\frac{t}{\tau_0{}^*} \right] \tag{45}$$

(the one-dimensional treatment of the transient response is discussed in Section 18). The expression in Eq. (45) neglects a small contribution from higher-order modes of decay.[22] These effective lifetimes are given for a slab geometry by

$$\frac{1}{\tau_{\text{eff}}} = \frac{1}{\tau} + \left\{ \frac{a}{s} + \tau \left[\frac{a}{L_{\text{a}}} \coth\left(\frac{a}{L_{\text{a}}} \right) - 1 \right] \right\}^{-1}$$

for the steady-state case,[21] and by

$$\frac{1}{\tau_0{}^*} = \frac{1}{\tau} + \frac{D_{\text{a}}}{\alpha_0{}^2}$$

for the transient case,[22] where a is half the slab thickness, s is the surface recombination velocity, and α_0 is the first root of

$$\frac{a}{\alpha_0} \tan \frac{a}{\alpha_0} = \frac{sa}{D_{\text{a}}}.$$

In the limit $s \to \infty$, the expressions become

$$\tau_{\text{eff}} = \tau[1 - (L_{\text{a}}/a) \tanh(a/L_{\text{a}})]$$

and

$$\tau_0{}^* = \tau[1 + (\pi L_{\text{a}}/2a)^2]^{-1}.$$

These two effective lifetimes are nearly identical, as shown in Fig. 11.

[21] R. Hirota, S. Tosima, and M. A. Lampert, *J. Phys. Soc. Japan* **18**, 535 (1963).

[22] R. Baron, O. J. Marsh, and J. W. Mayer, *J. Appl. Phys.* **37**, 2614 (1966).

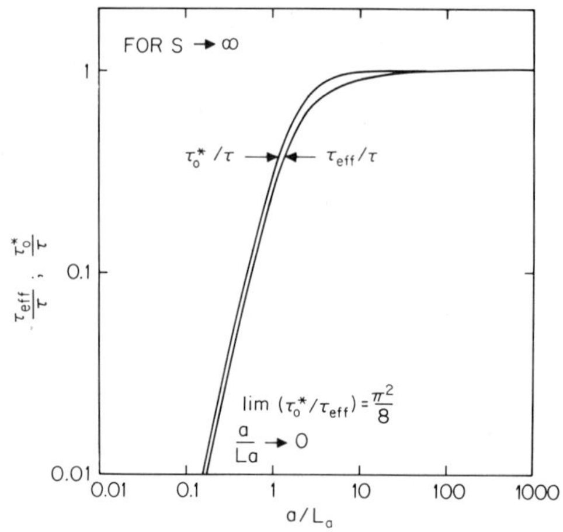

FIG. 11. The steady-state (τ_{eff}) and transient (τ_0*) effective lifetimes as a function of transverse sample dimension, showing the effect of an infinite surface velocity. (After Baron *et al.*[22])

Hirota *et al.* also consider a cylindrical geometry, and Baron *et al.* consider the high-level photodecay. The effect of nonplanar geometry (cylindrical and spherical) on both the steady state and transient cases has been treated by Lee and Baron.[22a]

III. Small Density of Deep Centers—Experiment

7. General Discussion

Double-injection behavior has been observed in a number of semiconductor materials. In many cases, the behavior was typical of either "thin" *p–i–n* junctions ($L/L_a \ll 1$), which have been described by Herlet and Spenke,[2] or negative-resistance structures characterized by deep levels (see Section 16). In this section, we discuss experimental results in devices whose behavior lies in the semiconductor and insulator regimes discussed in the preceding sections.

As seen in the previous discussion and in Section 21, the current–voltage characteristics are not sufficient to specify uniquely the particular regime of behavior. For example, a $J \propto V^2$ characteristic can result from the single-carrier space-charge-limited current regime, the Ashley–Milnes regime[10] (see Section 15), the semiconductor regime, or the high-level diffusion-dominated regime. In addition, $J \propto V^2$ behavior can be produced by poor contacts to the device (i.e., the contact drops dominate). A $J \propto V^3$ charac-

[22a] D. H. Lee and R. Baron, to be published.

teristic can result from diffusion effects in the semiconductor regime, from a low-level regime with a large density of deep centers (Section 11), or from poor contacts, as well as from the insulator regime.

A number of techniques can be used to supplement the I–V characteristics as a means of distinguishing the various regimes. The transient response (Section 18) to a voltage step provides a method of distinguishing between single-carrier and two-carrier effects, and may also indicate the presence of contact effects. Some confidence can be obtained by comparing experiment and theory if n_0 and p_0 are known and the high-level lifetime is measured independently (for example, by photoconductive decay or small-signal transient, Section 18). The length dependence (L^{-3} or L^{-5}) of the I–V behavior provides a good clue to the regime. Measurement of the field distribution can be used to distinguish unambiguously between the semiconductor and insulator regimes, even when diffusion effects cause the semiconductor regime to have an $I \propto V^3$ behavior. We will consider in detail only those experimental data which can be clearly identified with a particular double-injection regime. For comparison with the previous section, we will restrict our attention to cases where the number of deep centers is much smaller than the number of injected carriers.

8. Insulator Regime

The theory for the insulator regime has a beautiful simplicity, in that the current is proportional to $\tau V^3 L^{-5}$. It is natural to ask why eight years lapsed from the time the insulator regime was proposed[4] until the first experimental verification. The difficulty lay primarily in fabricating suitable samples. The sample and material requirements are the following: intrinsic material ($n_0 - p_0$ very small), absence of trapping centers, $L \gg L_a$, and injecting contacts. Although wide-band-gap materials (such as CdTe, ZnS) can be prepared in insulating form, there are large concentrations of traps and it is difficult to form both n^+ and p^+ contacts.

Silicon structures have been used to verify the predictions of the insulator regime.[23–26] To date, the most detailed investigations[23,24] have been made using p–i–n junctions fabricated by lithium-drift techniques. These results will be described in some detail below. Pell[27] and Mayer[28] have shown that the lithium-drift technique of producing p–i–n structures resulted in highly compensated intrinsic regions whose length (typically from 0.5 to 5 mm)

[23] R. Baron, J. W. Mayer, and O. J. Marsh, *Appl. Phys. Letters* **4**, 65 (1964).

[24] J. W. Mayer, R. Baron, and O. J. Marsh, *Phys. Rev.* **137**, A286 (1965).

[25] S. Okazaki and M. Hiramatsu, *Solid State Commun.* **5**, 475 (1967).

[26] S. Okazaki and M. Hiramatsu, *Japan. J. Appl. Phys.* **7**, 557 (1968).

[27] E. M. Pell, *J. Appl. Phys.* **31**, 291 (1960).

[28] J. W. Mayer, *J. Appl. Phys.* **33**, 2894 (1962).

could be controlled during the drift process. In this case, the lithium-doped n^+ region and the base p-type material bounding the intrinsic region serve as the injecting contacts. Extensive use of p–i–n junctions as nuclear-particle detectors has shown that trapping effects are negligible at room temperature.

In the insulator regime, the current is a function only of μ_n, μ_p, τ, and L; these quantities can be directly measured in a straightforward manner. The length can be determined from potential traverse or capacitance measurements, and the lifetime from either transient-pulse measurements (Section 18) or from the current decay after excitation by a brief, intense flash of light. When lifetime is measured, the sample must be biased into the V^3 portion of the current–voltage characteristic to ensure that the intrinsic region is in the high-level-lifetime regime.

a. Current–Voltage Characteristics

The forward-bias characteristics were measured on a number of p–i–n devices made by the lithium-drift technique. These characteristics were obtained by measuring the potential between two probes, one of which was placed on the n-type material near the n–i junction and the other on the p-type material near the p–i junction. This method was necessary in order to eliminate the metal-to-semiconductor contact potential drop present at high current levels in the current-carrying contacts. Figure 12 shows the log(current-density) versus log(voltage) relationship of a representative group of five units. At small voltages, i.e., in the regime where the number of injected carriers is less than the intrinsic carrier concentration n_i, the current density was ohmic in behavior. In this case, the sample lengths were long enough that the voltage drop at the junctions was small compared with that across the intrinsic region. The current levels agreed with the values predicted from $J = en_i(\mu_n + \mu_p)V/L$.

The current for units with $6 < L/L_a < 16$ generally exhibits a V^3 region followed by higher powers of voltage. This behavior follows qualitatively that predicted in Section 5a. The deviation from the predicted behavior results because the current (although rising faster than V^3) does not rise as rapidly as shown in Fig. 6. The most extreme case of this departure from theoretical prediction is shown by the behavior of sample 0-1-4 in Fig. 12. Here, although $L/L_a = 8.8$, for which Eq. (37) would predict a minimum slope of 4.6 for log J versus log V, the actual slope is very close to 3. This probably results because the sample is still in the transition region discussed in Section 5b. It is unfortunate that these measurements were not extended to higher current levels.

The simple theory may be tested by determining if the current is proportional to τL^{-5}. The current–voltage characteristics of 12 units which clearly showed a near V^3 component were analyzed by extrapolating the V^3

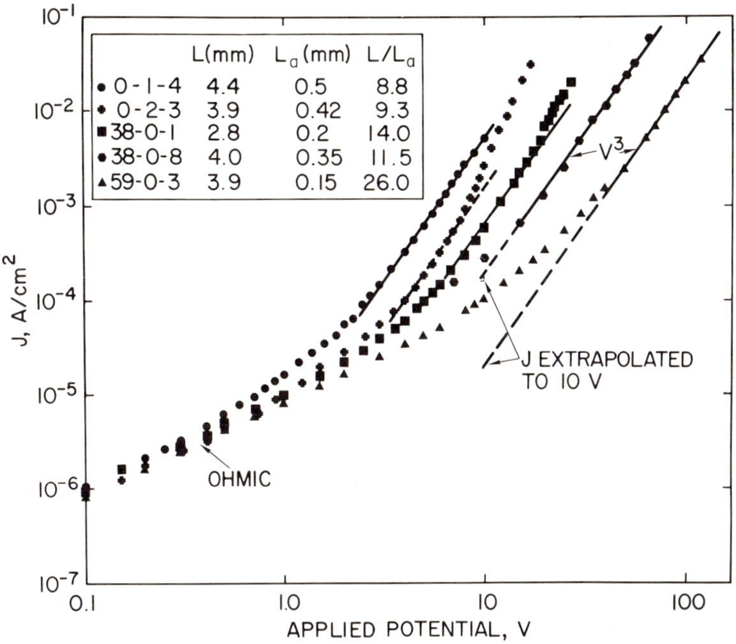

FIG. 12. The log(current-density) versus log(applied voltage) characteristics for five *p–i–n* junctions whose intrinsic lengths L and ambipolar diffusion lengths L_a are shown in the table. The solid lines represent a V^3 current dependence. (After Mayer *et al.*[24])

component of current to 10 V (see Fig. 12), and comparing the current with that predicted. In Fig. 13, which plots $J/\tau\mu_n\mu_p$ as a function of L^{-5} (solid points), there is only a weak correlation with sample length. A much better fit to the data was obtained by correcting L for carrier-diffusion effects in the vicinity of the contacts. A first-order correction was to take an effective length L_{eff} given by $L_{eff} = L - 2.2(2L_a)$, where the value 2.2 is chosen by an empirical fit. This is shown by the open points in Fig. 13. For further verification, two of the samples were cooled to 0, −75, −105, and −193°C, and their current–voltage characteristics and high-level lifetime were measured. For temperatures at which V^3 dependence could be obtained, the same empirical factor also gave a reasonable fit to the data. The use of the effective-length approximation is discussed in detail in Section 5.

b. Potential-Probe Measurements and Boundary Conditions

One of the crucial experimental tests of the theory of the insulator regime is the measurement of the field and carrier distributions. Current–voltage characteristics are good qualitative indications of double injection, but

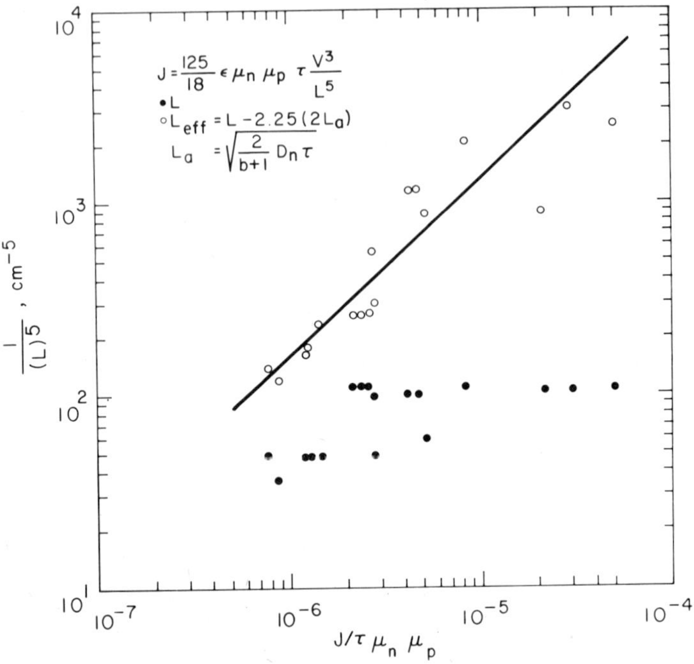

FIG. 13. Comparison of reciprocal of the fifth power of the sample length with the quantity $J/\tau\mu_n\mu_p$ at $V = 10$ V for 17 samples. The current values used were taken from the currents in the V^3 portion extrapolated to 10 V, as shown in Fig. 12. The closed points represent taking the sample length to be the measured length of the intrinsic region. The open points represent taking the sample length to be the effective length calculated from $L_{eff} = L - 2.2(2L_a)$. The solid line is the theoretical relation between L^{-5} and $J/\tau\mu_p\mu_n$ (at $V = 10$ V) of Eq. (17). (After Mayer *et al.*[24])

quantitative information can best be determined from $n(x)$ or $\mathscr{E}^{\dagger}(x)$. The mobile-carrier and space-charge distributions can be determined easily from probe techniques,[19,29,30] although infrared-absorption[31] measurements could also be used to find $n(x)$.

The interpretation of the measured probe potential is relatively straightforward in moderately doped semiconductor materials across which a voltage has been applied. In these cases, where the majority-carrier distribution is relatively unperturbed, the potential distribution is related directly to the electrostatic potential.[32–34] The standard measurement technique is to

[29] D. E. Sawyer, *Solid-State Electron.* **5**, 89 (1962).
[30] J. A. Swanson, *IBM J. Res. Develop.* **1**, 39 (1957).
[31] N. J. Harrick, *Phys. Rev.* **103**, 1173 (1956).
[32] J. R. Macdonald, *Solid-State Electron.* **5**, 11 (1962).
[33] J. R. Macdonald, *J. Chem. Phys.* **29**, 1346 (1958).
[34] J. R. Macdonald, *J. Chem. Phys.* **30**, 806 (1959).

use a fine-pointed metal probe connected to a high-impedance voltmeter.[29,35] Macdonald[34] points out that in the one-carrier system, the potential probe measured the quasi-Fermi level[36] potential of the mobile carriers, which is equivalent to measuring the electrostatic potential for the case where the majority-carrier density is unperturbed.

In the two-carrier case, where the density of injected carriers is greater than the density of thermally generated carriers, the analysis is more complicated, involving the separate quasi-Fermi levels of the holes and electrons. For this case, it is shown in the appendix that the voltage probe measures the weighted mean of the quasi-Fermi levels and not the electrostatic potential. Thus, the carrier concentrations can be obtained rather simply from the potential distribution by calculating

$$n = \left| \frac{J}{e\mu_p(b + 1)\mathscr{E}^\dagger} \right|,$$

where $\mathscr{E}^\dagger$ is defined in Section 5. Figure 14 illustrates a voltage traverse of a sample measured in the ohmic portion of the forward direction. The linear voltage versus distance data are an indication of the high degree of compensation. Lateral traverses are also indicated to show the uniformity of the sample.

When the samples were biased into the V^3 insulator regime, carrier injection from the junctions introduced a marked curvature in the potential distribution. Figure 15 illustrates the measured potential distribution in unit 0-2-3 for three applied voltages. The current–voltage characteristic for this unit can be seen in Fig. 12. There is a linear potential rise for the lowest voltage (curve A) and a marked distortion near the junctions with increasing voltage (curves B and C). A large potential step ΔV_p can be seen in the insert of Fig. 15 for curve C at the p–i junction, and a smaller step ΔV_n is evident near the n–i junction.

In the vicinity of the n–i and p–i junctions, where the carrier distribution is dominated by diffusion effects, $n(x) \propto \exp(-x/L_a)$. Carrier distributions near the n–i and p–i junctions are shown in Figs. 16 and 17 respectively, for two samples biased into the V^3 current regime. The exponential decrease in carrier concentration agrees closely with that predicted by the ambipolar diffusion length as calculated from the measured carrier lifetime. The well-defined exponential nature of the carrier distribution makes it possible to determine by extrapolation the carrier concentrations at the p–i and n–i junctions.

[35] G. L. Pearson, W. Read, and W. Shockley, *Phys. Rev.* **85**, 1055 (1952).

[36] See, for example, W. Shockley, "Electrons and Holes in Semiconductors," p. 308. Van Nostrand, Princeton, New Jersey, 1950.

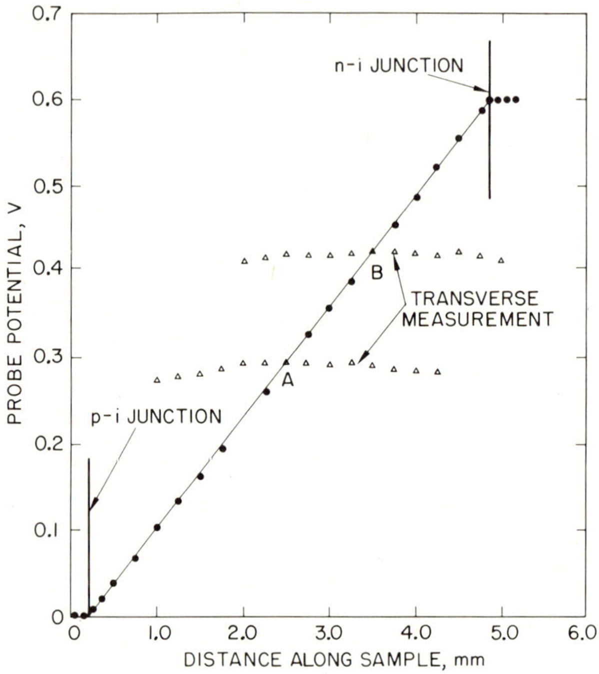

FIG. 14. Potential distribution along a *p–i–n* structure forward biased into the ohmic region. The points marked by triangles indicate transverse measurements made at points A and B. (After Mayer *et al.*[24])

As shown in Section 4c, the boundary conditions are given to first order by

$$p\text{–}i \quad \text{junction:} \qquad n(0) \approx JL_a/2\beta e\mu_p$$

and

$$n\text{–}i \quad \text{junction:} \qquad n(L) \approx JL_a/2\beta e\mu_n,$$

where $\mu_n/\mu_p \approx 2.8$ for silicon at room temperature.

These equations predict that the carrier concentration (for any J) near the *p–i* junction will be about three times greater than that near the *n–i* junction. Figure 18 shows the carrier concentration at the *p–i* and *n–i* junctions as a function of JL_a; the concentration values were determined by extrapolating $n(x)$ data similar to those shown in Figs. 16 and 17. The solid points represent the concentration at the *n–i* junction, the open points that at the *p–i* junction, and the solid and dashed lines the theoretical boundary values. The data agree qualitatively with the theoretical predictions, in that $n \propto J$ and $n_{p-i} > n_{n-i}$. However, in most cases, the values of n lie substantially below the predicted values.

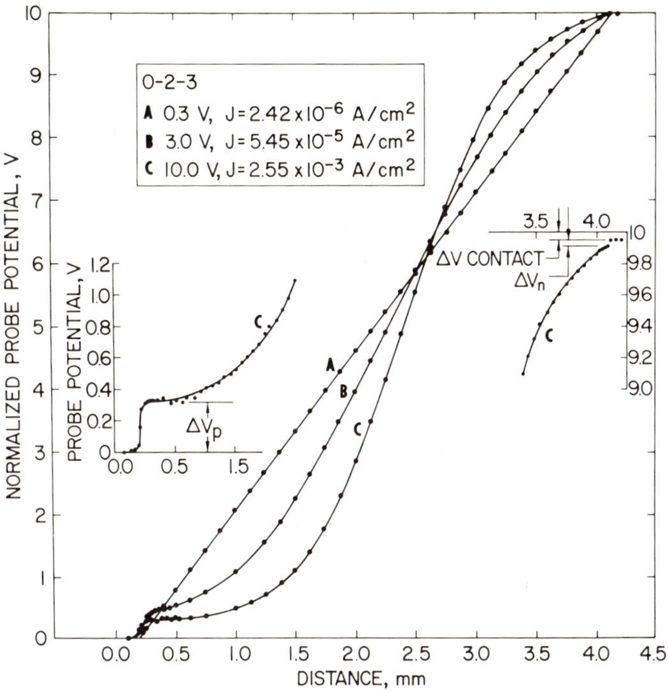

FIG. 15. The probe-potential distribution (normalized to 10 V) for unit 0-2-3 at three applied voltages. The *p–i* junction is on the left and the *n–i* junction is on the right. Curve A was taken at 0.3 V, which was in the ohmic portion of the current–voltage characteristic. The inserts represent the potential distribution of curve C near the junctions with an expanded voltage scale to show the voltage steps ΔV_p and ΔV_n. (After Mayer *et al.*[24])

c. Field and Carrier Distribution

The field and carrier distributions were examined in detail on eight units. Plots of $\mathscr{E}^\dagger(x)$ (E in Fig. 19) and $n(x)$ are shown for a representative unit in Figs. 19(a) and 19(b), respectively. The current–voltage characteristics for this unit are shown in Fig. 12.

The dashed line above the $\mathscr{E}^\dagger(x)$ points in Fig. 19(a) represents an exponential diffusion region using the value of L_a calculated from the measured lifetime. The line has been displaced above the data points for clarity. There is reasonable agreement between this line and the slope of the $\mathscr{E}^\dagger(x)$ points near the *n–i* and *p–i* junctions. It can be seen that the transition points between the diffusion and field-driven regions shift closer together with increasing current values. The shift in the transition point can also be seen in Fig. 17. This would lead to a change in L_{eff} and to deviations from a V^3 current law;

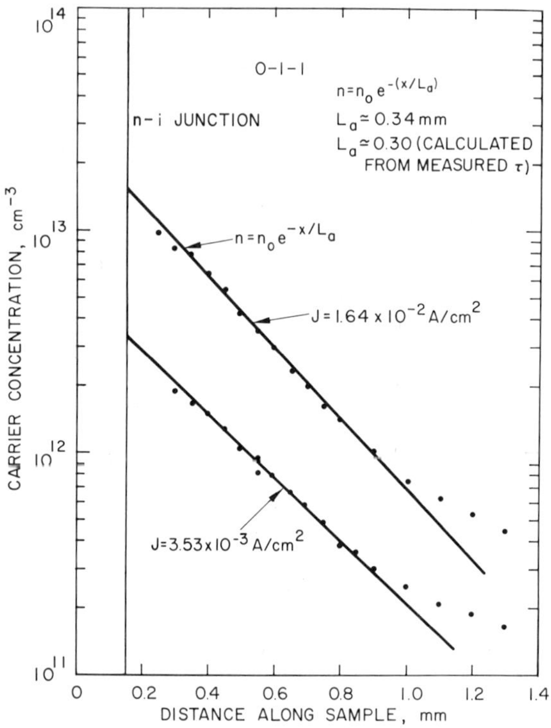

FIG. 16. Detail of the carrier concentration versus distance along sample 0-1-1 near the n–i junction, calculated from Eq. (150) and the measured probe potential. Here, n_0 is a constant of proportionality. (After Mayer *et al.*[19])

such a deviation can be seen in the current–voltage characteristics shown in Fig. 12 for this unit. Near the n–i and p–i junctions, the values of $\mathscr{E}^\dagger$ almost coincide. This results because the carrier distributions at the junctions are proportional to J, as shown in Fig. 18.

Figure 19(b) shows the carrier distribution derived from the $\mathscr{E}^\dagger(x)$ data in Fig. 19(a). It can be seen that the minimum carrier concentration n_{min} in the center of the intrinsic region is a factor of 1.5–50 times greater than the intrinsic carrier concentration ($n_i \approx 10^{10}$). The solid curves in Fig. 19(b) represent the theoretical curves that best fit the experimental data. As can be seen from Fig. 19(b), the theoretical curve is in reasonable agreement with the experimental points. Note that the deviations from the theoretical curves near the junctions may be caused by a change in the lifetime with distance; with this one exception, the form of the experimental curve is identical to that of the theoretical one.

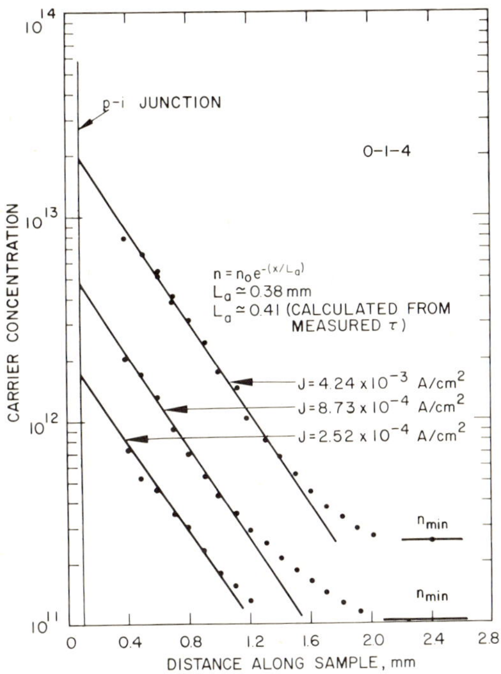

FIG. 17. Detail of the carrier concentration versus distance along sample 0-1-4 near the *p–i* junction, calculated from Eq. (150) and the measured probe potential. Here, n_0 is a constant of proportionality. (After Mayer *et al.*[19])

The theoretical treatment discussed above may be tested in a straightforward manner by plotting the current dependence of the carrier concentration n_0 in the center of the intrinsic region. Theory predicts[19] that n_{min} is proportional to $J^{2/3}$, in marked contrast to the junction regions, where $n \propto J$. Figure 20 shows n_{min} as a function of current density for nine *p–i–n* structures; the solid lines indicate a $J^{2/3}$ dependence. In most of the samples, $n_{min} \propto J^{2/3}$ over two decades in current.

Insulator-regime $J \propto V^3$ behavior has also been observed in n^+–π–p^+ silicon structures[25,26] where the π material had a resistivity of 10^4 ohm-cm. The *I–V* characteristics [Fig. 21(a)] show ohmic behavior followed by a V^3 regime. The variation in current density (measured at 10 V) as a function of *L* is shown in Fig. 22(b), where the solid line indicates a $J \propto L^{-5}$ dependence. For L/L_a ratios between 15 and 30, the V^3 portion of the current followed an L^{-5} dependence without an effective-length correction. Okazaki and Hiramatsu suggest that the properties of the injecting contacts (Au–Ga and Au–Sb) might suppress diffusion effects. Although p_0 was approximately

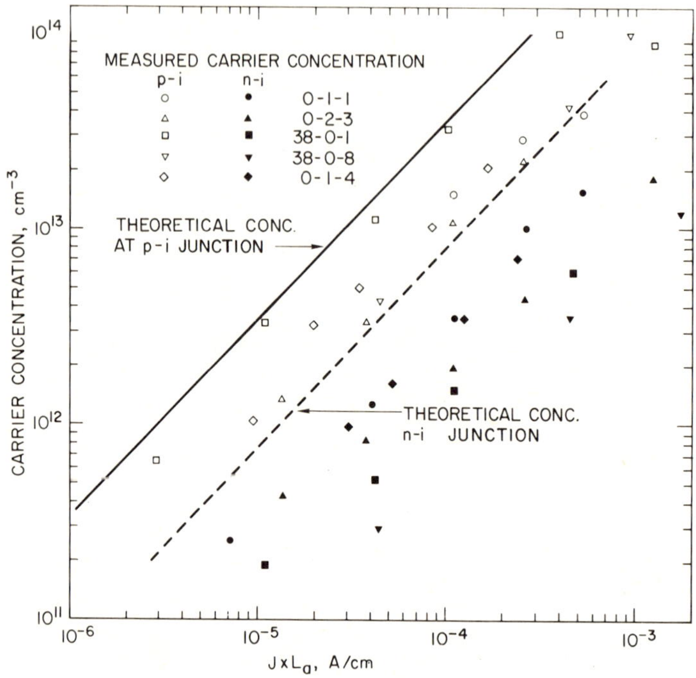

FIG. 18. The dependence upon JL_a of the carrier concentration at the *p–i* and *n–i* junctions for five units. The values of the carrier concentration were obtained by extrapolation to the junction of data similar to those shown in Figs. 16 and 17. The open points represent values near the *p–i* junction; closed points represent values near the *n–i* junction. The solid and dashed lines represent the concentrations at the *p–i* and *n–i* junctions as calculated from Eq. (25) and the corresponding equation for the *n–i* junction. (After Mayer *et al.*[24])

$10^{12}/cm^3$, the semiconductor regime was not observed because of the short length of the samples (less than 0.2 mm).

Logan *et al.*[37] have found $J \propto V^3$ behavior in GaP. However, the current density is a factor of 10^2 less than predicted, which can be attributed to the presence of trapping centers. The influence of deep levels is discussed in Section 8.

9. Semiconductor Regime

In many respects, samples suitable for studies of the semiconductor regime are easier to obtain than those in the insulator regime. In the case of silicon and germanium, materials with dopant concentrations of 10^{11} to $10^{13}/cm^3$ can be obtained. The first experimental verification of the V^2 semiconductor

[37] R. A. Logan, H. G. White, P. W. Foy, and C. J. Frosch, *Solid-State Electron.* 7, 473 (1964).

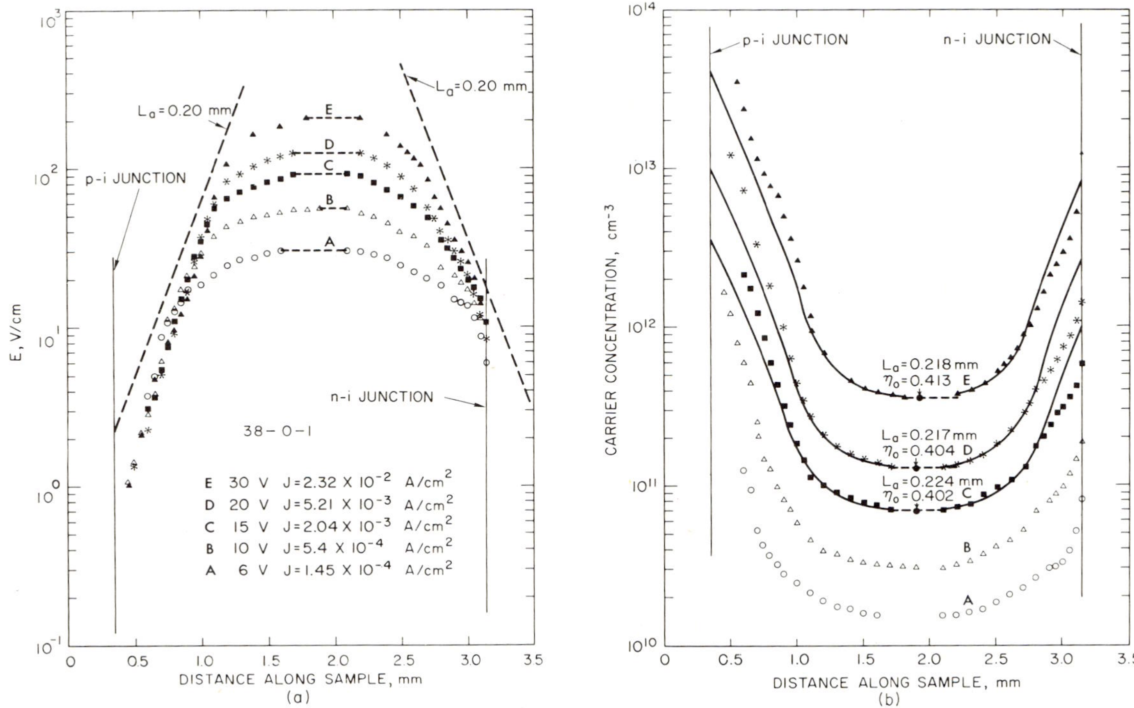

FIG. 19. (a) The measured distribution ($\mathscr{E}^{\dagger}$) for unit 38-0-1 for six values of the applied voltage. The dashed line indicates the variation in field calculated on the basis of the ambipolar diffusion length determined from lifetime measurements. The dashed line was shifted above the experimental points to avoid obscuring the data points. (b) The carrier distribution as calculated from Eq. (150) for the same values of applied voltage. The solid lines indicate the fit to the theoretical curves. (After Mayer *et al.*[24])

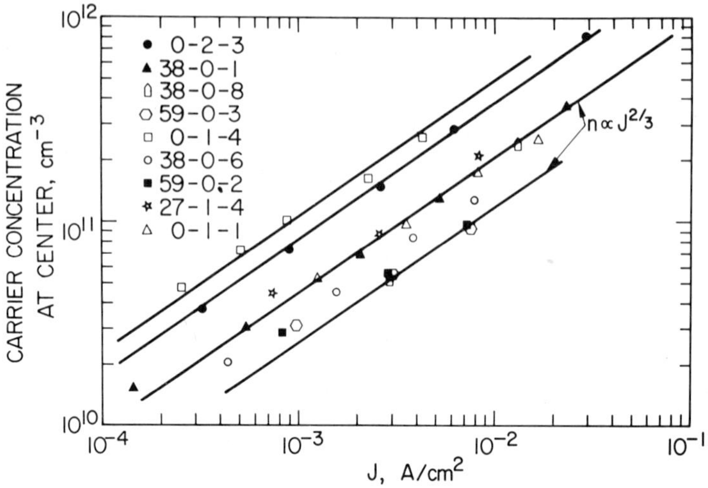

FIG. 20. The dependence of the minimum carrier concentration n_{min} on the current for nine samples. The solid lines indicate a $J^{2/3}$ dependence. (After Mayer *et al.*[24])

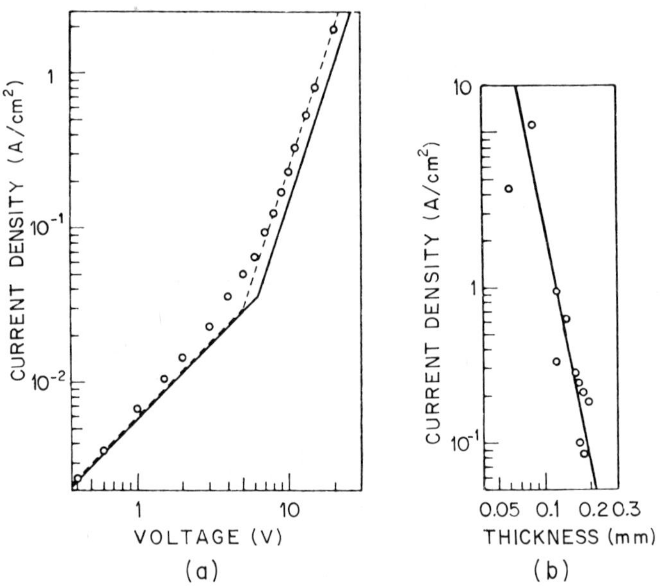

FIG. 21. (a) I–V characteristic of high-resistivity, p-type silicon with injection contacts at room temperature. Crystal thickness is 0.17 mm. Experimental points are shown by circles. The solid lines show an ohmic and $I \propto V^3$ dependence. The broken line is obtained from connecting the observed values. (b) Dependence of double-injection current density on crystal thickness. The points refer to the current density measured at 10 V. (After Okazaki and Hiramatsu.[25])

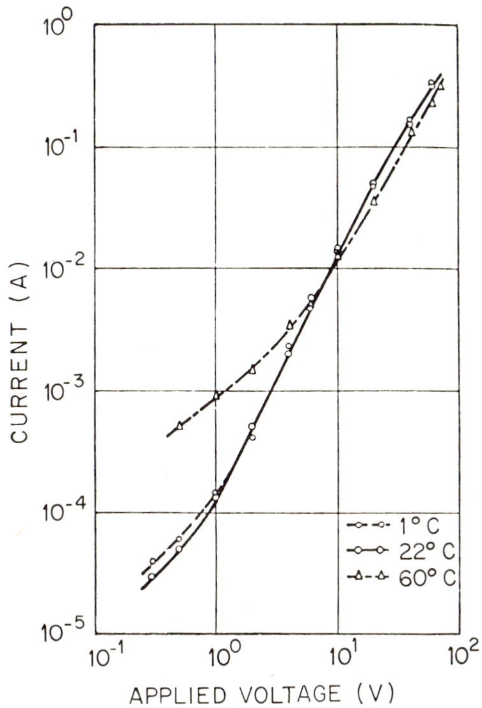

FIG. 22. The current–voltage characteristics of an n-type germanium sample at various temperatures. The sample was 0.97 cm long and 0.063 × 0.055 cm in cross section. (After Larrabee.[38])

regime was obtained by Larrabee,[38] using 1-cm-long germanium samples. As shown in Fig. 22, he obtained a reasonable fit to the square-law behavior at the three temperatures studied. At 60°C, the ohmic portion of the I–V characteristic had increased because of the increased number of carriers. In the square-law regime, the currents did not vary strongly with temperature. The lower magnitude of the 60°C curve was attributed to the decrease in carrier mobility at the higher temperatures.

More recently, Lee[38a] has studied the temperature dependence of double injection in long silicon p^+–π–n^+ samples. He found that the temperature dependence followed the prediction of Eq. (10) within 10% when measured values of $\mu_n(T)$, $\mu_p(T)$, and $\tau(T)$ were used.

[38] R. D. Larrabee, *Phys. Rev.* **121**, 37 (1961).

[38a] D. H. Lee, *J. Appl. Phys.* (to be submitted).

Deshpande[39] observed double-injection effects in long silicon p^+–p–n junctions. The resistivity of the starting material was about 6000 ohm-cm, which increased to 15,000–30,000 ohm-cm after the diffusion and annealing process steps. This increase was attributed to the formation of donor-type complexes during the temperature cycles. A typical I–V curve as shown in Fig. 23 exhibits an ohmic characteristic followed by a "break" to the V^2

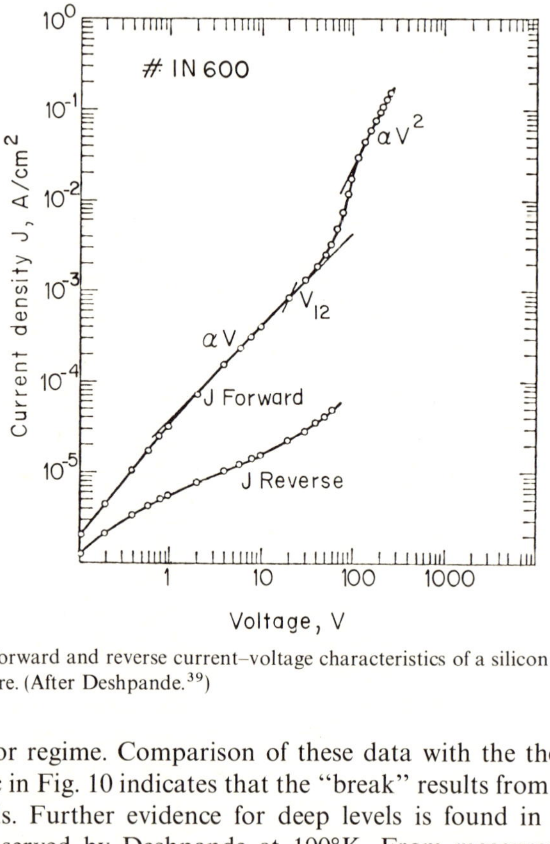

FIG. 23. The forward and reverse current–voltage characteristics of a silicon p–π–n device at room temperature. (After Deshpande.[39])

semiconductor regime. Comparison of these data with the theoretical I–V characteristic in Fig. 10 indicates that the "break" results from the presence of deep levels. Further evidence for deep levels is found in the negative resistance observed by Deshpande at 100°K. From measurements of the current in the ohmic region as a function of temperature, Deshpande found that the deep donor level has an activation energy between 0.30 and 0.34 eV. Based on etching studies, he concluded that these centers were introduced uniformly throughout the bulk.

In Deshpande's work, neither the lifetime nor the potential distribution was measured directly. In the following sections, we will discuss a more

[39] R. Y. Deshpande, *Solid-State Electron.* **9**, 265 (1966).

detailed set of measurements[40,41] on a series of p^+–π–n structures fabricated from high-resistivity float-zone-grown silicon. In these structures, the p^+ regions were formed by alloying aluminum and the n^+ regions by lithium or phosphorus diffusion. The sample lifetime was determined by biasing the sample into the V^2 portion of the current–voltage characteristic and then measuring the transient response to a voltage step or the current decay following excitation by a brief, intense flash of light (see Section 18). Potential-probe measurements in the low-voltage (ohmic portion of the current–voltage characteristics) regime were used to determine both p_0 and the uniformity of the material.

a. Current–Voltage Characteristics

The current–voltage characteristics of a representative selection of samples are shown in Fig. 24, in which log J versus log V characteristics are plotted for five samples. The heavy lines indicate a $J \propto V^2$ dependence. The closed points at 10 V are calculated from the measured parameters of τ, p_0, and L, using Eq. (10). The increasing deviation from theory for small L/L_a ratios (samples 2 and 3) is caused by diffusion from the contacts. The steeply rising current of sample 1 is typical of short p–i–n structures, in which diffusion dominates. Figure 25 shows the I–V characteristics of one sample (5-5, 8) fabricated from 10 kilohm-cm silicon (measured $p_0 = 7 \times 10^{11}$ cm^{-3}) which had been successively shortened and rediffused for new n–π contacts. The field distribution and effective-length correction for this sample is discussed in detail in the following sections.

In order to demonstrate the validity of $J \propto p_0 \tau V^2/L^3$, Fig. 26 shows $J/p_0\tau$ versus L^3 for a series of 19 samples with large L/L_a ratios. The values of J were determined from the V^2 component (as shown in Figs. 24 and 25) extrapolated to 10 V. The values of p_0 ranged from 10^{11} to 2×10^{12} cm^{-3}, and τ from 4 to 30 μsec. Over this range of parameters, the data show reasonably close agreement with the simple theory as shown by the solid line in Fig. 26.

b. Potential and Field Distributions

The theoretical treatment in Section 4 predicts that the field distribution should follow the relation $\mathscr{E} = [k(L - x)]^{1/2}$, with the maximum field strength near the p–π junction. Furthermore, if carrier-diffusion effects are dominant, the field and potential distributions should be proportional to $\exp(-x/L_a)$ near the n–π and p–π junctions. As in the case of the p–i–n structure, the voltage probe measures the weighted mean of the quasi-Fermi levels, and not the electrostatic potential.

[40] O. J. Marsh, J. W. Mayer, and R. Baron, *Appl. Phys. Letters* **5**, 74 (1964).

[41] J. W. Mayer, O. J. Marsh, and R. Baron, *J. Appl. Phys.* **39**, 1447 (1968).

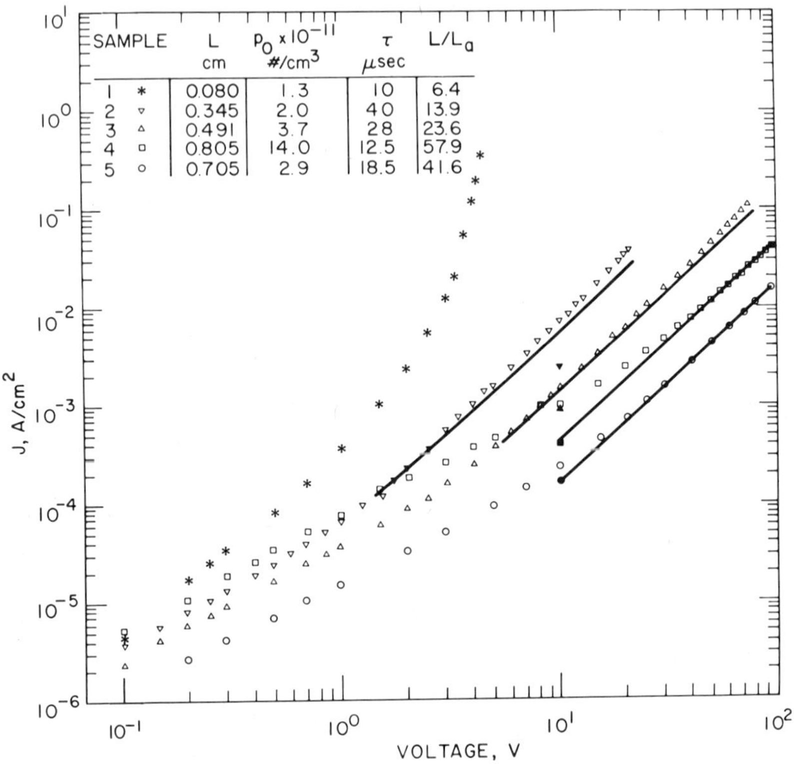

FIG. 24. The log current–log voltage characteristics of five silicon p–π–n diodes as a function of length. The solid lines indicate a V^2 current dependence. The solid data points at 10 V are the V^2 current components calculated from the measured parameters of τ, p_0, and L. (After Mayer *et al.*[41])

Figure 27 shows the potential variation with distance near the p–π and n–π junctions of a sample biased well into the V^2 regime. The contact voltage drop V_C and junction voltage V_J can be seen. As in the case of the p–i–n structure, V_J at the p–π junction is greater than at the n–π junction. These data were replotted in Fig. 28 as $\log(V - V_J - V_C)$ versus distance to show the influence of diffusion effects, which would lead to an exponential distribution proportional to $\exp(x/L_a)$. Because of the uncertainty of $V_J + V_C$, it is very difficult to determine the exact value of L_a from this plot. The existence of the diffusion region near the p–π junction is seldom as clear-cut as in the sample shown in Fig. 28. In general, however, the potential distribution changes very strongly with position in the region near the p–π junction.

A more useful procedure is to determine graphically the value of $\mathscr{E}^\dagger(x)$. A plot of these values for unit 5-5, 8 is shown in Fig. 29 for three different

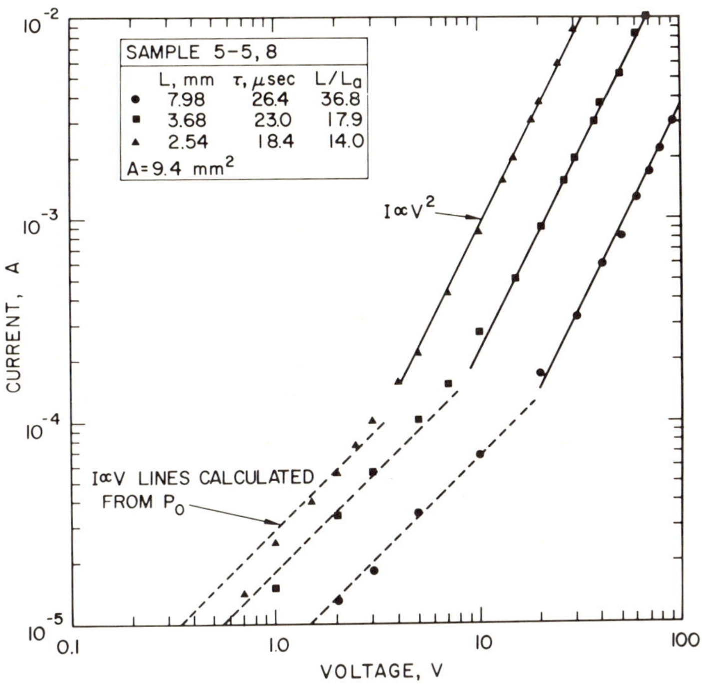

FIG. 25. The log current–log voltage characteristics of a sample which had been successively shortened and rediffused for new n–π contacts. The solid lines indicate a current proportional to V^2. (After Mayer *et al.*[41])

lengths of the unit and in Fig. 30 for one sample length, but three different applied voltages. The I–V characteristics of this unit are shown in Fig. 25. The heavy lines in Figs. 29 and 30 represent a fit to an $\mathscr{E} \propto x^{1/2}$ relation. It can be seen that this relation holds over more than half the sample lengths. The influence of diffusion can be seen near the n–π junction. It was found that in the region near the field maximum, the assumption that $n \approx p \gg p_0$ does not hold, so that the samples are not fully into the high-level regime. The hole concentrations are often 20–30% greater than the electron concentration; this is consistent with the long transition region discussed in Section 5c and probably accounts for the rounding of the field distribution on the p–π junction side of the field maximum, compared with the sharp corner predicted for high-level field distribution.[20] The data of Figs. 29 and 30 clearly show the asymmetrical nature of the field distribution and the tendency of $\mathscr{E}$ to follow an $x^{1/2}$ relation in the central region.

The $\mathscr{E} \propto x^{1/2}$ dependence is shown more strikingly in Fig. 31, which is a plot of $\mathscr{E}^2$ versus x for two bias voltages on one sample. The fit to Lampert's prediction can be seen to hold over most of the sample length.

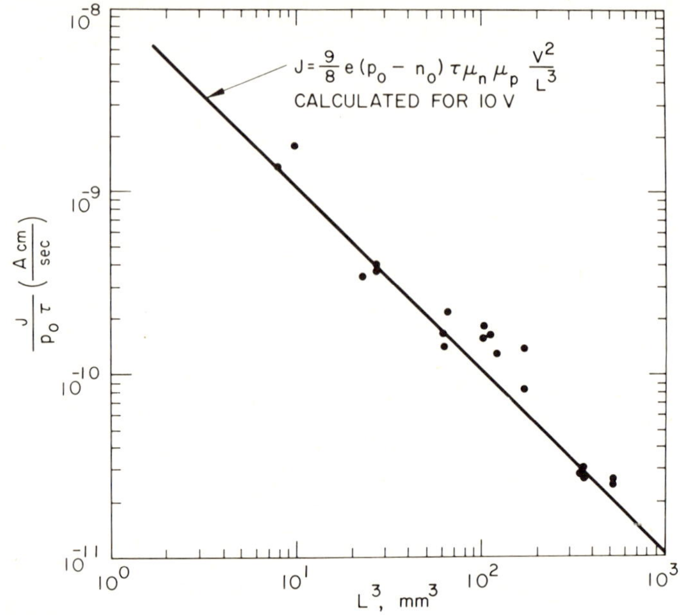

FIG. 26. The quantity $J/p_0\tau$ plotted versus L^3 for a series of samples with large L/L_a ratios. Here, J was determined from the V^2 component extrapolated to 10 V as shown in Fig. 24 by the straight lines. The straight line in this plot represents the theoretical relation between the plotted quantities. (After Mayer *et al.*[41])

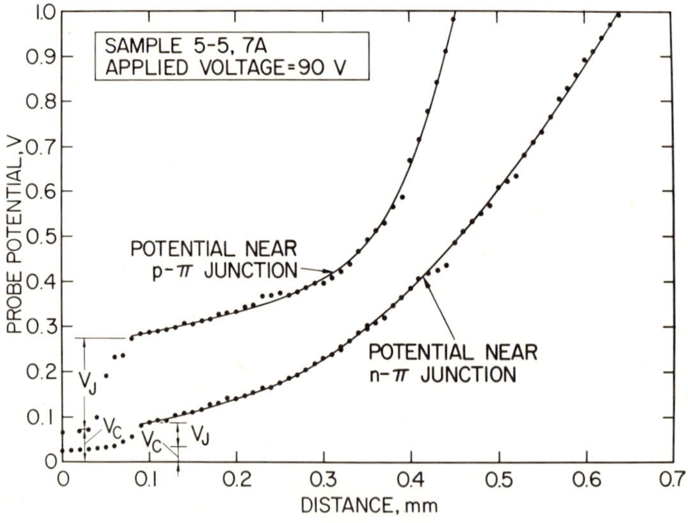

FIG. 27. The measured probe potential as a function of distance for a sample biased well into the semiconductor regime. The potentials near the n–π and p–π junctions are superposed for clarity. The contact voltage drop V_C and junction voltage V_J are indicated. (After Mayer *et al.*[41])

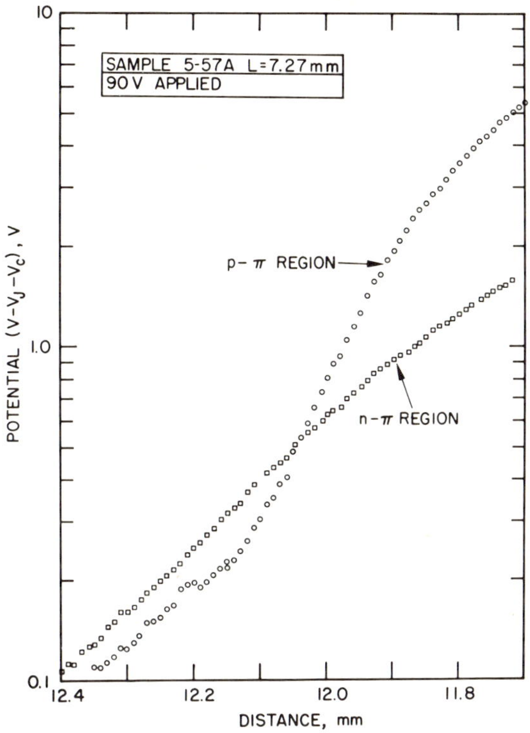

FIG. 28. Data of Fig. 27 replotted as $\log(V - V_J - V_C)$ versus distance to show the exponential nature of the potential distribution near the contacts. (After Mayer *et al.*[41])

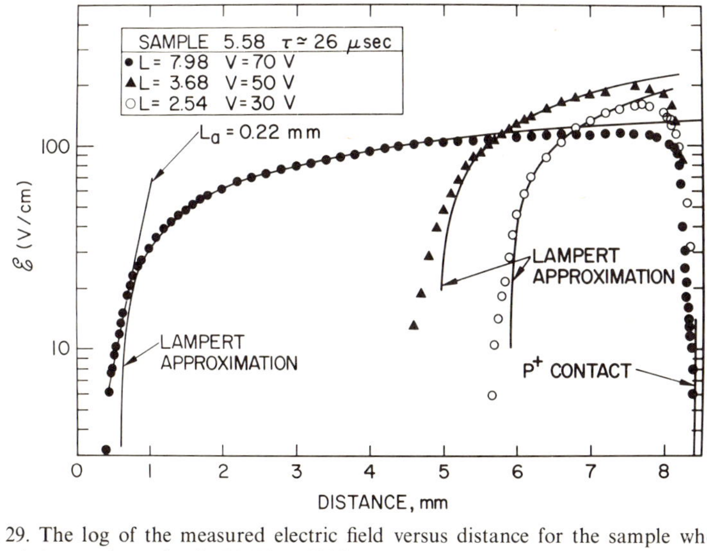

FIG. 29. The log of the measured electric field versus distance for the sample whose I–V characteristics are shown in Fig. 25. The solid lines represent the Lampert (simple semiconductor regime) approximation to the field distribution. In the longest sample, the solid straight line near the n^+–π junction (which occurs at the left end of each of the curves) represents a field distribution proportional to $\exp(x/L_a)$. The p^+–π junction is on the right. (After Mayer *et al.*[41])

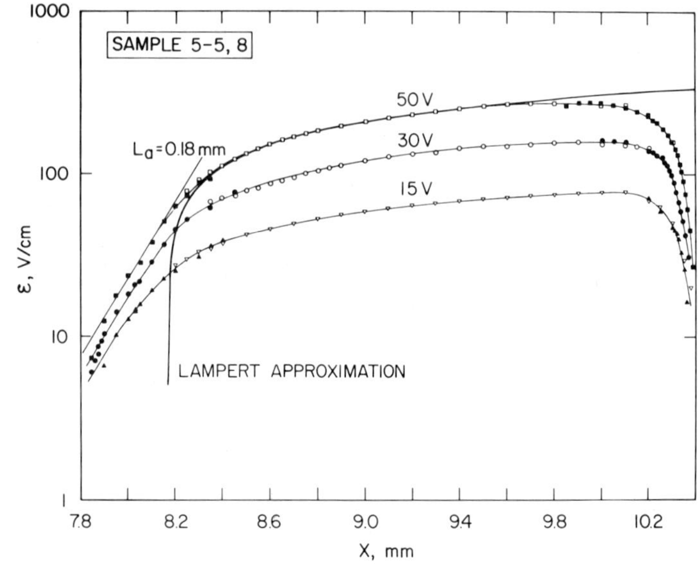

FIG. 30. The log of the measured field versus distance as a function of applied voltage for the 2.54-mm-long sample shown in Figs. 25 and 29. The heavy curve represents the Lampert (simple semiconductor regime) approximation for the 50-V curve. The straight line near the n^+–π junction (on the left) represents a field distribution proportional to $\exp(x/L_a)$. The p^+–π junction is on the right. (After Mayer *et al.*[41])

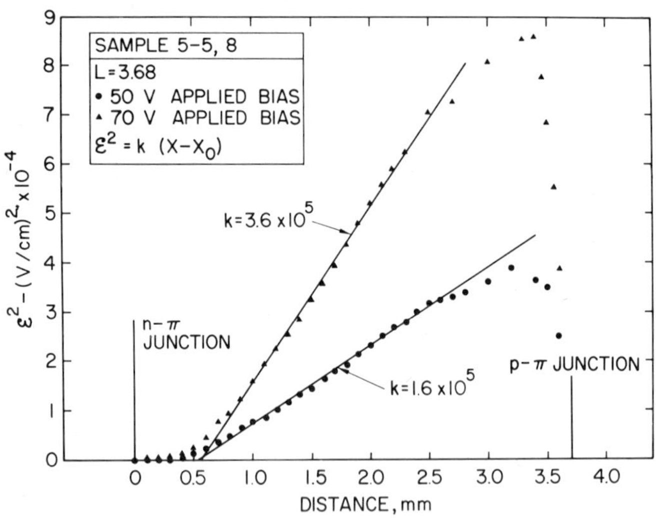

FIG. 31. The square of the measured electric field versus distance for two values of applied voltage for the sample shown in Figs. 25 and 29. The straight lines represent the theoretical curves for the Lampert (simple semiconductor regime) approximation. (After Mayer *et al.*[41])

It has been shown that the experimental results obtained with silicon p–π–n structure are in reasonable agreement with the predictions of the theory.

1. The current–voltage characteristics are in agreement with the simple theory (Section 4a), since $J \propto V^2$ and the magnitude of J is given by the expression $J = (9/8)e\mu_p\mu_n p_0 \tau V^2/L^3$ to within a factor of two at worst.
2. The electric field distribution follows the general shape predicted by Baron.[20] That is, the field approximates a distribution of $\mathscr{E} = [k(x - x_0)]^{1/2}$ in the central region, and has a transition to diffusion-dominated exponential regions near the junctions. This transition is abrupt at the end near the p–π junction and is gradual at the end near the n–π junction.

IV. Large Numbers of Deep Centers—Theory

10. General Discussion

As has been noted previously, δp_{R} cannot be neglected in Poisson's equation when the number of recombination centers is large. The presence of this term, which generally varies with both position and injection level, greatly complicates the problem. With few exceptions, results have been obtained in this area only when the quasineutrality approximation can be made. That is, the term $(\varepsilon/e)\, d\mathscr{E}/dx$ can be neglected in Poisson's equation. There has been slow progress toward the inclusion of space charge. For the case of partially filled deep centers and no thermal carriers, Ashley[10a] succeeded in deriving the threshold voltage for the onset of current flow. Lampert and Schilling, using the regional approximation method, in Chapter 1 of this volume, obtain the entire current–voltage characteristic for the two cases of partially filled deep centers and completely filled deep centers. It is inherent in the method that the sample be divided into regions which either satisfy the quasineutrality approximation or are dominated by space charge (trapped space charge in the case of partially filled centers and free space charge in the case of completely filled centers).

Even in the limit of the quasineutrality approximation, a considerable variation in behavior is possible. Among these is the current–controlled negative resistance first predicted by Lampert and Rose.[5] We shall discuss some of the many distinct behavior regimes which can occur under the quasineutrality approximation.

Before proceeding further, we introduce a very useful variable, the ratio

$$u = \delta p/\delta n. \tag{46}$$

In terms of this variable, the recombination equations (5) take on a very simple form. Using (5) and the quasineutrality form of Poisson's equation, i.e.,

$$0 = \delta p - \delta n + \delta p_{\mathrm{R}}, \tag{47}$$

it is easily shown that

$$\delta n = \frac{p_3(u - u_0)}{(1 - u)(\beta + u)}, \tag{48}$$

$$\delta p = \frac{p_3 u(u - u_0)}{(1 - u)(\beta + u)}, \tag{49}$$

$$\delta p_{\mathrm{R}} = \frac{p_3(u - u_0)}{\beta + u}, \tag{50}$$

and

$$r = \frac{[-p_2 + (N_{\mathrm{R}} + p_2 + n_2)u - n_2 u^2](u \quad u_0)p_3 c_n}{(1 - u)(\beta + u)^2}. \tag{51}$$

The constants are defined as follows:

$$n_2 \equiv n_0 + n_1, \qquad p_2 \equiv p_0 + p_1,$$

$$n_3 \equiv \beta p_{\mathrm{R},0} + p_2 + \beta n_2, \qquad p_3 \equiv n_{\mathrm{R},0} + p_2 + \beta n_2,$$

$$u_0 \equiv n_3/p_3. \tag{52}$$

Here, and in the remainder of Part IV, β is used in its second meaning of $\beta = c_n/c_p$. Note that u_0 is the thermal equilibrium value of u, since $\delta n = \delta p = \delta p_{\mathrm{R}} = r = 0$ for $u = u_0$. Note also that δn, δp, and r approach ∞ as $u \to 1$ (the high-level limit). Thus, u is confined to lie between u_0 and 1, and cannot cross either limit. For simplicity, we will confine the discussion to the case where $u_0 < 1$.[41a] The choice of $u_0 < 1$ physically means that $\tau_{p,0} < \tau_{n,0}$, since $u = \delta p/\delta n \equiv \tau_p/\tau_n$. We will also limit the discussion to the case that $N_{\mathrm{R}} \gg p_2$, n_2, and p_2/u_0 (which corresponds exactly to the requirement that $n_{\mathrm{R},0} \gg p_0$, $p_{\mathrm{R},0} \gg n_0$, and $\beta p_{\mathrm{R},0} \gg p_0$). This allows us to simplify some of the above expressions to

$$r \approx \frac{N_{\mathrm{R}} u(u - u_0) n_{\mathrm{R},0} c_n}{(1 - u)(\beta + u)^2}$$

$$n_3 \approx \beta p_{\mathrm{R},0}, \qquad p_3 \approx n_{\mathrm{R},0}. \tag{53}$$

Having $r(u)$, it remains to evaluate either dJ_n/dx or dJ_p/dx as a function of u. Because J_n and J_p are nearly proportional to J in many cases, the leading

[41a] This does not limit the generality of this treatment, as the case where $u_0 > 1$ can be obtained merely by exchanging all n's and p's.

terms do not contribute to the derivative, and it is necessary either to find some linear combination in which these terms cancel (as was done for the high-level case), or to calculate $J_n(u)$ or $J_p(u)$ to sufficiently high order.[41b] We will now consider several classes of cases where this is possible.[41c]

First, we will consider the class of cases which occur at "higher" injection levels, where both n_0 and p_0 can be considered small with respect to the injected carrier densities. For an insulator, of course, this may include all injection levels. We will then consider the class of cases which occur at "lower" injection levels, where either n_0 or p_0 is relatively large compared with the respective injected carrier density. The physical interpretation of what happens during the transition from "lower" injection levels is discussed at some length. Finally, it is shown that the Ashley–Milnes regime[10] is not dependent on the quasineutrality condition, and the effects of space charge in this regime are discussed.

11. Thermal Densities Small

If n_0 and p_0 are sufficiently small, the current equation becomes

$$J = e\mu_p(b + u)\,\delta n\mathscr{E}\,; \tag{54}$$

when solved for $\delta n\mathscr{E}$ and substituted in the approximate recombination equation for this case,

$$r = \mu_n \frac{d}{dx}(\delta n\mathscr{E}), \tag{55}$$

Eq. (54) yields the following master equation for this class of cases:

$$\frac{K}{J} = -\frac{(1 - u)(\beta + u)^2}{u(u - u_0)(b + u)^2}\frac{du}{dx}, \tag{56}$$

where we have set $K = eN_{\mathrm{R}}n_{\mathrm{R},0}c_n/b$. When $u(x)$ is determined by integrating Eq. (56), $\mathscr{E}(x)$ is found from

$$\mathscr{E} = \frac{J}{e\mu_p n_{\mathrm{R},0}}\frac{(1 - u)(\beta + u)}{(u - u_0)(b + u)} \tag{57}$$

and integrated to give the current–voltage characteristic. As it stands, Eq. (56) is still formidable; however, it consists of a number of factors linear in u. It is is assumed, for example, that $u \ll \beta$ (or $u \gg \beta$), a much simpler expression results. If the constants u_0, β, and b are sufficiently well separated, a number of distinct behavior regimes result—much as with the high-level regimes.

[41b]It is the difficulty of doing this that has prevented the inclusion of space charge.

[41c]R. Baron, unpublished work, 1967–1969.

a. $b \gg 1 \gg \beta \gg u_0$

Consider first the sequence of regimes that occur when b, β, and u_0 satisfy the above condition. The condition $\beta \gg u_0$ is equivalent to $n_{\mathrm{R},0} \gg p_{\mathrm{R},0}$, so that the centers are mostly occupied with electrons. At the highest injection levels, $u \approx 1$, and Eq. (56) becomes

$$\frac{K}{J} = -\frac{(1-u)}{b^2}\frac{du}{dx}, \tag{58}$$

which integrates to

$$\frac{b^2 Kx}{J} = \frac{(1-u)^2}{2}; \tag{59}$$

which satisfies the boundary condition $u = 1$ ($\mathscr{E} = 0$) at $x = 0$. Combining this with Eq. (57) gives

$$\mathscr{E} = \left[\frac{2JN_{\mathrm{R}}c_n}{e\mu_p\mu_n n_{\mathrm{R},0}}x\right]^{1/2}, \tag{60}$$

which integrates to

$$V = \left[\frac{8}{9}\frac{JL^3}{e\mu_p\mu_n\tau n_{\mathrm{R},0}}\right]^{1/2}, \tag{61}$$

where the high-level lifetime $\tau = 1/N_{\mathrm{R}}c_n$ has been substituted. This is just the semiconductor regime, and is more easily recognizable if it is turned around, i.e.,

$$J = \frac{9}{8}e\mu_p\mu_n\tau n_{\mathrm{R},0}\frac{V^2}{L^3}. \tag{62}$$

Note that the material has been effectively turned into an n-type semiconductor of doping $n_{\mathrm{R},0}$, as this is the number of electrons promoted from the deep levels to the conduction band. Alternatively, $n_{\mathrm{R},0}$ is the amount of fixed charge in the crystal. This approximation is valid for $\delta n \gg n_{\mathrm{R},0}$. A lower limit to this regime may be found by setting the minimum value of δn (at $x = L$) to $n_{\mathrm{R},0}$, so that the regime is valid for

$$J \gtrsim 2b^2 KL. \tag{63}$$

As the injection level falls, u decreases until it satisfies $\beta \gg u \gg u_0$, and Eq. (56) becomes

$$\frac{K}{J} = -\frac{\beta^2}{u^2 b^2}\frac{du}{dx}, \tag{64}$$

which integrates to

$$\frac{Kx}{J} = \frac{\beta^2}{ub^2}. \tag{65}$$

Here, we neglect the details of the boundary condition at $x \approx 0$. Including the details of the boundary conditions requires a regional approach similar to that of Lampert (see Chapter 1), where the present region must be joined at some $x = x_1$ to a region $0 \leqslant x < x_1$ which satisfies the high-level solution given in Eq. (59). Thus, neglecting this region corresponds to assuming $x_1 \ll L$. Therefore, Eq. (65) does not actually imply $u \to \infty$ at $x = 0$, since Eq. (65) is not valid over the region $0 \leqslant x < x_1$.

Finding $\mathscr{E}(x)$ and integrating as above, we obtain

$$V = L^2/2\mu_p\tau\beta. \tag{66}$$

Note that V is independent of current. In addition, note that for the present case of $n_{\mathrm{R},0} \gg p_{\mathrm{R},0}$, $\tau\beta \approx \tau_{p,0}$, the low-level hole lifetime. This is the Lampert threshold voltage. The negative resistance occurs during the transition between these two regimes. If x_1 is allowed to be arbitrary, the shape of the current–voltage characteristic in the negative-resistance region can be determined. Although this shape is shown in Fig. 32, we will only derive the distinct regimes of behavior here (i.e., those regimes where a very simple solution fills almost the entire length of the sample). Once again, there is a lower limit to the current, this time determined by the condition that $\delta n > p_{\mathrm{R},0}$ everywhere, given by

$$J \gtrsim \frac{b^2 p_{\mathrm{R},0}}{\beta n_{\mathrm{R},0}} KL. \tag{67}$$

It should be noted that the condition on this sequence of regimes, that $\beta \gg u_0$, is equivalent to $n_{\mathrm{R},0} \gg p_{\mathrm{R},0}$.

As the injection level falls still further, u approaches u_0 until it satisfies $(u - u_0) \ll u_0 \approx u$. Thus, we may set $u = u_0$ except where terms of the form $(u - u_0)$ occur, and Eq. (56) becomes

$$\frac{K}{J} = -\frac{\beta^2}{(u - u_0)u_0 b^2} \frac{du}{dx}, \tag{68}$$

which integrates to

$$\frac{Kx}{J} = -\frac{\beta^2}{u_0 b^2} \ln\left(\frac{u - u_0}{u_0}\right). \tag{69}$$

The choice of the integration constant $\ln u_0$ corresponds to the upper limit

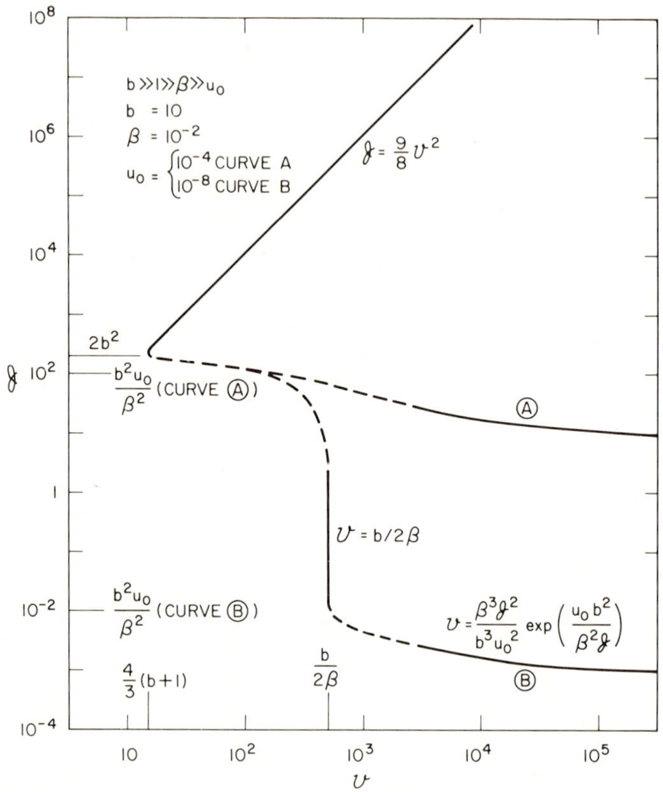

FIG 32. The current–voltage characteristic for a sample satisfying $b \gg 1 \gg \beta \gg u_0$ (Section 11a) for two values of u_0. The full lines represent the distinct regimes discussed in the text. The dashed lines represent transition regions. (After Baron.[41c])

of validity of $u \lesssim 2u_0$ or $\delta n \lesssim p_{\mathrm{R},0}$. Finding $\mathscr{E}(x)$ and integrating gives

$$V = \frac{J^2 \beta n_{\mathrm{R},0}}{e\mu_n b^2 p_{\mathrm{R},0}^2 K}\left[\exp\left(\frac{u_0 b^2 KL}{\beta^2 J}\right) - 1\right]. \tag{70}$$

This last regime is that discussed by Keating.[42,43] The resulting current–voltage characteristic is shown in Fig. 32 in terms of a reduced current and voltage:

$$\mathscr{J} = J/J^*, \qquad \mathscr{V} = V/V^*,$$

[42] P. N. Keating, *Solid State Commun.* **1**, 210 (1963).
[43] P. N. Keating, *Phys. Rev.* **135**, A1407 (1964).

where

$$J^* = KL,$$

$$V^* = \frac{KL^2}{e\mu_p n_{R,0}} = \frac{L^2}{\mu_n \tau}.$$

These reduced variables were chosen because the resulting expressions are very simple, with Eqs. (62), (66), and (70) becoming, respectively,

$$\mathscr{J} = (9/8)\mathscr{V}^2, \tag{71}$$

$$\mathscr{V} = b/2\beta, \tag{72}$$

and

$$\mathscr{V} = (\beta^3 \mathscr{J}^2/b^3 u_0^2)[\exp(u_0 b^2/\beta^2 \mathscr{J}) - 1]. \tag{73}$$

Note that the constant-voltage regime does not appear a distinct regime unless the more stringent condition that $p_{R,0} \ll \beta n_{R,0}$ is satisfied. The regime characterized by Eq. (73) is valid for currents no higher than $\mathscr{J} = b^2 u_0/\beta^2$ (see Fig. 32). Thus, the minimum possible value of the exponent is 1, and there is no point in expanding the exponential term.

The characteristic shown in Fig. 32 is, of course, not valid at arbitrarily low injection levels. Eventually, either the quasineutrality condition, or the condition that n_0 and p_0 are small, breaks down, and some other behavior dominates. The behavior in this event will be discussed later. We will now consider the current–voltage characteristics which result from other assumptions about the relative values of b, β, and u_0.

b. $b \gg 1 \gg u_0 \gg \beta$

This case differs from the previous one in that $u_0 \gg \beta$, or, equivalently, $p_{R,0} \gg n_{R,0}$, so that the centers are mostly occupied by holes. If this condition is to be satisfied along with $1 \gg u_0$, it is necessary to have a very small value of β. The high-level regime is the same as in Section 11a. The constant-voltage regime can no longer exist as a distinct regime. The next regime is again where $u \approx u_0$, so that Eq. (56) becomes

$$\frac{K}{J} = -\frac{u_0}{(u - u_0)b^2}\frac{du}{dx}. \tag{74}$$

This is functionally the same as Eq. (68), but the constants are different, leading to[43a]

$$\mathscr{V} = (u_0 \mathscr{J}^2/b^3)[\exp(b^2/u_0 \mathscr{J}) - 1]. \tag{75}$$

[43a] In the remaining discussion, we will give only the form of Eq. (56) and the resulting current–voltage characteristic, as the procedure for determining the characteristic is clear from the preceding discussion.

This again is one of the cases considered by Keating.[43] Note that the knee of the current–voltage characteristic occurs at $\mathscr{J} = 2b^2$ (Fig. 33), so that the minimum value of the exponent in Eq. (75) is $1/2u_0$, which is always large. Thus, we again do not consider expanding the exponential.

However, it must be noted that this regime [and thus the expression in Eq. (75)] is valid only for $\mathscr{J} \ll 2b^2$. If Eq. (75) is evaluated at $\mathscr{J} = 2b^2$, we obtain $\mathscr{V} = 4u_0 b \exp(1/2u_0)$, which can be very large indeed. Thus Eq. (75) will not be valid until extremely high values of $\mathscr{V}$ are reached. Accuracy may be improved considerably if it is realized that a large contribution to $\mathscr{V}$ comes from the integration of $\mathscr{E}$ from 0 to x_1, in which region Eq. (74) is not valid. By restricting the integration of $\mathscr{E}$ to the range x_1 to L, and using the value of $x_1 = \mathscr{J}L/2b^2$ (determined from the condition that the high-level

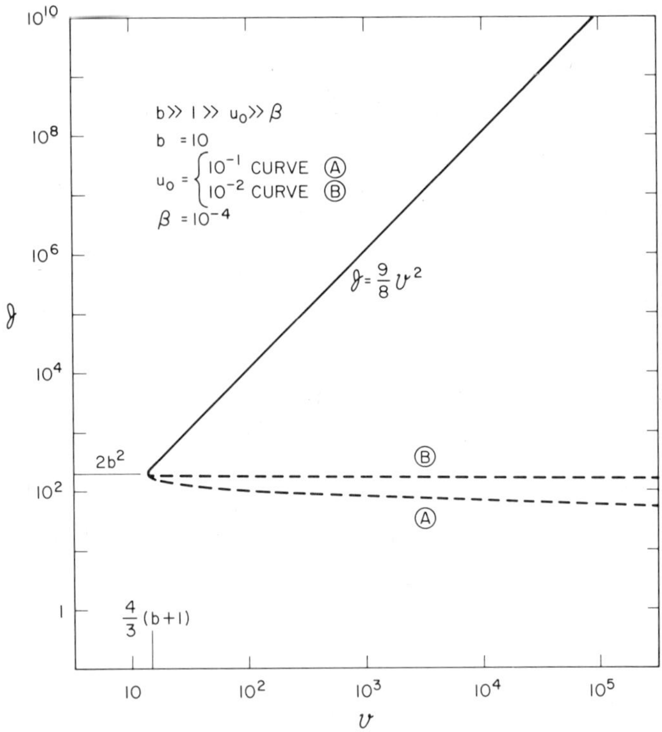

FIG 33. The current–voltage characteristic for a sample satisfying $b \gg 1 \gg u_0 \gg \beta$ (Section 11b) for two values of u_0. The full line is the high-level semiconductor regime. As mentioned in the text, the low-level regime does not appear because it occurs at a very high voltage. (After Baron.[41c])

regime stops when $\delta n = n_{R,0}$), we obtain

$$\mathscr{V} = \frac{u_0 \mathscr{J}^2}{b^2}\left[\exp\left(\frac{b^2}{u_0 \mathscr{J}} - \frac{1}{2u_0}\right) - 1 + \frac{1}{3u_0}\right]. \tag{75'}$$

The correction discussed above appears in the exponent. The term $1/3u_0$ is the approximate voltage across the high-level region.[43b] It is actually Eq. (75′) which is plotted in Fig. 33, as the region of validity of Eq. (75) does not even appear on the figure.

c. $1 \gg b \gg \beta \gg u_0$

We now return to the case of a recombination center initially occupied primarily by electrons[43c] and consider what happens when $b \ll 1$. This causes a transition from electrons carrying most of the current at low injection levels to holes carrying most of the current at higher injection levels. For this case, the high-level regime is again the same, except it is now valid down to

$$\mathscr{J} \gtrsim 2. \tag{76}$$

The two regimes for u which satisfy $\beta \gg u \gg u_0$ and $u \approx u_0$ occur exactly as in Section 11a. However, a new regime where $1 \gg u \gg \beta$ must now be inserted. For this regime,

$$\frac{K}{J} = -\frac{1}{(b+u)^2}\frac{du}{dx}, \tag{77}$$

which leads to

$$\mathscr{V} = 1/2 \tag{78}$$

for the current–voltage characteristic. This regime is valid for

$$\mathscr{J} \gtrsim b. \tag{79}$$

Physically, this new regime is made necessary because hole conduction dominates the high-level regime (since $n \approx p$ and $b \ll 1$), whereas electron conduction dominates below the lower limit of the regime described by Eq. (78), since then $b\,\delta n \gg \delta p$. The resulting current–voltage characteristic is shown in Fig. 34.

[43b] A similar correction can be found for Eq. (73) in the previous section. However, the region of usefulness of Eq. (73) is far greater and the correction is not as necessary.

[43c] A more general statement, covering both this and the complementary case where $u_0 \gg 1$, is that most of the centers are in the charged, as opposed to the neutral, state.

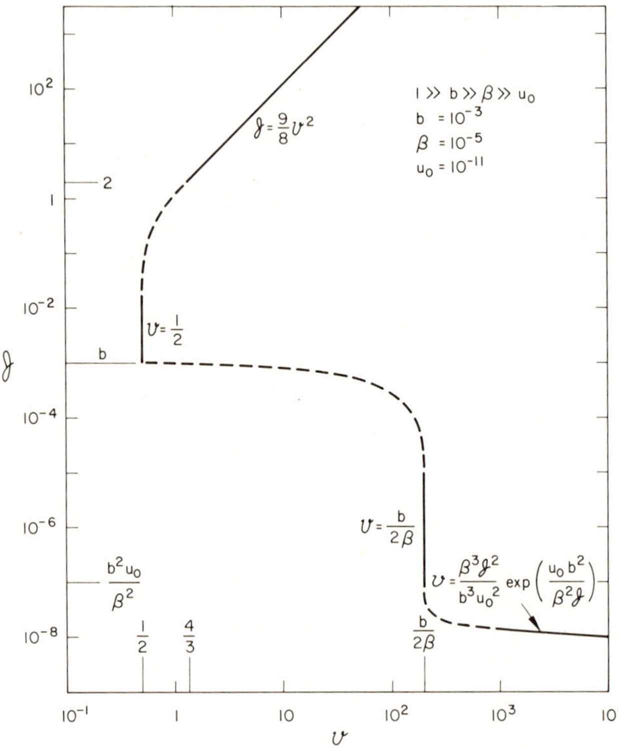

FIG. 34. The current–voltage characteristic for a sample satisfying $1 \gg b \gg \beta \gg u_0$ (Section 11c). The full lines represent the distinct regimes discussed in the text. Dashed lines represent transition regions. (After Baron.[41c])

d. $1 \gg \beta \gg b \gg u_0$

The previous case is now extended to still smaller values of b. A new regime is uncovered when holes are allowed to dominate the conduction to still lower levels of injection. The high-level regime and the $\mathscr{V} = \frac{1}{2}$ regime are the same as in Section 11c except that the $\mathscr{V} = \frac{1}{2}$ regime is now valid only down to

$$\mathscr{J} \gtrsim \beta. \tag{80}$$

A new regime appears for $\beta \gg u \gg b$, where Eq. (56) becomes

$$\frac{K}{J} = -\frac{\beta^2}{u^4}\frac{du}{dx}, \tag{81}$$

and leads to

$$\mathscr{J} = (125/243)\beta\mathscr{V}^3, \tag{82}$$

or

$$J = \frac{125}{243} e\mu_p\mu_n{}^2\beta\tau^2 n_{\mathrm{R},0} \frac{V^3}{L^5} \tag{83}$$

for the current–voltage characteristic. This regime is valid for

$$\mathscr{J} \gtrsim 3b^3/\beta^2. \tag{84}$$

The ratio of this value of $\mathscr{J}$ to the lower limit for the next higher regime is $3(b/\beta)^3$, so that, even for $b/\beta \gtrsim 1$, this regime is valid over a wide range of currents. The remaining two regimes valid at lower injection levels are the same as in Section 11a. Note that the constant-voltage regime $\mathscr{V} = b/2\beta$ now is at a lower voltage than the upper constant-voltage regime $\mathscr{V} = 1/2$. The current–voltage characteristics are shown in Fig. 35. The slight amount of negative resistance may be an artifact of the approximation.

It should be stressed that the $J \propto V^3$ regime predicted above (which, to the authors' knowledge, has not been reported previously) has absolutely no relation to the insulator regime. The present case assumes quasineutrality, and space charge does not enter the calculations. In addition, the field and density distributions are asymmetrical, demonstrating the greater similarity to the semiconductor regime.

The physical interpretation of this $J \propto V^3$ regime is not completely clear at present; in addition, the requirement on the material parameters is not easily satisfied. Even the complementary case $u_0 \gg b \gg \beta \gg 1$, which yields the same I–V curve, is not easily satisfied. However, consider the case described in the following section.

e. $\beta \gg 1 \gg b \gg u_0$

We extend the previous case by allowing $\beta \gg 1$. The condition that $u_0 \ll 1$ is maintained by the fact that the centers are almost completely filled with electrons. The result for this case is identical to that in Section 11d, except that the constant-voltage regime $\mathscr{V} = \frac{1}{2}$ is missing. Thus, the appearance of the current–voltage characteristic is as given in Fig. 36. It should be possible to realize the complementary case $u_0 \gg b \gg 1 \gg \beta$ in materials such as GaAs or InSb, which have large values of b.

f. Classification of Regimes by Lifetime Behavior

The progression from regime to regime may be classified by the manner in which the hole and electron lifetimes τ_p and τ_n are varying. These lifetimes

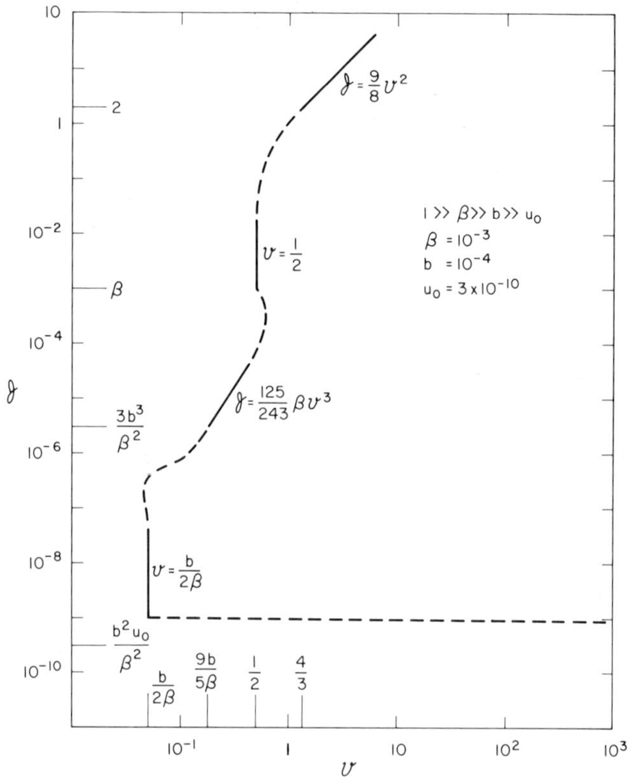

FIG. 35. The current–voltage characteristic for a sample satisfying $1 \gg \beta \gg b \gg u_0$ (Section 11d). The full lines represent the distinct regimes discussed in the text. Dashed lines represent transition regions. The lowest distinct regime does not appear because it occurs at a very high voltage. (After Baron.[41c])

are defined by the relations

$$r \equiv \frac{\delta n}{\tau_n} \equiv \frac{\delta p}{\tau_p}. \tag{85}$$

Thus, from Eqs. (53), (48), and (49),

$$\tau_n = \frac{\beta + u}{u N_R c_n} \tag{86}$$

and

$$\tau_p = \frac{\beta + u}{N_R c_n}. \tag{87}$$

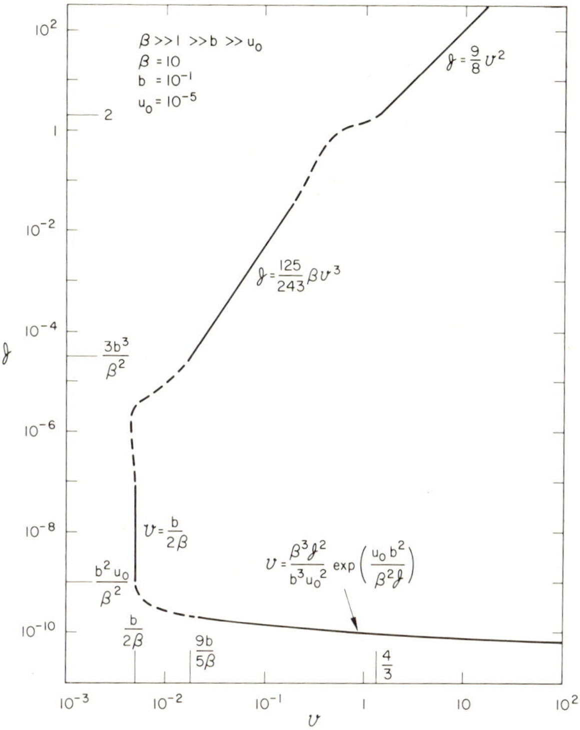

FIG. 36. The current–voltage characteristic for a sample satisfying $\beta \gg 1 \gg b \gg u_0$ (Section 11e). The different regimes are discussed in the text. (After Baron.[41c])

Consider the case in Section 11a at the lowest injection level, where $u - u_0 \ll u_0$; τ_n and τ_p are constant and equal to the thermal equilibrium values. In the next higher regime (where $\mathscr{V} = b/2\beta$), the condition $\beta \gg u \gg u_0$ is satisfied. In this regime, τ_p remains constant and equal to its thermal equilibrium value, while τ_n is decreasing in proportion to $1/u$ toward its high-level value. The final regime discussed in Section 11a is the high-level regime, where $u \lesssim 1$. Now, both τ_n and τ_p have become constant and equal to $\tau = (\beta + 1)/N_R c_n$, the common high-level value. The intermediate case, where $1 \gg u \gg \beta$, does not appear as a distinct regime in the current–voltage characteristic, but has the same $\mathscr{J} = (9/8)\mathscr{V}^2$ dependence as the high-level regime. This occurs because τ_n is constant and equal to its high-level value and the current is carried primarily by electrons ($b \gg 1$) in both this and the high-level regime. Thus, this intermediate case is distinguished from the high-level case by the fact that τ_p is varying, with δp adjusting to this—however, this does not affect the current–voltage characteristic.

In the case in Section 11b, $\beta \ll u_0$, so that the condition $\beta \gg u \gg u_0$ can never be satisfied, and the $\mathscr{V} = b/2\beta$ regime does not exist. In both these cases, the negative-resistance regime is a transition case where one region of the sample is in the high-level regime and the remainder of the sample is in a lower regime. The negative resistance occurs as the region in the high-level regime grows to fill the length of the sample.

In the case in Section 11c, the intermediate regime mentioned above (where τ_p is changing and $\tau_n = \tau$) now leads to a distinct regime. This new regime ($\mathscr{V} = \frac{1}{2}$) exists to provide the change from current carried mostly by electrons at low levels of injection ($u \ll b$) to current carried mostly by holes at higher levels of injection ($u \gg b$).

In the case in Section 11d, the $\mathscr{J} \propto \mathscr{V}^3$ regime occurs because it is now possible to have a situation where the current is carried mostly by holes, τ_p is constant, and τ_n is varying.

Finally, in the case in Section 11e, the $\mathscr{V} = \frac{1}{2}$ regime is missing because it is no longer possible to satisfy the condition necessary for its existence, i.e., that hole current dominates and that τ_n is constant while τ_p varies.

12. Hole Thermal Density Not Small

We will now consider what happens when $p_0 \gg \delta p$, or, more strictly, when

$$\frac{d(p_0\mathscr{E})}{dx} \gg \frac{d(\delta p\mathscr{E})}{dx}. \tag{88}$$

The recombination equation then becomes

$$r = -\mu_p p_0\, d\mathscr{E}/dx, \tag{89}$$

because we neglect the term on the right-hand side of Eq. (88). In addition, we limit ourselves to considering cases where b is not too small, so that $b\,\delta n \gg \delta p$ at all times (remember that $u = \delta p/\delta n < 1$ has already been assumed). We will reverse the order of discussion from Section 11 and proceed from low to high injection levels.

At the lowest injection levels, $b\,\delta n \ll p_0$, and the current is carried entirely by the thermal holes. The behavior is then just that of an ordinary ohmic resistor and there are no injection effects, i.e.,

$$J = e\mu_p p_0 V/L. \tag{90}$$

Once the injection level has risen to the point where $b\,\delta n \gg p_0$, the current equation may be written as

$$J = e\mu_n\,\delta n\mathscr{E}. \tag{91}$$

As in the previous section, a "master" equation may be obtained from Eqs. (89), (91), and (53). It is now more convenient to use the dependent variable $v = u - u_0$, and we obtain

$$\frac{K''}{J} = \frac{[(\beta + u_0) + v^2](1 - v)[(\beta + u_0) + v]^2}{v^3(u_0 + v)} \frac{dv}{dx}, \tag{92}$$

where

$$K'' = \frac{ebn_{R,0}^2 N_R c_n}{p_0}, \tag{93}$$

and we have assumed, as in the previous section, that $1 \gg u_0$ and $N_R \gg p_2, n_2$. We have also assumed that $N_R u_0 \gg p_2$ (or, equivalently, that $\beta p_{R,0} \gg p_0$). When Eq. (92) is integrated to yield $v(x)$, $\mathscr{E}(x)$ is found from

$$\mathscr{E} = \frac{J}{e\mu_n n_{R,0}} \frac{(1 - v)(\beta + u_0 + v)}{v}, \tag{94}$$

and is then integrated to give the current–voltage characteristic. As before, various regimes of behavior are exhibited, depending on how the value of v compares with the constants u_0, $\beta + u_0$, and $(\beta + u_0)^{1/2}$.

a. $v \ll u_0 \ll 1$

If $v \ll u_0$, so that both τ_n and τ_p are constant and equal to their thermal equilibrium values, Eq. (92) reduces to

$$\frac{K''}{J} = \frac{(\beta + u_0)^3}{u_0 v^3} \frac{dv}{dx}, \tag{95}$$

which integrates to

$$\frac{K''(L - x)}{J} = \frac{(\beta + u_0)^3}{2u_0 v^2}. \tag{96}$$

Note that the boundary condition is now that v is large near $x = L$, rather than near $x = 0$ as before.[43d] This is required to keep the sign of Eq. (96) correct, i.e., both sides positive. Physically, this means that the easy injecting contact is the n–π junction at $x = L$, the bulk being a lightly doped p-type semiconductor. In the previous case (Section 11), the bulk was converted to an n-type semiconductor by the recombination traffic, so that the easy injecting contact there was the p–"v" junction at $x = 0$. Thus, in the present

[43d]As before, the exact details of how v is "large" near $x = L$ are not specified in this first approximation. The adjustment to a more exact boundary condition is achieved in a thin layer adjacent to $x = L$.

case,

$$\mathscr{E} \approx \frac{J}{e\mu_n n_{\mathrm{R},0}} \frac{\beta + u_0}{v}$$

$$= \left[\frac{2u_0 N_{\mathrm{R}} c_n J(L - x)}{(\beta + u_0)e\mu_p\mu_n p_0}\right]^{1/2}, \tag{97}$$

which, when integrated and solved for J, yields

$$J = \frac{9}{8} e\mu_n\mu_p\tau_{n,0} p_0 \frac{V^2}{L^3}, \tag{98}$$

where we have substituted the low-level electron lifetime $\tau_{n,0}$ for $(\beta + u_0)/u_0 N_{\mathrm{R}} c_n \equiv 1/p_{\mathrm{R},0} c_n$ [see Eq. (86)].

In terms of our reduced variables, Eq. (98) becomes

$$\mathscr{J} = \frac{9}{8} \frac{p_0(\beta + u_0)}{n_{\mathrm{R},0} u_0} \mathscr{V}^2, \tag{99}$$

or

$$\mathscr{J} = \frac{9}{8} \frac{p_0}{p_{\mathrm{R},0}} \mathscr{V}^2$$

for $\beta \gg u_0$. This regime is similar to the lowest injection level regime of Section 11 in that both τ_n and τ_p are constant and equal to their thermal equilibrium values. The difference, of course, is that $p \approx p_0$ in the present case, whereas $p \approx \delta p$ previously.

This regime is the one investigated by Ashley and Milnes.[10] Note that our derivation was done using the quasineutrality assumption. It was not necessary to invoke space charge. Actually, as we shall show in Section 15, where this regime is discussed in greater detail, it is not necessary to invoke any form of Poisson's equation in deriving the current–voltage characteristic for this regime.

b. $u_0 \ll v \ll (u_0 + \beta) \ll 1$

This condition, of course, is possible only if $\beta \gg u_0$ (equivalently, $n_{\mathrm{R},0} \gg p_{\mathrm{R},0}$), so that deep centers are initially filled mostly with electrons. The variable v satisfying this condition is equivalent to $p_{\mathrm{R},0} \ll \delta p_{\mathrm{R}} \ll n_{\mathrm{R},0}$, so that τ_p remains constant and equal to its thermal equilibrium value while τ_n is decreasing toward its high-level value. If it is satisfied, Eq. (92) reduces to

$$\frac{K''}{J} = \frac{(\beta + u_0)^3}{v^4} \frac{dv}{dx}. \tag{100}$$

This leads to

$$J = \frac{8}{9} e\mu_n \left(\frac{\mu_p p_0 n_{R,0}}{N_R c_n} \right)^{1/2} \frac{V^{3/2}}{L^2} \tag{101}$$

or

$$\mathscr{J} = \frac{8}{9} \left(b \frac{p_0}{n_{R,0}} \right)^{1/2} \mathscr{V}^{3/2} \tag{102}$$

for the current–voltage characteristic. As in Section 12a, v (and thus δn and δp) is large near $x = L$.

In the previous regime (Section 12a), the current–voltage characteristic could be changed from $\mathscr{J} \propto \mathscr{V}^2$ to $\mathscr{J} \propto \mathscr{V}^{3/2}$ by fields large enough to produce hot carriers and field-dependent mobilities. However, the current will remain proportional to p_0, so that it is still possible to distinguish between the two regimes (i.e., that of Section 12a with field-dependent mobilities and that of Section 12b with constant mobilities), because the present regime has $J \propto p_0^{1/2}$.

c. $(u_0 + \beta) \ll v \ll (u_0 + \beta)^{1/2} \ll 1$

With $(u_0 + \beta) \ll v$, τ_n has become constant and equal to its high-level value, while τ_p is now increasing toward its high-level value. The condition that $v \ll (\beta + u_0)^{1/2}$ is related to the divergence of J rather than directly to the recombination kinetics, and thus (as shown below) to the condition on δp and p_0. Under this condition, Eq. (92) becomes

$$\frac{K''}{J} = \frac{(\beta + u_0)}{v^2} \frac{dv}{dx}, \tag{103}$$

which leads to

$$J = e\mu_n n_{R,0} V/L \tag{104}$$

or

$$\mathscr{J} = b\mathscr{V} \tag{105}$$

for the current–voltage characteristic. This ohmic behavior occurs because $\delta n \approx n_{R,0}$ (a constant) over this regime and the current is carried primarily by electrons. The integration of Eq. (103) is necessary only to arrive at $v(x)$ and $\delta p(x)$.

There is no point in extending this chain of regimes to larger values of v ($\approx u$), since large values of v lead to a breakdown of the condition on δp and p_0. Referring back to Eq. (88), we can evaluate these expressions in terms of v

and find that

$$\frac{d(\delta p\mathscr{E})}{dx}\bigg/\frac{d(p_0\mathscr{E})}{dx} = \frac{n_{\mathrm{R},0}}{p_0}\,\frac{v^2}{(\beta + u_0) + v^2}. \tag{106}$$

Thus, in order to satisfy Eq. (88), we must have that $v^2/(\beta + u_0) \ll p_0/n_{\mathrm{R},0} \ll 1$ (since $p_0/n_{\mathrm{R},0} \ll 1$ by previous assumption).

13. Transition from p_0 Not Small to p_0 Small

The transition out of the series of regimes discussed in Section 12 occurs before Eq. (88) becomes invalid near the n–π junction (the easy injecting contact). The physical situation is the following: both junctions are injecting. Thus, the sample must be divided into two regions. The region nearest the n–π junction behaves according to Section 12, with $d(p_0\mathscr{E})/dx$ dominant. The region nearest the p–π junction behaves according to Section 11, with $d(\delta p\mathscr{E})/dx$ dominant. At low injection levels, the region nearest the p–π junction is very narrow, and the series of regimes discussed in Section 12 occurs. At high injection levels, the region nearest the n–π junction is very narrow, and the series of regimes discussed in Section 11 occurs. The transition between the two series of regimes occurs as the region near the p–π junction spreads to fill the length of the sample.

In Fig. 37, we have reproduced Fig. 32, showing the sequence of regimes discussed in Section 11a, where we assumed $b \gg 1 \gg \beta \gg u_0$. The sequence of regimes discussed in Section 12 is also shown, for several values of $p_0/n_{\mathrm{R},0}$. These regimes and their transition points are summarized in Table I. It is

TABLE I

Current–Voltage Regimes Discussed in Section 12 and Their Transition Points

Regime $\mathscr{J}$–$\mathscr{V}$ characteristic	$\mathscr{V}_{\text{transition}}$	$\mathscr{J}_{\text{transition}}$
$\mathscr{J}_0 = \dfrac{p_0}{n_{\mathrm{R},0}}\mathscr{V}$		
	$\mathscr{V}_{01} = \dfrac{8}{9}\dfrac{u_0}{\beta + u_0}$	$\mathscr{J}_{01} = \dfrac{8}{9}\dfrac{u_0}{\beta + u_0}\dfrac{p_0}{n_{\mathrm{R},0}}$
$\mathscr{J}_1 = \dfrac{9}{8}\dfrac{p_0}{n_{\mathrm{R},0}}\dfrac{\beta + u_0}{u_0}\mathscr{V}^2$		
	$\mathscr{V}_{12} = \left(\dfrac{8}{9}\right)^4\left(\dfrac{u_0}{\beta + u_0}\right)^2\dfrac{bn_{\mathrm{R},0}}{p_0}$	$\mathscr{J}_{12} = \left(\dfrac{8}{9}\right)^7\left(\dfrac{u_0}{\beta + u_0}\right)^3\dfrac{b^2 n_{\mathrm{R},0}}{p_0}$
$\mathscr{J}_2 = \dfrac{8}{9}\left(\dfrac{bp_0}{n_{\mathrm{R},0}}\right)^{1/2}\mathscr{V}^{3/2}$		
	$\mathscr{V}_{23} = \left(\dfrac{9}{8}\right)^2\dfrac{bn_{\mathrm{R},0}}{p_0}$	$\mathscr{J}_{23} = \left(\dfrac{9}{8}\right)^2\dfrac{b^2 n_{\mathrm{R},0}}{p_0}$
$\mathscr{J}_3 = b\mathscr{V}$		

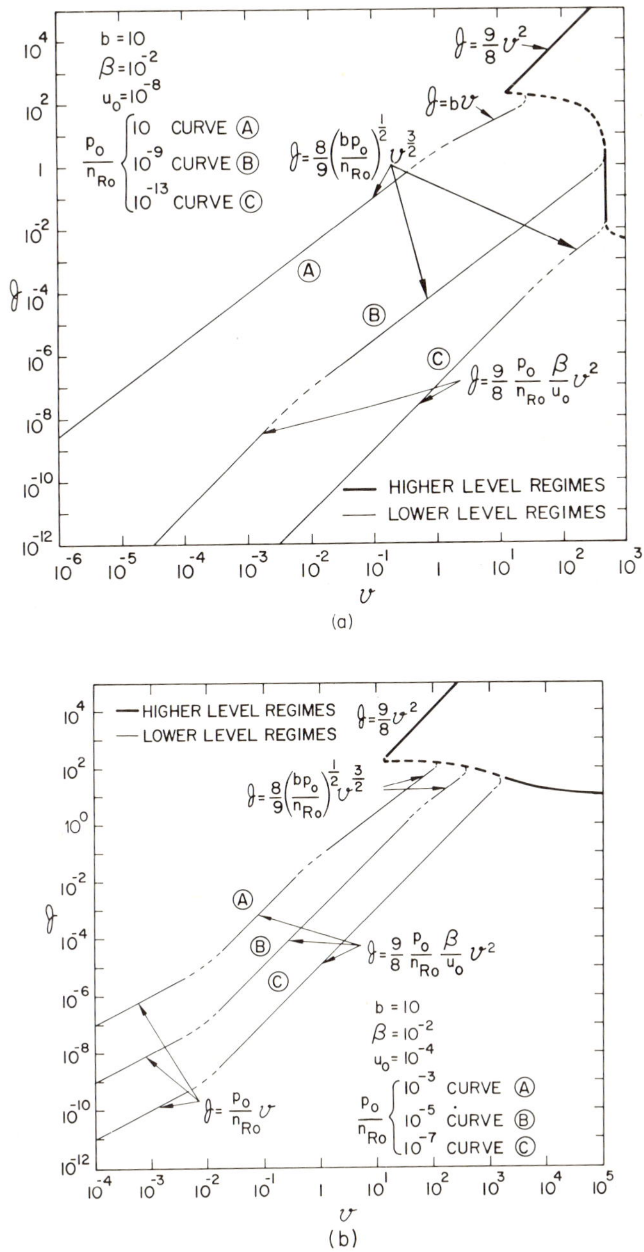

FIG. 37. The current–voltage characteristics for a sample satisfying $b \gg 1 \gg \beta \gg u_0$, showing the manner in which the lower-level regimes discussed in Section 12 intersect the higher-level regimes discussed in Section 11 for several values of u_0 and $p_0/n_{R,0}$. (After Baron.[41c])

easy to show from the appropriate inequalities that it is not possible for the $\mathscr{J} \propto \mathscr{V}^2$ regime of Section 12 to intersect the $\mathscr{V} = b/2\beta$ regime of Section 11. Thus, the $\mathscr{J} \propto \mathscr{V}^{3/2}$ regime of Section 12 will always appear as a distinct regime if the quasineutrality assumption is valid. The $\mathscr{J} \propto \mathscr{V}$ regime of Section 12, on the other hand, can never appear as a distinct regime at the same time the current–voltage characteristic exhibits any appreciable negative resistance. This is because its current–voltage characteristic passes very close to the knee of the higher-level curve. Thus, the appearance of this regime corresponds to the electronic transformation of a *p*-type to an *n*-type semiconductor at injection levels so low that high-level injection is small compared with the resulting ohmic conduction. If this is to occur, it is necessary that $\mathscr{V}_{23} < \frac{4}{3}b$, or $n_{\mathrm{R},0} \lesssim p_0$. Thus, as could be inferred from the disappearance of the negative resistance, the "many-deep-levels" behavior is changing to a "few-deep-levels" behavior.

We shall show in Section 15 that it is possible to have a low-level $\mathscr{J} \propto \mathscr{V}^2$ regime without imposing the quasineutrality assumption. The behavior can be quite different in progressing to higher injection levels if the space charge is not small.

14. Electron Thermal Density Not Small

If b is not too small and the material is not too intrinsic (so that $n_0 \gg p_0$), the behavior will be ohmic at low injection levels, i.e.,

$$J = e\mu_n n_0 V/L. \tag{107}$$

As the injection level increases, δp will become $> p_0$ first, but it will not cause a change in the behavior, since $J \approx J_n$. With a further increase in the injection level, δn will become $> n_0$, and the conditions for the regimes discussed in Section 11 will be valid. Thus, the ohmic characteristic, Eq. (107), will be valid until it intersects one of the regimes discussed in Section 11.

If b is very small, the situation where $\delta p \gg bn_0 \gg b\,\delta n$ (even though $\delta n \gg \delta p$) can exist. In this case, $J \approx J_p$, but the recombination equation can be approximated by

$$r = \mu_n n_0\, d\mathscr{E}/dx. \tag{108}$$

This case will not be discussed in detail; however, the development would parallel that given in Section 12. At low enough injection levels that $\tau_p \approx \tau_{p,0} = 1/c_p n_{\mathrm{R},0}$ (i.e., $\delta p_{\mathrm{R}} \ll n_{\mathrm{R},0}$, or, equivalently, $v \ll \beta + u_0$), Eq. (108) may be written as

$$\frac{\delta p}{\tau_p} = \frac{J}{e\mu_p \tau_p \mathscr{E}} = \mu_n n_0 \frac{d\mathscr{E}}{dx}, \tag{109}$$

which yields a current–voltage characteristic of

$$J = \frac{9}{8} e\mu_p \mu_n n_0 \tau_{p,0} \frac{V^2}{L^3}. \tag{110}$$

This is just a variation of the Ashley–Milnes regime discussed in Section 12. Note that in Section 12, it was necessary to have $v \ll u_0$ (i.e., $\delta p_R \ll p_{R,0}$, $\tau_n \approx \tau_{n,0}$) to have the $J \propto V^2$ regime, whereas here it is necessary only to have $v \ll u_0 + \beta$ (i.e., $\delta p_R \ll n_{R,0}$, $\tau_p \approx \tau_{p,0}$).

15. Ashley–Milnes Regime

We will now consider the Ashley–Milnes regime[10] in greater detail, directing our attention to the question of space charge and its effects. The following basic assumptions are necessary for the existence of this regime:

(i) The current is carried by only one carrier (e.g., electrons), so that

$$J = e\mu_n \, \delta n \mathscr{E}. \tag{111}$$

(ii) The lifetime for this carrier is constant, independent of injection level and position, i.e., $|\delta p_R| \ll p_{R,0}$, so that $\tau_n \approx \tau_{n,0} = 1/c_n p_{R,0}$.

(iii) The density of injected carriers of the other type (holes) is so small that $d(p_0 \mathscr{E})/dx \gg d(\delta p \mathscr{E})/dx$.

In order to satisfy these conditions, it is necessary that the hole lifetime τ_p be small enough, since we require that $\delta p/\delta n \approx u_0 = \beta p_{R,0}/n_{R,0} = \tau_{p,0}/\tau_{n,0} \ll b$. With these assumptions, the recombination equation can be written as

$$\frac{\delta n}{\tau_{n,0}} = \frac{J}{e\mu_n \tau_{n,0} \mathscr{E}} = -\mu_p p_0 \frac{d\mathscr{E}}{dx}, \tag{112}$$

which is completely equivalent to Eqs. (94) and (95) in Section 12, and, of course, also yields Eq. (98) as a current–voltage characteristic.

Note that nothing was assumed about the space charge, and that Poisson's equation was not used in deriving Eq. (112). It is, of course, true that Poisson's equation must be satisfied. The necessary space charge builds up on the recombination centers without in any way affecting the distribution of mobile carriers, the field, or the current–voltage characteristic. Because of the form of Eq. (112), the ratio of the space charge to δn is easily calculated, giving

$$\left(\frac{\varepsilon}{e} \frac{d\mathscr{E}}{dx}\right) \Big/ \delta n = -\frac{\varepsilon}{e\mu_p p_0 \tau_{n,0}} = -\frac{\Theta_{p,0}}{\tau_{n,0}}, \tag{113}$$

where $\Theta_{p,0} = \varepsilon / e\mu_p p_0$ is the low-level dielectric relaxation time (equivalent to what Lampert calls the ohmic dielectric relaxation time for $p_0 \gg n_0$).

Thus, Poisson's equation becomes

$$\delta p - \delta n + \delta p_{\mathrm{R}} = -\frac{\Theta_{p,0}}{\tau_{n,0}}\delta n, \tag{114}$$

as long as Eq. (112) is valid. Unless $b \gg 1$, a consequence of the basic assumptions is that $u_0 \ll 1$, so that $\delta p \approx u_0\, \delta n$ may be neglected in Eq. (114). The subsequent discussion will be confined to this assumption ($u_0 \ll 1$), in which case, we may solve Eq. (114) for δp_{R}, obtaining

$$\delta p_{\mathrm{R}} = \left(1 - \frac{\Theta_{p,0}}{\tau_{n,0}}\right)\delta n. \tag{115}$$

If $\Theta_{p,0}/\tau_{n,0} \ll 1$, the space charge is a small fraction of the injected charge [see Eq. (113)] and the quasineutrality assumption is valid; here, the space charge of the electrons is compensated by an equal and opposite charge on the recombination centers ($\delta p_{\mathrm{R}} \approx \delta n$). This is the case discussed in Section 12.

If $\Theta_{p,0}/\tau_{n,0} \gg 1$, the space charge is very large and resides mainly on the recombination centers [$\delta p_{\mathrm{R}} \approx -(\Theta_{p,0}/\tau_{n,0})\,\delta n$]. It must be reemphasized that the space charge adjusts to match the field distribution, which is determined solely by the recombination kinetics. This affects the limit on the validity of assumption (ii), and the behavior at higher injection levels.

If $\Theta_{p,0}/\tau_{n,0} \ll 1$, assumption (ii) is equivalent to $\delta n \ll p_{\mathrm{R},0}$ (or $v \ll u_0$), and the behavior at higher injection levels is as discussed in Section 12, with a transition to either a $J \propto V^{3/2}$ or a $J \propto V$ behavior and an eventual transition to the higher-level behavior discussed in Section 11. Note also in this case that $\delta p_{\mathrm{R}} > 0$, so that τ_p tends to increase and τ_n tends to decrease with increasing injection level. Thus, the tendency is toward the higher-level condition of equal lifetimes.

If $\Theta_{p,0}/\tau_{n,0} \gg 1$, assumption (ii) is equivalent to $(\Theta_{p,0}/\tau_{n,0})\,\delta n \ll p_{\mathrm{R},0}$, so that the assumption becomes invalid at a much lower injection level. In addition, $\delta p_{\mathrm{R}} < 0$, so that τ_p tends to decrease and τ_n tends to increase as the injection level increases. Thus, we have a transition to trap-free, single-carrier SCLC (with τ_n essentially infinite and τ_p essentially zero), which is similar to the trap-filled-limit transition found in pure single-carrier behavior. The voltage at which this transition takes place may be estimated by setting $|\delta p_{\mathrm{R}}| = p_{\mathrm{R},0}$ at $x = L/2$ (the point at which the trap-free, single-carrier, SCLC-dominated region extends halfway across the sample) and obtaining

$$V_{\mathrm{TFL}} \approx \frac{e p_{\mathrm{R},0} L^2}{3\varepsilon} \tag{116}$$

or

$$\mathscr{V}_{\mathrm{TFL}} \approx \frac{b p_{\mathrm{R},0}^2}{3 N_{\mathrm{R}} p_0} \frac{\tau_{n,0}}{\Theta_{p,0}}, \tag{117}$$

which, for $u_0 \ll \beta$, reduces to

$$\mathscr{V}_{\text{TFL}} \approx \frac{bn_{\text{R},0}}{3p_0} \frac{u_0^2}{\beta^2} \frac{\tau_{n,0}}{\Theta_{p,0}}.$$

The trap-free, SCLC regime, in turn, has a transition to the higher-level double-injection regimes (discussed in Section 11) as the hole injection from the p–π junction spreads into the sample, forcing the lifetimes the other way again (as also discussed by Lampert and Schilling in Chapter 1).

The ratio of current in the Ashley–Milnes regime to that for the trap-free, single-carrier case is just

$$\frac{J_{\text{AM}}}{J_{\text{SC}}} = \frac{\tau_{n,0}}{\Theta_{p,0}}, \tag{118}$$

so that whenever $\Theta_{p,0}/\tau_{n,0} \gg 1$, $J_{\text{AM}} \ll J_{\text{SC}}$, and there can indeed be a trap-filled-limit transition as discussed above. It should also be noted that J_{AM} is larger than the single-carrier SCLC which would flow if the center behaved as a shallow trap.

Several possible I–V characteristics are given in Fig. 38. The different regimes discussed above are indicated on the figure. Note particularly the trap-filled-limit transition to trap-free SCLC. It may occur at a lower voltage than the threshold voltage for the quasineutrality solution (Fig. 38a), or it may occur and provide a "threshold" voltage even when the quasineutrality threshold voltage ($\mathscr{V} = b/2\beta$ regime) does not occur (Fig. 38b).

V. Large Numbers of Deep Centers—Experiment

16. General Discussion and Results

The existence of negative-resistance behavior due to deep levels has been confirmed in Si, Ge, GaAs, InSb, and CdS. There have been more than 30 papers published concerning this phenomenon. For germanium, Stafeev and co-workers[44–46] predicted and presented evidence for negative resistance in long diodes. Others have discussed effects of plastic deformation,[47,48] impact

[44] A. A. Lebedev, V. I. Stafeev, and V. M. Tuchkevich, *Zh. Tekhn. Fiz.* **1**, 2131 (1956) [*English Transl.*: Soviet Phys.–Tech. Phys. **1**, 2071 (1957)].

[45] V. I. Stafeev, *Fiz. Tverd. Tela* **1**, 841 (1959) [*English Transl.*: *Soviet Phys.–Solid State* **1**, 763 (1959)].

[46] V. I. Stafeev, *Fiz. Tverd. Tela* **3**, 2513 (1961) [*English Transl.*: *Soviet Phys.–Solid State* **3**, 1829 (1962)].

[47] L. J. Van Ruyven and W. H. Th. Adriaens, *Phys. Letters* **3**, 109 (1962).

[48] O. Ryuzan, *J. Phys. Soc. Japan* **16**, 2177 (1961).

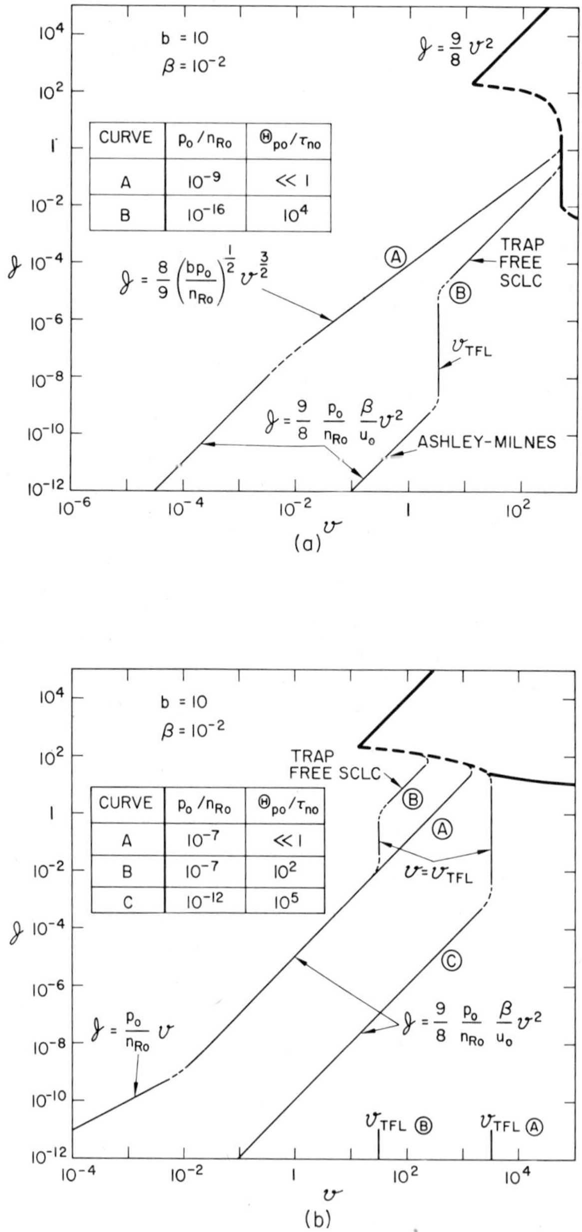

FIG. 38. The current–voltage characteristics for $u_0 = 10^{-8}$ (a) and $u_0 = 10^{-4}$ (b), showing the effect of large space charge on the way in which the transition between the lower-level and higher-level regimes takes place. (After Baron.[41c])

ionization,[49] irradiation-induced centers,[50] and diffused structures.[9,12] Light emission coupled with the breakup from a low-level state has been observed in GaAs[51,52] and CdS.[53] In InSb, the negative-resistance effect is attributed to the formation of high-conductivity filaments[54–56]; in some cases,[56] the postbreakdown region ($I \propto V^3$) is in agreement with a recombination-limited-current model. The influence of filaments[57,58] on negative resistance is discussed in detail by Barnett in Chapter 3 of this volume. It should only be mentioned that the possibility of the existence of filament formation must always be considered in the analysis of negative-resistance data, particularly in structures whose transverse dimensions are large compared with the length.

The greatest number of experimental investigations have been carried out in silicon. In 1962, Holonyak[9] described gold- and cobalt-doped silicon *p–i–n* structures and showed that light lowered the threshold voltage, and that the prenegative-resistance region had an $I \propto V^n$ characteristic with $n \approx 2$. Although single-carrier, space-charge-limited behavior and certain regimes of double-injection theory[59] predict $I \propto V^2$ behavior, other experimental results[13,16,60] indicate that the situation can be more complex in these structures. In particular, strong oscillations have been noted in gold-,[15,16,60] cobalt-,[14,61] and radiation-defect-compensated[62,62a] silicon structures. The frequency of these oscillations depends on many factors, such as the density of deep centers, length of the *i* region, level of illumination, and applied bias voltage. The exact mechanism has not yet been defined, although two models

[49] T. Wada, Y. Fukuoka, and T. Arizumi, *J. Phys. Soc. Japan* **25**, 165 (1968).

[50] A. B. Gerasimov and S. M. Ryvkin, *Fiz. Tverd. Tela* **7**, 657 (1965) [*English Transl.*: *Soviet Phys.–Solid State* **7**, 526 (1965)].

[51] K. Weiser and R. S. Levitt, *J. Appl. Phys.* **35**, 2431 (1964).

[52] W. P. Dumke, *in* "Radiative Recombination in Semiconductors" (7th Intern. Conf.), p. 157. Dunod, Paris and Academic Press, New York, 1965.

[53] C. W. Litton and D. C. Reynolds, *Phys. Rev.* **133**, A536 (1964).

[54] I. Melngailis and R. H. Rediker, *J. Appl. Phys.* **33**, 1892 (1962).

[55] I. Melngailis and R. H. Rediker, *Proc. IRE* **50**, 2428 (1962).

[56] J. E. Nordman and H. Kvinlaug, *J. Appl. Phys.* **39**, 3244 (1968).

[57] A. M. Barnett and A. G. Milnes, *J. Appl. Phys.* **37**, 4215 (1966).

[58] B. K. Ridley, *Proc. Phys. Soc.* (*London*), **82**, 954 (1963).

[59] V. V. Osipov and V. I. Stafeev, *Fiz. Tekhn. Poluprov.* **1**, 1795 (1968) [*English Transl.*: *Soviet Phys.–Semicond.* **1**, 1486 (1968)].

[60] Yu. A. Bykovskii, K. N. Vinogradov, and V. V. Zuev, *Fiz. Tekhn. Poluprov.* **1**, 1559 (1968) [*English Transl.*: *Soviet Phys.–Semicond.* **1**, 1295 (1968)].

[61] B. G. Streetman, M. M. Blouke, and N. Holonyak, *Appl. Phys. Letters* **11**, 200 (1967).

[62] F. M. Berkovskii and R. S. Kasymova, *Fiz. Tverd. Tela* **8**, 1985 (1966) [*English Transl.*: *Soviet Phys.–Solid State* **8**, 1580 (1967)].

[62a] B. G. Streetman, N. Holonyak, H. V. Krone, and W. D. Compton, *Appl. Phys. Letters* **14**, 63 (1969).

predict behavior which follows the experimental results: (1) a periodic fluctuation of carrier concentration between the boundaries and center of the "i" region,[16] or (2) an increase and then decrease of the lifetime as the injection level increases.[62]

There have been two general approaches to the study of negative resistance (aside from oscillations): investigation of the prebreakdown and threshold region,[11,63,64] and of the postbreakdown region. In the latter case, since the samples will be in the high-lifetime state, relatively long samples are required to prevent the domination of the postbreakdown I–V characteristics by diffusion effects.[65–67] Only a few examples of the experimental data will be discussed in detail. In part, this decision is based on the fact that the theory is sufficiently complex that there has been quantitative agreement between theory and experiment in only a few limiting cases.

Some of the features of negative-resistance behavior are illustrated[68] in Figs. 39 and 40, which represent the current–voltage characteristics of silicon p–π–n structures similar to those discussed in Section 9. The room-temperature characteristics (not shown) were in agreement with the high-level semiconductor-regime behavior expected for a sample with a small density of deep centers. Two different types of device characteristics were observed at lower temperatures, as shown in Figs. 39 and 40; these types can be classified broadly by the existence or nonexistence of negative resistance. In the latter case (samples A and B), the current at low voltages is limited by the potential drop at the junctions. Subtracting the junction potential drop from the applied voltage leads to an ohmic characteristic (shown by the dashed line).

The current–voltage characteristics of sample C were representative of long ($L > 4$ mm) structures with high-level lifetimes between 1 and 100 μsec. The large density of deep centers responsible for the negative resistance apparently was introduced during high-temperature process steps[39,68] (as was also found in short structures[65,66]). The current between the two conduction states increases by a factor of about 10^3 at 200°K and 10^5 at 80°K. The high-conduction-state current ($I \propto V^2$) is in agreement with the predictions of Eq. (62), with the value of $n_{R,0}$ taken as the value of p_0 at room temperature. These results, along with more detailed measurements[69] of the effect of light on the population of recombination centers, are consistent with the following model. A set of deep donor levels N_D lies in the lower half of the energy gap

[63] J. L. Wagener and A. G. Milnes, *Solid-State Electron.* **8**, 495 (1965).
[64] E. Schibli and A. G. Milnes, *Solid-State Electron.* **10**, 97 (1967).
[65] K. Shohno, *Japan. J. Appl. Phys.* **3**, 114 (1965).
[66] K. Shohno, *Japan. J. Appl. Phys.* **5**, 358 (1966).
[67] T. Fukami and K. Homma, *Japan. J. Appl. Phys.* **2**, 535 (1963).
[68] J. W. Mayer, R. Baron, and O. J. Marsh, *Appl. Phys. Letters* **6**, 38 (1965).
[69] O. J. Marsh, R. Baron, and J. W. Mayer, *Appl. Phys. Letters* **7**, 120 (1965).

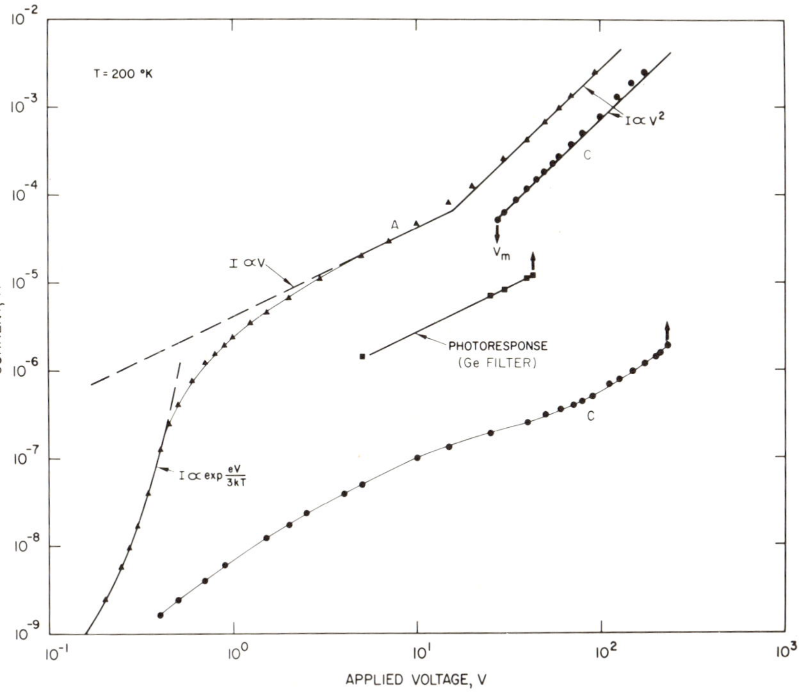

FIG. 39. Current–voltage characteristics of two silicon p–π–n diodes at 200°K. The extrapolated ohmic component of current obtained by subtracting the junction drop is shown by a dashed line for sample A, which did not exhibit negative resistance. The influence of germanium-filtered light on the low-level characteristic and the threshold voltage of sample C is shown. (After Mayer *et al.*[68])

and is partly compensated by a set of shallow acceptor levels N_A (so that $N_D > N_A$). The resulting holes on the donor levels are fully ionized into the valence band at room temperature, so that $p_0 = N_A$. At sufficiently low temperatures, the donor states are occupied by $p_R \approx N_A$ holes and $n_R \approx (N_D - N_A)$ electrons in thermal equilibrium, and current is carried by the now small number of thermal holes, p_0 ($p_0 \ll N_A$). As the rate of carrier injection is increased by raising the forward bias, the donor states which act as recombination centers fill with electrons and the device has a $J \propto V^2$ behavior once the filling is complete.

Some of the most detailed investigations of the low-level characteristics (excluding oscillations) have been carried out by Milnes and co-workers.[10,10a,63,64] In their early work,[10] they predicted (the Ashley–Milnes model, treated in Section 15), and confirmed with gold-doped germanium samples, the influence of variation in the ratio of occupied and unoccupied recombination

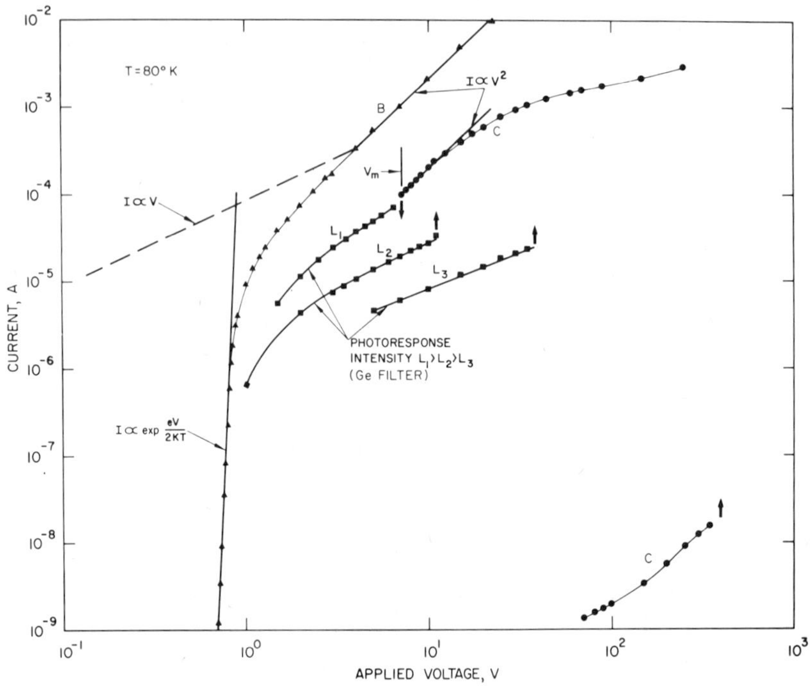

FIG. 40. Current–voltage characteristics of two silicon p–π–n diodes at 80°K. The dashed line was obtained by subtracting the junction drop from the applied voltage and indicates the ohmic current–voltage characteristic of sample B. The influence of germanium-filtered light on the low-level characteristic and the threshold voltage of sample C is shown. (After Mayer *et al.*[68])

centers. The low-level current–voltage characteristic consists of an ohmic followed by a square-law regime. In this and other work,[10a,63] they use theory and experimental results to determine capture cross sections of deep impurities.[70] In the most recent work,[64] the theory (and experiment) was extended to include the effects of gradients in the density of the deep impurities.

The low-level I–V characteristics of indium-doped p–i–n silicon structures[63] illustrate some of the features of the Ashley–Milnes model. The p–i–n devices were fabricated from silicon containing 2×10^{16} atoms/cm^3 of indium (deep acceptor, $\sigma_p^- > \sigma_n^0$) and compensated with about 4×10^{13} atoms/cm^3 of shallow donors. The I–V characteristics shown in Fig. 41 show three distinct regions. The ohmic and V^2 regions follow the Ashley–Milnes theory, and the $J \propto V^{2.8}$ regime results from the influence of field-dependent mobilities and capture cross sections. In the low-level regime, the current is proportional

[70] E. Shibli and A. G. Milnes, *Mater. Sci. Eng.* **2**, 229 (1968).

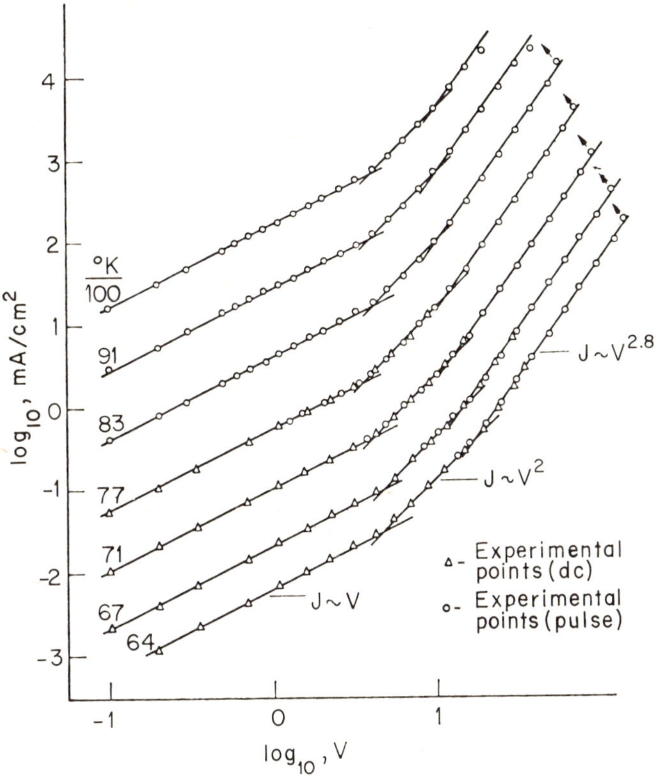

FIG. 41. Current–voltage characteristics of a 15.9-mil-thick indium-doped ($2 \times 10^{16}/cm^3$) silicon device as a function of temperature. Circles indicate experimental points taken with pulse techniques, triangles are experimental points taken with direct current. (After Wagener and Milnes.[63])

to the thermal hole density, as shown in Fig. 42, where the straight lines correspond to an impurity with an energy level similar to indium in silicon (0.158 eV above the valence band).[63] The experimental points correspond to the current density as a function of current density at constant voltage in the ohmic, V^2, and high-field regimes. The transition between the ohmic and V^2 regimes was found to vary as the square of the sample length, as predicted. At this point, the electron transit time equals its lifetime, so that the value of the capture cross section $\sigma_n{}^0$ can be determined. The values of the breakdown voltage V_B as a function of sample length L at temperatures of 64 and 77°K are shown in Fig. 43. For the shorter devices, the $V_B \propto L^2$ dependence follows the predicted behavior. For longer samples, V_B occurs in the high-field region, where the cross sections are field dependent. In the treatment by

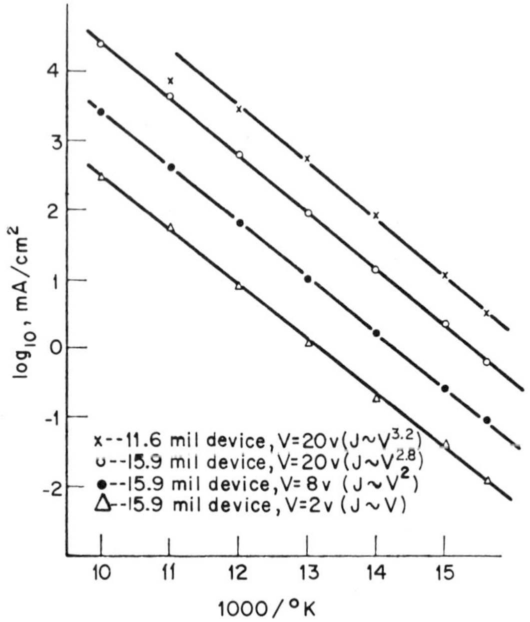

FIG. 42. Current as a function of temperature at constant voltage for the different regions of the characteristics shown in Fig. 41. The straight lines correspond to an energy level 0.158 ± 0.002 eV above the valence band. (After Wagener and Milnes.[63])

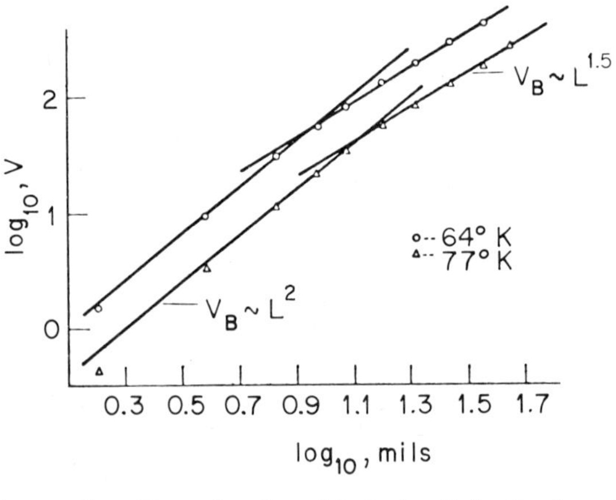

FIG. 43. Breakdown voltage V_B as a function of device length for the characteristics shown in Fig. 41. (After Wagener and Milnes.[63])

Wagener and Milnes,[63] the theory was extended to include high-field effects, and the theoretical predictions are in agreement with experimental results.

In general, however, the experimental picture is less clear-cut. Figure 44 shows the current–voltage characteristics of gold-doped silicon p–v–n structures in which the thickness of the v region was 0.2–0.4 cm.[60] These samples also exhibited current-oscillation effects in the prebreakdown regions. It is obvious that further experimental investigations are required to test some of the theoretical predictions outlined in the previous section. The difficulty may well lie in fabricating suitable structures with a known concentration of deep levels. It would also be desirable to clarify the physical mechanisms

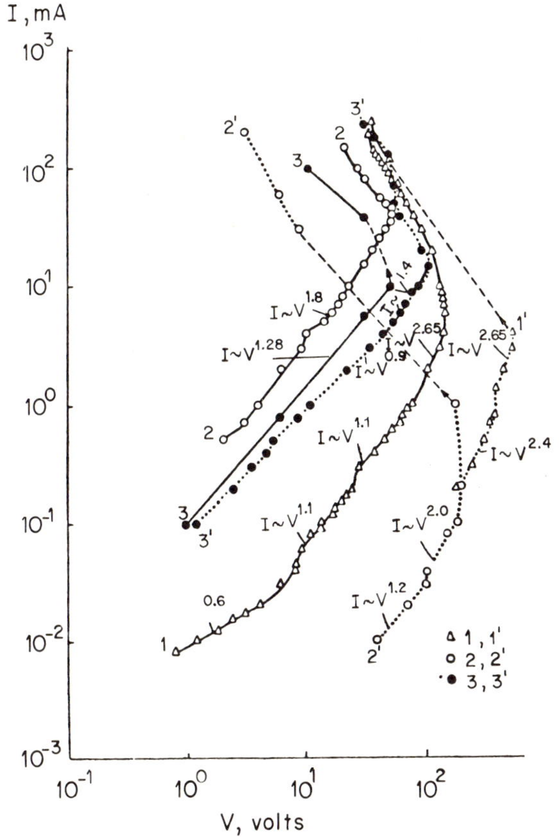

FIG. 44. Current–voltage characteristics of gold-doped silicon p–v–n structures. (1, 1′)T = 300 and 83°K, respectively; sample No. 1, $R_{T=300°K}$ = 200 kilohms. (2,2′)T = 300 and 83°K, respectively; sample No. 2, $R_{T=300°K}$ = 25 kilohms. (3,3′)T = 300 and 83°K, respectively; sample No. 3, $R_{T=300°K}$ = 4 kilohms. (After Bykovskii *et al.*[60])

responsible for the oscillatory behavior. One can anticipate that this may well be achieved before all the features of the low-level behavior are delineated.

VI. Transient and Small-Signal AC Response

17. General Discussion

The investigation of the transient response is a very powerful tool both in determining the carrier lifetime and in distinguishing between double injection and other modes of behavior. In this section, we discuss the response to a voltage step, the small-signal ac response, noise properties, and the comparison of double injection with single-carrier, space-charge-limited currents.

For the time-dependent case, the defining equations (1) and (2) must be replaced by the time-dependent current equations

$$J_n = e\mu_n\left[(n_0 + \delta n)\mathscr{E} + \beta\frac{dn}{dx}\right], \tag{119a}$$

$$J_p = e\mu_p\left[(p_0 + \delta p)\mathscr{E} - \beta\frac{dp}{dx}\right], \tag{119b}$$

$$J = J_n + J_p + \varepsilon\frac{\partial\mathscr{E}}{\partial t}, \tag{119c}$$

(where $\varepsilon\,\partial\mathscr{E}/\partial t$ is the displacement current and β again stands for kT/e), and the time-dependent particle-conservation equations

$$\frac{\partial n}{\partial t} = -r_n + \frac{1}{e}\frac{\partial J_n}{\partial x}, \tag{120a}$$

$$\frac{\partial p}{\partial t} = -r_p - \frac{1}{e}\frac{\partial J_p}{\partial x}, \tag{120b}$$

$$\frac{\partial p_{\mathrm{R}}}{\partial t} = r_p - r_n. \tag{120c}$$

Poisson's equation remains unchanged. It can easily be shown from these equations that $\partial J/\partial x \equiv 0$ when the displacement current is included in the definition of J.

18. Step Response

The transient response can be described most easily by considering the transient current response to a voltage step. Two limiting cases can be distinguished: the "small-signal" case and the "large-signal" case. In the "small-signal" case, the diode is initially biased in one of the high-level regimes and the final state is in the same high-level regime. We have used this definition because the results of the small-signal approximation remain valid over the range of voltage steps for which the above is true. In the "large-signal" case, the voltage is such that the initial and final states are in different regimes. Specifically, we will consider the case where the diode is initially unbiased and the final state is in the high-level semiconductor regime.

a. Small-Signal Case

Let us consider a long structure which is entirely in the high-level semiconductor regime.[22] As for the steady-state case, this allows us to assume $\tau_n = \tau_p = \tau$ and to neglect diffusion and space-charge terms. Using the same linear combinations and assumptions as for the high-level steady-state case (Section 4a), the equations reduce to

$$J = e\mu_p(b + 1)n\mathscr{E} + \varepsilon\frac{\partial \mathscr{E}}{\partial t} \tag{121}$$

and

$$\frac{(b + 1)}{\mu_n}\frac{\partial n}{\partial t} = -\frac{(b + 1)n}{\mu_n \tau} + (n_0 - p_0)\frac{\partial \mathscr{E}}{\partial x}. \tag{122}$$

The steady-state solutions are, of course, those given in Section 4a.

Consider what happens when a voltage step from an initial voltage V_i to a final voltage V_f is applied, where both V_i and V_f fall in the semiconductor regime. There will first be a capacitive surge because of the geometrical capacitance of the sample. As the time constant of this surge approaches zero for an ideal voltage step, it may be neglected. At the end of this surge, an additional uniform field of magnitude $\Delta\mathscr{E} = (V_f - V_i)/L$ will exist in the sample. Charge will then move around and reduce the displacement current to zero. This last will occur in a time of the order of the local dielectric relaxation time $\Theta' \approx \varepsilon/(b + 1)ne\mu_p$. If double injection is to be observed, it is necessary that $\Theta' \ll \tau$. This may easily be seen by reference to Lampert and Rose.[5] They show that if double injection is to be observed (in an n-type semiconductor), it is necessary that $\Theta_{\Omega,n} < \tau$ (otherwise, single-carrier, space-charge-limited currents will be observed) and that $t_p < \tau$ (otherwise, ohmic currents will be observed), where $\Theta_{\Omega,n} = \varepsilon/(n_0 - p_0)e\mu_n$ is their "ohmic" dielectric relaxation time and $t_p = L^2/V\mu_p$ is the hole transit time. The maximum value of Θ' may be written as $\Theta'_{max} = \Theta_{\Omega,n}(t_p/\tau)$, from which $\Theta'_{max} < \Theta_{\Omega,n} < \tau$ immediately follows. Thus, high-level double injection implies $\Theta' < \Theta'_{max} \ll \tau$.

It is then possible to choose a time t_1 such that $\Theta' \ll t_1 \ll \tau$. At this time t_1, the density distribution n has not had time to change appreciably by means of recombination, since $t_1 \ll \tau$. In addition, n has not been changed appreciably by the relaxation (movement) of the space charge to accommodate the added field. This last follows because the change in n due to the relaxation process is at most of the order of the total density of space charge in the sample, and this has been assumed to be much less than n. Then, if we define "immediately after the step" to mean such a time t_1, the following conclusions may be drawn: the current J_1 present immediately after the voltage step is given by

$$J_1 = J_i(V_f/V_i). \tag{123}$$

The subscripts i, 1, and f indicate, respectively, the initial value of the quantity before the onset of the voltage step, the value of the quantity immediately after the onset of the pulse ($t = t_1$), and the final value of the quantity after steady state is established. Since the density at $t = t_1$ has not had time to change, the density distribution may be obtained from Eqs. (9)–(11):

$$n_1 = n_i = 3\mu_n\tau(n_0 - p_0)V_i/[4(b + 1)L^{3/2}x^{1/2}]. \tag{124}$$

In addition, the displacement current at $t \geqslant t_1$ vanishes, and the field may be calculated from Eqs. (121), (123), and (124) to give

$$\mathscr{E}_1 = J_1/e\mu_p(b + 1)n_1 = \mathscr{E}_f. \tag{125}$$

The fact that the field does not change after the initial capacitive surge is the key to the entire problem. Since $\partial\mathscr{E}/\partial t \equiv 0$ for the time range of interest, the displacement current may be eliminated from Eq. (121). It also allows replacing $\mathscr{E}$ by $\mathscr{E}_f$ in Eqs. (121) and (122). Furthermore, by evaluating the continuity equation (122) at $t = \infty$, where $\partial n/\partial t = 0$ and $n = n_f$, Eq. (122) becomes

$$0 = -\frac{(b + 1)n_f}{\mu_n\tau} + (n_0 - p_0)\frac{\partial\mathscr{E}_f}{dx}.$$

This may be used to eliminate the term $(n_0 - p_0)\,\partial\mathscr{E}_f/\partial x$ in Eq. (122), leading to

$$\partial n/\partial t = -(n - n_f)/\tau. \tag{126}$$

This has the solution

$$n(x, t) = n_f(x)\left[1 - \frac{n_f - n_i}{n_f}e^{-t/\tau}\right], \tag{127}$$

$$= n_f(x)\left[1 - \frac{V_f - V_i}{V_f}e^{-t/\tau}\right], \tag{128}$$

so that the density distribution remains constant in shape and varies in time only by an exponential factor.

The variation of the current with time can now be determined easily from Eqs. (121) and (128) to be

$$J(t) = J_f\left[1 - \frac{V_f - V_i}{V_f} e^{-t/\tau}\right]. \tag{129}$$

The response to a rectangular voltage pulse of magnitude ΔV on top of a dc bias of V_i is shown in Fig. 45 for $\Delta V \ll V_i$.

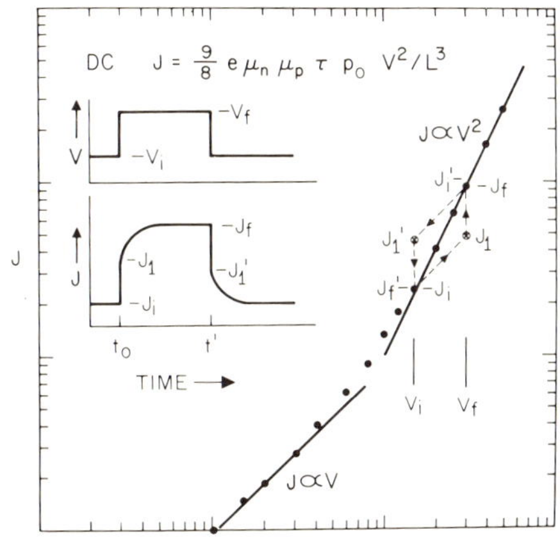

FIG. 45. A schematic representation of the small-signal current response to a voltage pulse, along with the dynamic path on the I–V characteristic.

In Fig. 45, the path of the transient current–voltage characteristic is shown schematically by the heavy arrows. For the initial part of the pulse, with the current rising from J_i to J_1, the diode acts as an ohmic resistor—the carrier density changes negligibly and the increased current is due to the increased field. In the latter part of the pulse, as the current rises from J_1 to J_f, the density increases exponentially with a time constant equal to τ at a constant voltage and field. At t', when the voltage pulse ends, the corresponding transient for a negative voltage step occurs—an ohmic drop in the current from J_i' to J_1' followed by an exponential decay from J_1' to $J_f' = J_i$.

An almost identical result can be derived for the insulator regime,[22] giving

$$n(x, t) = n_f(x)\left[1 - \frac{V_f^2 - V_i^2}{V_f^2} e^{-t/\tau}\right], \tag{130}$$

and

$$J(t) = J_f\left[1 - \frac{V_f^2 - V_i^2}{V_f^2} e^{-t/\tau}\right]. \quad (131)$$

Of course, $n_f(x)$ is now the appropriate function for the insulator regime.

It is clear from the above that it is the time independence of $\mathscr{E}(x)$ (for $t \geqslant t_1$) that makes possible the simple result that the decay is exponential with a time constant equal to the lifetime. This time independence of $\mathscr{E}(x)$ depends ultimately on the fact that $n_i(x)$ and $n_f(x)$ differ only by a constant factor. This will be true as long as the "pure" $J \propto V^2$ or $J \propto V^3$ case holds. However, if there is a transition from one regime to another, or if the inequality $L_a \ll L$ is not satisfied and the diffusion term in the continuity equation must be considered, $n_i(x)$ and $n_f(x)$ will no longer differ only by a constant factor, and the analysis breaks down. Note that nowhere has it been assumed that $\Delta V \ll V_i$, as is normally necessary in discussing a small-signal case. Instead, the only restriction is that V_i and V_f are in the same regime. Thus, ΔV could be much greater than V_i if the regime extended over a wide range of voltage.

To illustrate these conclusions, experimental results for a silicon p–π–n structure exhibiting double-injection behavior are given below. This sample is a bar of transverse dimensions 1.9 × 0.6 mm and length (π region) 3.0 mm, and exhibits $J \propto V^2$ behavior (Lampert semiconductor regime[5]) at room temperature. The response of the current to a voltage pulse from 45 to 55 V and back again is shown in Fig. 46. This response is as predicted, as can be seen by comparing this with the result shown in Fig. 45 for the $\Delta V \ll V_i$ case.

The tail of the response shown in Fig. 46 is reproduced in Fig. 47(b) along with the response to a light flash in Fig. 47(a). The light flash was filtered through a $\frac{1}{4}$-in. thick silicon filter to produce reasonably penetrating light, and had a decay time of about 1.5 μsec. These decays are plotted in Fig. 47(c)

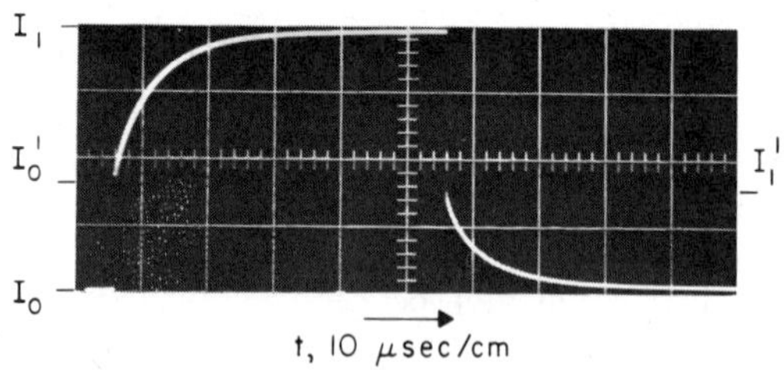

FIG. 46. Current response to a voltage pulse of a silicon p–π–n sample biased in the semiconductor regime. The initial and final voltages are similar to those shown in Fig. 45. (After Baron *et al.*[22])

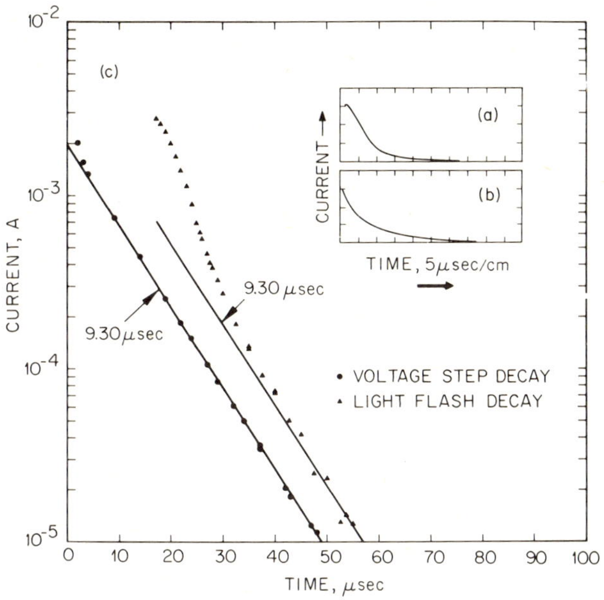

FIG. 47. (a) Current response to a flash of penetrating light with sample biased in semiconductor regime. (b) Tail of current response shown in Fig. 46. (c) Plot of responses in (a) and (b) on semi-log scale. (After Baron *et al.*[22])

on a semilog scale. They both tend to a final time constant of 9.30 μsec. The deviation from exponential decay during the first part of the light-flash response is caused by the excitation of higher-order decay modes with shorter time constants.[22] Because the light-flash decay would have to be followed down to very low levels in order to determine accurately the high-level lifetime, the voltage-step decay is a more practical method of determining τ. The lifetime as measured from the decay of a voltage step for a series of silicon *p–π–n* and *p–i–n* structures[41,24] gave values which allowed consistent calculations of the current from the appropriate formula given in Part II.

b. Large-Signal Case

The change in behavior introduced when the initial and final states of the system are not in the same regime is illustrated in Fig. 48, which shows the response of the current to a voltage step from zero bias to a number of final voltages in the double-injection regime. The response is now more complicated, being slower and exhibiting an inflection point in the current response. As is evident in Fig. 48, the inflection point moves to shorter times with increased bias.

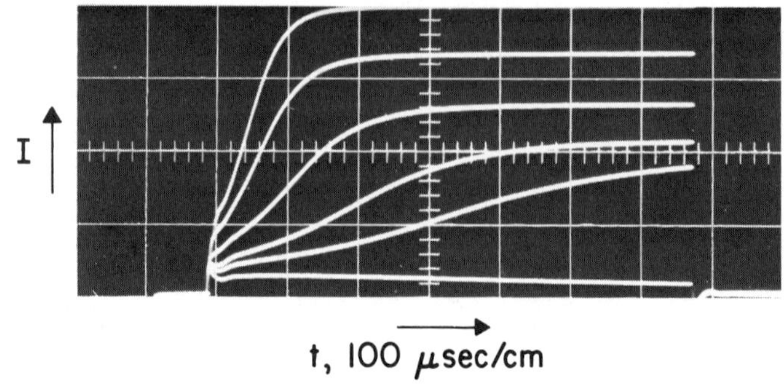

FIG. 48. Current response of a silicon p–π–n structure to a voltage step from zero bias to a number of final voltages. All but the lowest of the final voltages bias the sample into the semiconductor regime. (After Baron *et al.*[22])

Dean[71,71a] has considered a limiting case for a semiconductor using the following assumptions: (i) The low-level dielectric relaxation time is much less than the transit time, which, in turn, is much less than the lifetime; (ii) diffusion and trapping effects are negligible[71b]; and (iii) the lifetime is constant.

Dean then follows the drift of the minority carriers injected into the sample. His assumptions allow him to consider that the region behind the leading edge of the minority carriers is in the semiconductor regime. The assumed short dielectric relaxation time allows majority carriers to flow into this region and relax the space charge of the injected minority carriers. In front of the leading edge there are no injected minority carriers, and this region exhibits ohmic behavior.

Dean finds that the field in the modulated region is proportional to $x^{1/2}$ at all times, with of course a uniform field in the unmodulated region. The time dependence of the current is given by

$$J = J_0[1 - \tfrac{2}{3}(t/t_r)]^{-1/2} \qquad \text{for} \qquad t \leqslant t_a, \tag{132}$$

where J_0 is the ohmic current in the absence of injected carriers, $t_r = L^2/\mu_a V$ is the ambipolar transit time, and $t_a = (5/6)t_r$ is the arrival time of the leading edge at the end of the sample. At $t = t_a$, the electric field is the same as the final field, and the carrier distribution is proportional to the final distribution. For the reason given in the previous section, Dean finds that the subsequent rise of the current is exponential with a time constant equal to the

[71] R. H. Dean, *Appl. Phys. Letters* **13**, 164 (1968).
[71a] R. H. Dean, *J. Appl. Phys.* **40**, 585 (1969).
[71b] The influence of traps on the large signal response has also been considered.[71c]
[71c] R. H. Dean, *J. Appl. Phys.* **40**, 596 (1969).

lifetime. One consequence of his analysis is that there is a cusp in the time derivative of the current at $t = t_a$.

Dean has verified his analysis in room-temperature measurements on p-type germanium. He has measured the electric field as a function of position and time with probes, and has found good agreement with the predicted $x^{1/2}$ dependence behind the leading edge. His oscilloscope traces of the current and its time derivative are shown in Fig. 49 (for two different time scales). The upper traces are the current and the lower traces are the time derivative. The cusp in the derivative occurs at $(9/11)t_r$, which agrees reasonably well with the predicted value of $(5/6)t_r$. The measured value of the current at time t_a is $(3/2)J_0$, which also agrees with the theory.

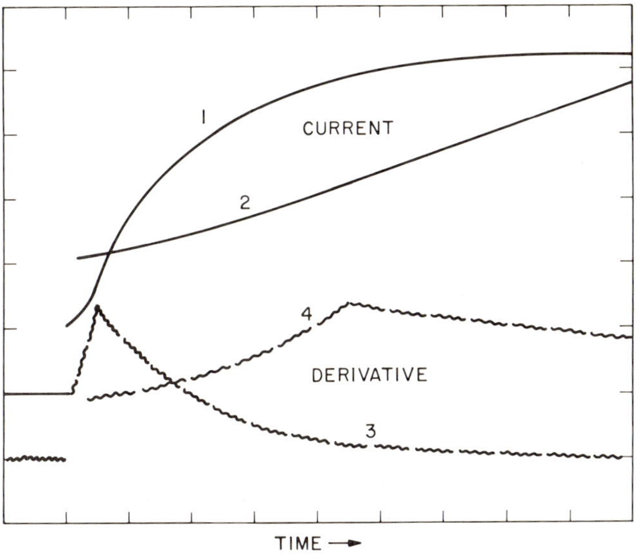

FIG. 49. Oscilloscope traces of the current (traces 1 and 2) and its time derivative (traces 3 and 4) following the application of a 100-V step (from zero bias) to a 1.0 × 1.0 × 2.0 mm germanium p–π–n structure. The horizontal scale is 20 μsec/div for traces 1 and 3, and 2 μsec/div for traces 2 and 4. Traces 1 and 2 are the current (25.4 mA/div, 12.7 mA/div). Traces 3 and 4 are the derivative (1.0 mA/μsec/div). The cusp occurs at approximately 9 μsec. (After Dean.[71])

19. AC Small-Signal Impedance

At times, it is useful to know the small-signal impedance of double-injection devices. For example, an understanding of noise in double-injection devices (discussed in the next section) requires this knowledge. In addition, because double-injection devices exhibit inductive behavior, it is necessary to determine the Q in order to evaluate possible device applications.

a. Semiconductor Regime

Baron[41c] has calculated the small-signal impedance for the limiting case of the semiconductor regime—with the usual assumptions concerning high level and neglect of diffusion. The starting points are Eqs. (121) and (122), and all variables are assumed to be of the form

$$J = J_0 + J_1 e^{j\omega t}, \qquad n = n_0 + n_1 e^{j\omega t},$$

etc. The variables with zero subscripts are easily shown to be the dc solutions. The terms proportional to $e^{j\omega t}$ are collected, and higher-order terms are neglected, thus linearizing the equations. The resulting equation for $\mathscr{E}_1(x)$ is

$$\frac{d\mathscr{E}_1}{dx} + (1 + j\omega t)\left[\frac{1}{2x} + j\omega\frac{\Theta}{2(xL)^{1/2}}\right]\mathscr{E}_1 = \frac{3}{4}I_1\frac{R_0}{L^2}\left(\frac{L}{x}\right)^{1/2}(1 + j\omega t),$$

where $R_0 = V_0/I_0$ and $\Theta = (3/2)C_0R_0$. Here, C_0 is the geometrical capacitance, so that Θ is of the order of the average high-level dielectric relaxation time. Since $V_1 = \int_0^L \mathscr{E}_1\, dx$, this leads to

$$Z_1 = \frac{V_1}{I_1} = 3R_0(1 + j\omega\tau)\int_0^1 dz' \int_0^{z'} dz \times \left\{ z'\left(\frac{z}{z'}\right)^{(1+j\omega\tau)} \exp[j\omega\Theta(1 + j\omega\tau)(z - z')]\right\} \tag{134}$$

for the small-signal impedance, where τ is, as usual, the high-level lifetime and $z = (x/L)^{1/2}$.

(1) $\omega\Theta \ll 1$. In the limit that $\omega\Theta \ll 1$, this may be evaluated by expanding in powers of $\omega\Theta$, giving

$$Z_1 \approx R_0\frac{1 + j\omega\tau}{2 + j\omega\tau}\left[1 - j\omega\Theta\frac{3}{4}\frac{1 + j\omega\tau}{3 + j\omega\tau}\right] \tag{135}$$

for the first two terms. To find an equivalent circuit, it is easiest to calculate $Y_1 = 1/Z_1$ and expand to give

$$Y_1 \approx \frac{1}{R_0[1 + \frac{3}{4}(\Theta/\tau)]} + \frac{1}{R_0 + j\omega\tau R_0} + \frac{9}{8}j\omega C_0 + \frac{1}{\frac{4}{3}(\tau/\Theta)R_0 + \frac{4}{9}j\omega(\tau/\Theta)\tau R_0}. \tag{136}$$

This gives the equivalent circuit shown in Fig. 50(a), which reduces to that in Fig. 50(b) if the terms of order $\omega\Theta$ are neglected. Note that $(\Theta/\tau) \ll 1$ (as shown in Section 18a). The real and imaginary parts of the parallel admittance of the circuit in Fig. 50(b) are shown in Fig. 51. These curves exhibit the following characteristics:

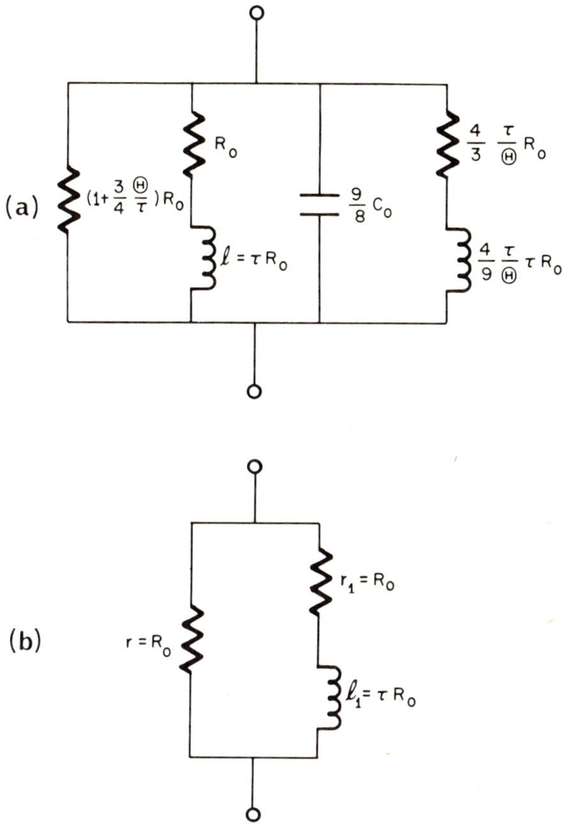

FIG. 50. (a) Equivalent circuit for a diode biased in the semiconductor regime (Section 19). (b) Simplified equivalent circuit for the case when terms of order $\omega\Theta$ in the impedance are neglected. (After Baron.[41c])

Real Part of Admittance. For $\omega\tau \ll 1$, the conductance is given by $G_1 = 2/R_0 = dI_0/dV_0$. That is, it is equal to the dc differential conductance. For $\omega\tau \gg 1$ (but $\omega\Theta \ll 1$), $G_1 = 1/R_0$. This follows because the carrier density does not have time to change, and thus the device behaves as an Ohmic resistor.

Imaginary Part of Admittance. For both $\omega\tau \gg 1$ and $\omega\tau \ll 1$, the susceptance B_1 goes to zero,[72] and thus ceases to contribute to the impedance, and

[72] This is true, of course, only in the approximation of Fig. 50(b) where the terms in Eq. (135) involving $\omega\Theta$ are neglected. The term $\frac{9}{8}j\omega C_0$ in Eq. (136), which, of course, does not go to zero for increasing ω, can be neglected, since the assumption $\omega\Theta \ll 1$ implies $\omega C_0 \ll 1/R_0$. At higher frequencies ($\omega\Theta \gg 1$), this capacitive term becomes dominant, as discussed in Section 19a(2).

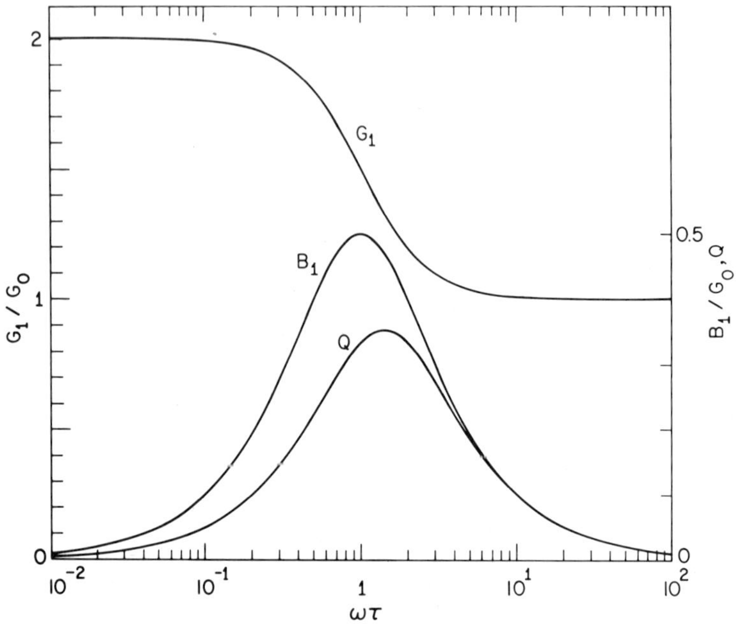

FIG. 51. The small-signal conductance, susceptance, and Q of a structure biased into the high-level semiconductor regime. (After Baron.[41c])

the device impedance is pure resistive. This occurs at low frequencies because the carrier concentration follows the varying field, and occurs at high frequencies because the carrier concentration cannot follow the field. The susceptance is large only at frequencies near $\omega\tau = 1$, where the lag in the carrier concentration behind the field produces an appreciable reactive component in the admittance.

Q Values. A similar argument leads to the $Q = |B_1|/G_1$ peaking at $\omega\tau = \sqrt{2}$ (near $\omega\tau = 1$). Note that the peak value is only $Q = 0.35$. This low value of Q arises because the value of $\omega l = R_0$ at $\omega\tau = 1$.

It should also be noted that the equivalent circuit shown in Fig. 50(b) has a transient response to a small voltage step identical to that derived in Section 18a.

Experimental Results. Bilger *et al.*[73] have measured the small-signal admittance as a function of frequency, as well as the transient response to a step voltage of a silicon p–π–n structure. When biased in the forward direction, this structure exhibited ohmic behavior followed by an $I \propto V^2$ region,

[73] H. R. Bilger, D. H. Lee, M-A. Nicolet, and E. R. McCarter, *J. Appl. Phys.* **39**, 5913 (1968).

with a transition voltage of ≈ 8 V. The current response to a voltage step is shown in Fig. 52. The real and imaginary parts of the admittance for several values of the bias are shown in Fig. 53. The solid lines in these figures are the theoretical curves which have been least-squares fitted to the data, adjusting the values of r, r_1, and l_1 of Fig. 50(b). The values of these parameters, as found by the various methods, are in reasonable agreement. The values found by Bilger *et al.* for a bias of 30 V are shown in Table II. The values of r and r_1

TABLE II

EQUIVALENT CIRCUIT PARAMETERS OF A DOUBLE-INJECTION DIODE BIASED AT $V_0 = 30.0$ V, $I_0 = 1.60\ \mu$A, $R_0 = 18.8$ kilohms[a]

	r (kilohms)	r_1 (kilohms)	l_1 (H)	τ (μsec)
Step response	18.1	18.5	0.70	38
Re(Y_1) [$=G_1$]	17.8	16.8	0.69	41
Im(Y_1) [$=B_1$]	—	17.9	0.63	37

[a] Data from Bilger *et al.*[73]

should be compared with their theoretical value: $r = r_1 = R_0 = V_0/I_0 =$ 18.8 kilohms. The fact that r and r_1 are indeed nearly equal, as expected, is related to the fact that the height of the initial step in the current transient (Fig. 52) is one half of the final value (for $\Delta V \ll V_i$).

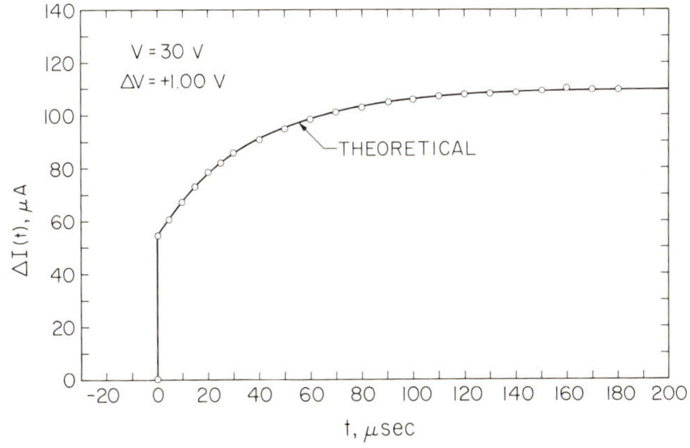

FIG. 52. Current response of a silicon p–π–n diode biased in the semiconductor regime to a voltage step. The solid line is a least-squares fit of $\Delta i(t) = \Delta i(0) + \Delta i(\infty)\ (1 - \exp t/\tau)$. The values of the parameters are given in Table II. (After Bilger *et al.*[73])

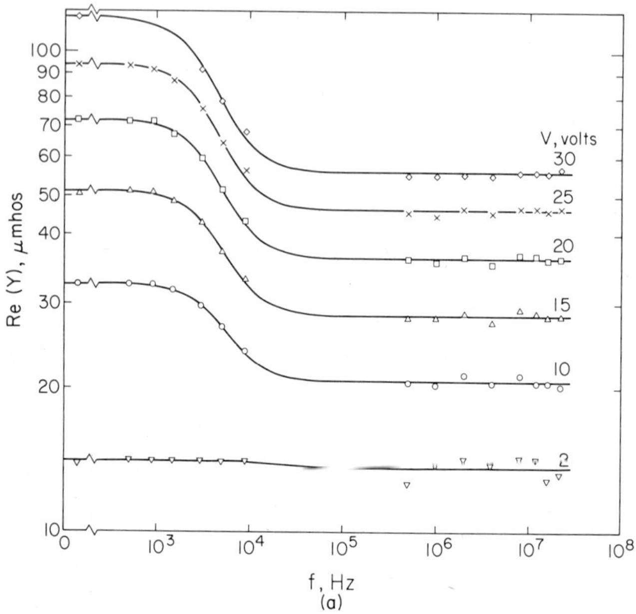

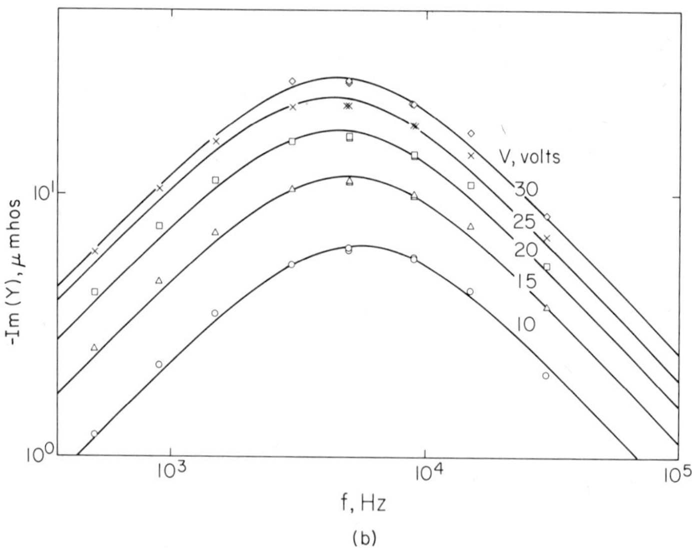

FIG. 53. (a) Real part of the diode (Fig. 52) admittance versus frequency. (b) Imaginary part of the diode admittance versus frequency. The solid lines are least-squares fits to the data. The fitting function is that derived from the equivalent circuit of Fig. 50 [$\omega C = \mathrm{Im}(Y)_{f\to\infty}$ was first subtracted from Im(Y)]. (After Bilger *et al.*[73])

(2) $\omega\Theta \gg 1$. In the limit that $\omega\Theta \gg 1$, the integral of Eq. (134) may be expanded in powers of $1/\omega\Theta$, which leads to

$$Y_1 \approx j\omega C_0 + \frac{4}{3}\frac{1}{R_0} \tag{137}$$

for the first two terms in the expression for the admittance. Since the space-charge distribution does not have time to change at these frequencies, the capacitance becomes the geometrical capacitance. It is easily seen from

$$\omega C_0 \Big/ \left(\frac{4}{3}\frac{1}{R_0}\right) = \frac{1}{2}\omega\Theta \gg 1$$

that the capacitance term is now dominant.

The higher value of the conductance arises because the time-varying field is now spatially independent, having insufficient time to relax to a $x^{1/2}$ dependence. Thus, the added field is not small in regions where n is large. A simple derivation is as follows. The average conduction current due to the time-varying field $\mathscr{E}_1$ is given by

$$I_1 = (A/L)\int_0^L e\mu_p(b+1)n_0\mathscr{E}_1\,dx\,.$$

This is essentially equivalent to adding the velocities of all the carriers and dividing by the length. Now,

$$n_0 = \frac{J_0}{e\mu_p(b+1)\mathscr{E}_0} = \frac{J_0}{e\mu_p(b+1)\frac{3}{2}(V_0/L)(x/L)^{1/2}}.$$

Therefore,

$$I_1 = \frac{4}{3}\frac{A}{L}\frac{J_0\mathscr{E}_1 L^{3/2}}{V_0}L^{1/2} = \frac{4}{3}\frac{I_0\mathscr{E}_1 L}{V_0},$$

and thus

$$Y_1 = \frac{I_1}{V_1} = \frac{4}{3}\frac{1}{R_0}.$$

b. Insulator Regime

The direct derivation of the small-signal impedance for the insulator regime has proved intractable to date. However, as seen in the previous section, it is possible to obtain the dominant term in the small-signal impedance (for $\omega\Theta \ll 1$) from the transient response. This results in

$$Y_1 = \frac{1}{R_0}\frac{3 + j\omega\tau}{1 + j\omega\tau},$$

which corresponds to the equivalent circuit shown in Fig. 54. The change in the value of r_1 corresponds to the fact that the height of the initial step in the current transient is now 1/3 of the final value (for $\Delta V \ll V_i$). This leads to a maximum value of $Q = 1/\sqrt{3}$ at $\omega\tau = \sqrt{3}$.

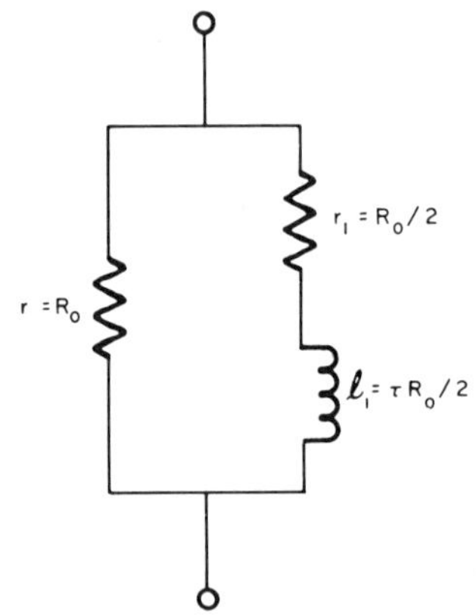

FIG. 54. Equivalent circuit for a structure biased in the insulator regime for $\omega\Theta \ll 1$. (After Baron.[41c])

c. *Diffusion-Dominated Regime*

Nordman and Greiner[74] have calculated the small-signal impedance of a long *p–i–n* diode in which the injection level is sufficiently high that the carrier density in the *i* region is dominated by the diffusion term in the continuity equation. This is the case discussed in Section 4c. However, they limit their calculations to small enough injection levels that the junctions can be considered to be perfectly injecting ($J_n = 0$ at the *p–i* junction and $J_p = 0$ at the *n–i* junction). They also neglect displacement currents, so that their results are limited to frequencies low enough that $\omega\Theta \ll 1$ is satisfied. For this regime, as we saw in Section 4c, the dc voltage across the *i* region is independent of the current. The total voltage includes the voltage across the junctions, which is current dependent.

The calculation follows that given above for the semiconductor regime; however, the small-signal boundary conditions must be developed. For this the voltage across the junction is a useful parameter, and Nordman and

[74] J. E. Nordman and R. A. Greiner, *IEEE Trans. Electron Devices* **ED-10**, 171 (1963).

Greiner set for the n–i junction (using the notation of this chapter),

$$\begin{aligned} n(0) = p(0) &= n_i \exp(V_J/\beta) = n_i \exp[(V_{J,0} + V_{J,1}e^{j\omega t})/\beta] \\ &\approx n_i \exp(V_{J,0}/\beta + n_i(V_{J,1}/\beta)e^{j\omega t} \exp(V_{J,0}/\beta) \\ &= n_0(0) + n_1(0)e^{j\omega t} \end{aligned}$$

for $V_{J,1}/\beta \ll 1$. Using the condition (for this junction) that $J_p = 0$ allows the derivation of a relation between J_1 and $V_{J,1}$, and the solution then proceeds in a straightforward manner.

The impedance of the device consists of the impedance of the i region in series with the impedance of the junctions. Nordman and Greiner's result for the junction impedance agrees with the usual results for a p–n junction, corresponding approximately to an equivalent circuit of a resistor in parallel with a capacitor for $\omega\tau \lesssim 2$. Their result for the i-region impedance involves integrals which can only be evaluated numerically. The expressions are too involved to give here, but the results correspond approximately to an equivalent circuit of a resistor in parallel with an inductor. This equivalent circuit is approximately valid for all frequencies considered here (i.e., $\omega\Theta \ll 1$) when $L/L_a \lesssim 1$. However, for $L/L_a \gg 1$, the equivalent circuit is useful only at lower frequencies, where $\omega\tau \ll 1$. Thus, their result for the i region corresponds roughly to our result for the semiconductor regime. The total impedance for the device also corresponds approximately to an equivalent circuit of a resistor in parallel with an inductor for low frequencies ($\omega\tau \lesssim 1$) and long i regions ($L/L_a \gtrsim 1$). For $L/L_a < 1$, the junction impedance dominates, and the equivalent circuit becomes roughly a resistor in parallel with a capacitor. As mentioned above, Nordman and Greiner do not treat the very high injection case, where p–n junctions exhibit inductive behavior, which would lead to inductive behavior for the p–i–n device, even for $L/L_a < 1$.

One important difference is Nordman and Greiner's result for the Q of the device reproduced here in Fig. 55. Note that for large values of L/L_a (their d/L in the figure), large values of Q are obtainable; however, they peak at frequencies where $\omega\tau \ll 1$. The occurrence of high values of Q for this regime can be understood qualitatively from the following argument. As we have seen for the semiconductor and insulator regimes, the transient response consists of an ohmic step followed by an inductive rise. The higher power law for the dc I–V characteristic in the insulator regime makes the ohmic step in the current a smaller fraction of the total step, and thus a higher Q results. Extending this argument to the diffusion-dominated regime, where the I–V characteristic is very steep, one expects a still smaller ohmic step and higher Q. Since larger values of L/L_a for this regime mean that the junction drop is a smaller proportion of the voltage, these larger values of L/L_a lead to steeper

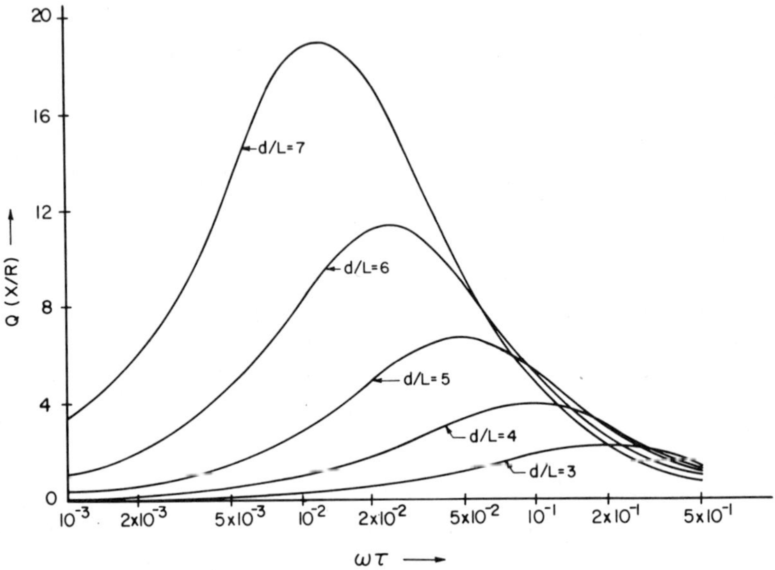

FIG. 55. Plots of Q of a p–i–n structure biased in the constant-voltage, diffusion-dominated regime. Note that d/L in this figure is L/L_a in the notation of this chapter. (After Nordman and Greiner.[74])

I–V curves, smaller values of the ohmic step, and higher values of Q, in agreement with the results of Nordman and Greiner.

Material restrictions set an upper limit on L/L_a and thus on the Q. A lower limit on the injection level is set by the requirement that n is everywhere $\gg n_i$, and an upper limit on the injection level is set by the requirement that the junctions do not wash out [i.e., $n(0) \ll p_p, n_n$].[74a] Nordman and Greiner show that both of these conditions can be satisfied only for L/L_a less than some critical value. This critical value of L/L_a is ~ 3 for germanium and ~ 7 for silicon. The corresponding peak values of Q are ~ 2.3 for germanium and ~ 19 for silicon, as seen in Fig. 55. In addition, these upper limits are attainable only at the highest allowable injection levels, corresponding to total junction voltages of ~ 0.4 V for germanium and ~ 0.8 V for silicon.

20. NOISE

Noise measurements on double-injection devices can give further insight into the high-frequency behavior of the charge carriers. It is clear from the preceding sections that changes in the mobile hole and electron concentra-

[74a] Nordman and Greiner did not consider that the upper limit on the injection level may be set by the transition to the high-current $J \propto V^2$ regime (beginning of nonperfect injection).

tions can take place only by way of recombination mechanisms. Thus, for times short compared with the lifetime, or for frequencies large enough so that $\omega\tau \gg 1$, the electron and hole concentrations become decoupled. Under these high-frequency conditions, the electrons and holes behave separately as ohmic resistors. Thus the noise arises only from Nyquist noise in these separate resistors and the mean-square fluctuation in the current is given by

$$\langle i^2 \rangle = 4kTG_1(\omega)\,\Delta f, \tag{138}$$

where $G_1(\omega)$ is the small-signal conductance at the measurement frequency.

Lee and Nicolet[75] have measured $I_{eq} = \langle i^2 \rangle / 2e\,\Delta f$ for the double-injection (semiconductor-regime) structure whose admittance is given in Fig. 53, over the temperature range from 140 to 350°K. They measured I_{eq} from 60 kHz to 22 MHz and found that the noise spectra exhibited a constant level above a certain frequency. This frequency is typically 1 MHz and depends on the excess noise present at lower frequencies (this excess noise was attributed to generation–recombination noise). In this frequency range (1–22 MHz), the condition $1/\tau \ll \omega \ll 1/\Theta$ is satisfied. Lee and Nicolet find close agreement between their results and Eq. (138), as shown in Fig. 56, where their plot of I_{eq} versus $G_1(\omega)$ is reproduced. Their results are

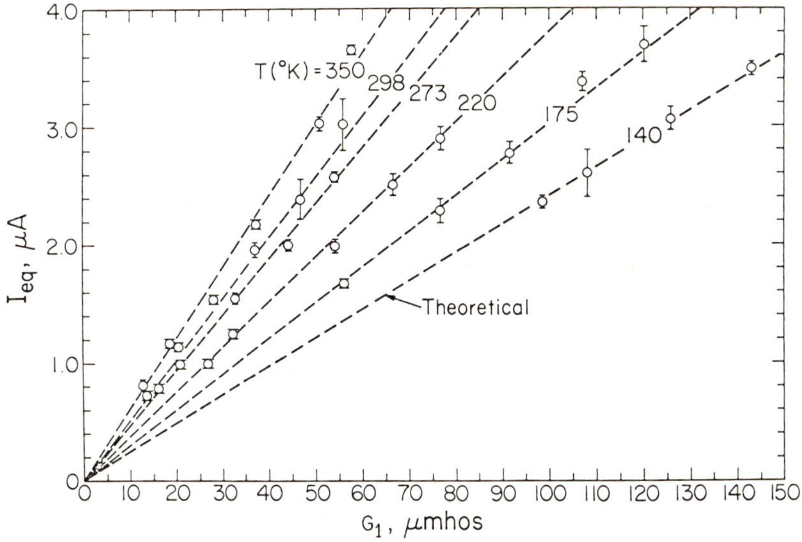

FIG. 56. High-frequency equivalent noise current I_{eq} versus high-frequency conductance $g = 1/r$ for the diode of Figs. 52 and 53 at temperatures from 140 to 350°K. The dashed theoretical lines have slopes of $2kT/e$. (After Lee and Nicolet.[75])

[75] D. H. Lee and M-A. Nicolet, *Phys. Rev.* **184**, 806 (1969).

also consistent with previous work on the noise properties of double-injection structures.[73,75a,76,77]

21. Comparison with Single-Carrier Injection

Injection effects can be used to measure such material parameters as carrier mobilities and recombination time constants. However, the current–voltage characteristic, when treated as a separate entity, may lead to erroneous conclusions. Single-carrier, space-charge-limited currents (SCLC), double injection, or a mixture of the two[78,79] can lead to power-law I–V characteristics. In addition, quite stringent requirements are imposed on the contacts or junctions if "pure" single- or two-carrier injection is to be observed. Marsh and Viswanathan,[79] in some cases, found currents in excess of single-carrier behavior in silicon p^+–π–p^+ structures. These excess currents were identified with double-injection behavior caused by electron injection. Negative resistance and current instabilities, caused by hole injection at high fields, have also been noted[79a] in n–π–n silicon diodes. Pulse measurement of the transient characteristics represents a very convenient method of determining the mode of operation.

In single-carrier, space-charge-limited conditions, the trap-free current density, neglecting diffusion, is $J_{\mathrm{SCL}} = (9/8)\varepsilon\mu V^2/L^3 = (9/8)J_0$ and the transient current density to a voltage step[80,81] is

$$J(t) = \frac{1}{2}J_0\left(1 - \frac{t}{2t_r}\right)^{-2}, \qquad t < t_1, \tag{139}$$

where the transit time of the leading edge of the injected space charge is given by $t_1 = 0.79t_r$, where $t_r = L^2/\mu V$. The initial step in the current is to a value $J_0/2$ [Fig. 57(a)] and then the current rises monotonically to a value of approximately $1.2J_{\mathrm{SCL}}$ at time t_1. The current then decays to its steady-state value at $t \approx 2t_1$ for the case of no trapping. If trapping centers are present, the current decays below the J_{SCL} value [Fig. 57(b)] with times characteristic of the trapping time τ_c. Lemke[82] has shown that at sufficiently low voltages, the injection current decreases exponentially with a time constant of τ_c.

[75a] M-A. Nicolet, H. R. Bilger, and E. R. McCarter, *Appl. Phys. Letters* **9**, 434 (1966).
[76] S. T. Liu, S. Yamamoto, and A. van der Ziel, *Appl. Phys. Letters* **10**, 308 (1967).
[77] F. Driedonks, R. J. J. Zijlstra, and C. T. J. Alkemade, *Appl. Phys. Letters* **11**, 318 (1967).
[78] U. Büget and G. T. Wright, *Solid-State Electron.* **10**, 199 (1967).
[79] O. J. Marsh and C. R. Viswanathan, *J. Appl. Phys.* **38**, 3135 (1967).
[79a] A. K. Hagenlocher and W. T. Chen, *IBM J. Res. Dev.* **13**, 533 (1969).
[80] W. Helfrich and P. Mark, *Z. Physik* **166**, 370 (1962).
[81] A. Many and G. Rakavy, *Phys. Rev.* **126**, 1980 (1962).
[82] H. Lemke, *Phys. Status Solidi* **16**, 413 (1966).

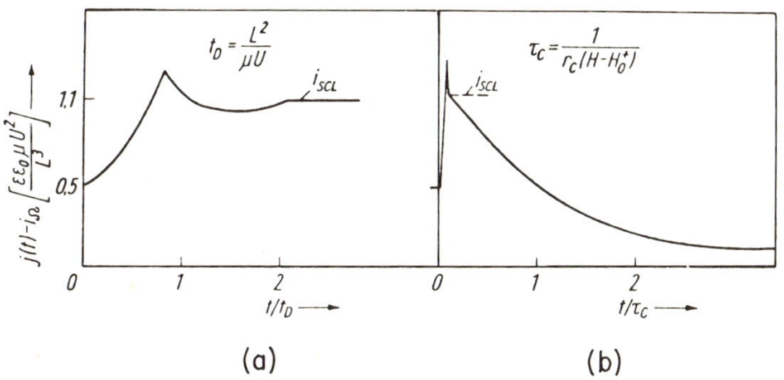

FIG. 57. Theoretical transient response of single-carrier, SCLC structures to a voltage step. (a) Trap-free case; t_D of the figure is t_r of the text. (b) With traps present and having a characteristic trapping time τ_C. (After Lemke.[82])

Lemke and Müller[83] have obtained quantitative agreement with the theoretical predictions using p^+–p–p^+ samples. Figure 58(a) is a sampling oscilloscope trace of the injection current response to a voltage step, in which the cusp at 30 nsec can be seen. The height of the cusp J_1 is about 20% greater than J_{SCL}. The time dependence of the current pulse for $t < t_1$ was in good agreement with the theoretical expression, Eq. (139). Carrier-mobility values determined from t_1 in a number of samples were between 430 and 460 cm^2 V^{-1} sec^{-1}, which is in good agreement with hole-mobility values of 420 (Zerbst and Heywang[84]) and 475 (Ludwig and Watters[85]) cm^2 V^{-1} sec^{-1}. In samples in which trapping was present, the current exhibited a decay similar to that shown in Fig. 58(b) (note the difference in time scale).

Space-charge-limited currents of holes in silicon were investigated by Marsh and Viswanathan.[79] In general, they found close agreement between theory and experimental values. Under some conditions, however, excess currents were found, as shown in Fig. 59. The two characteristics shown in Fig. 59 correspond to the two polarities of the dc bias measured at the button. The lower curve (button positive) shows ohmic behavior at low voltages with a transition to square-law (single-carrier) behavior. The excess current (button negative) was found to be a double-injection effect due to injection of electrons from the surface adjacent to the button. This can be deduced from the transient response to a voltage step, as shown in Fig. 60. Figure 60(a) shows the response to a positive step; the current values are in agreement with SCL predictions. Note that the time scale is too long to

[83] H. Lemke and G. O. Müller, *Phys. Status Solidi* **24**, 127 (1967).

[84] M. Zerbst and H. Heywang, *Z. Naturforsch.* **11a**, 608 (1956).

[85] G. W. Ludwig and R. L. Watters, *Phys. Rev.* **101**, 1699 (1956).

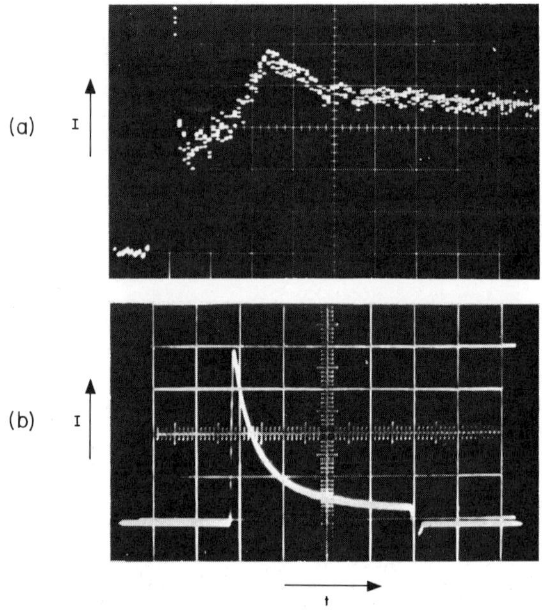

FIG. 58. Transient current response of SCLC of a silicon p–π–p sample of about 0.064 cm thickness doped with deep donors (Fe) acting as traps. (a) Sampling oscilloscope trace of the initial part of the time dependence; horizontal scale, 10 nsec/cm. (b) Oscilloscope trace at longer times, showing the effects of trapping. Horizontal scale, 2 μsec/cm. (After Lemke and Müller.[83])

permit identification of a current cusp. Figure 60(b) is the response to a negative voltage step, the polarity in which excess currents were seen. The appearance of a slowly rising component is characteristic of double-injection effects.

These results indicate that dc measurements of current–voltage characteristics should be supplemented with transient analysis. The difference in the characteristic time response between two-carrier and single-carrier phenomena permits a clear distinction between the different regimes.

VII. Appendix—Interpretation of Potential-Probe Measurements

22. Probe Measurements

In Section 8, potential-probe measurements were used to determine the field and carrier distributions. Justification for this method is given below.

It has been shown by Shockley[36] that the current equations can be written in terms of the quasi-Fermi levels ϕ_n and ϕ_p for electrons and holes. Following Shockley, ϕ_n and ϕ_p are defined in terms of the carrier

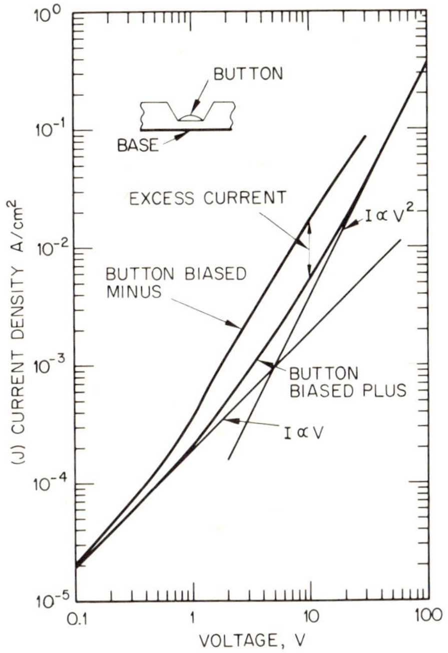

FIG. 59. Current density versus voltage of a typical SCL silicon p–π–p sample for both polarities of bias of the button with respect to the base. (After Marsh and Viswanathan.[79])

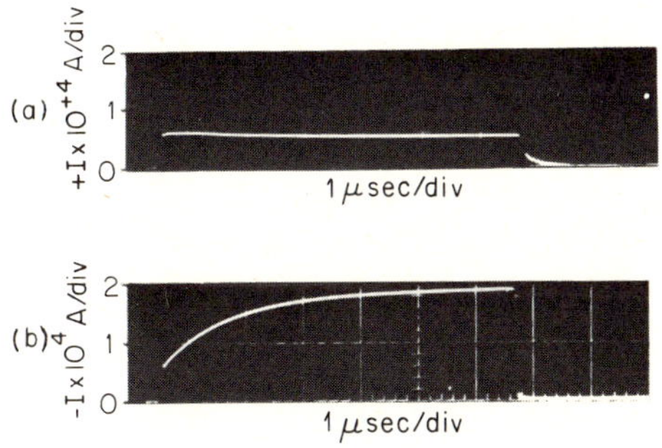

FIG. 60. The current transient response of a silicon p–π–p sample (a) to a positive voltage step on the button (see Fig. 59) with positive dc bias (in the $I \propto V^2$ regime) applied to the button, (b) to an equal negative voltage step with equal negative dc bias applied to the button. The latter case (b) represents the case where "excess" currents were observed. (After Marsh and Viswanathan.[79])

concentrations n and p by

$$n = n_i \exp[(\psi - \phi_n)/\beta], \qquad \phi_n = \psi - \beta \ln(n/n_i);$$
$$p = n_i \exp[(\phi_p - \psi)/\beta], \qquad \phi_p = \psi + \beta \ln(p/n_i); \tag{140}$$

where ψ is the electrostatic potential and is defined to coincide with the Fermi level under equilibrium conditions for intrinsic material, and $\beta = kT/e$.

The electron and hole current densities $\mathbf{J}_n$ and $\mathbf{J}_p$ can be written as

$$\mathbf{J}_n = -e\mu_n n \nabla\phi_n, \qquad \mathbf{J}_p = -e\mu_p p \nabla\phi_p. \tag{141}$$

The total current density $\mathbf{J}$ is the sum of $\mathbf{J}_p$ and $\mathbf{J}_n$ and is therefore written

$$\mathbf{J} = -e\mu_p p \nabla\phi_p - e\mu_n n \nabla\phi_n. \tag{142}$$

If a metal probe is placed in contact with a semiconductor, the two quasi-Fermi levels ϕ_n and ϕ_p merge into one level[85a] $\bar{\phi}$ in the metal, as shown schematically in Fig. 61, where y denotes the direction perpendicular to the surface. On the semiconductor side near the contact, the carrier densities differ from the values in the bulk; the change of $n(y)$ and $p(y)$ is reflected in the sharp variation of ϕ_n and ϕ_p near the probe contact. Here, $\tilde{\phi}_n$ and $\tilde{\phi}_p$ denote the values of the quasi-Fermi level, and $\tilde{n}$ and $\tilde{p}$ the carrier densities, for the bulk of the material far away from the probe.

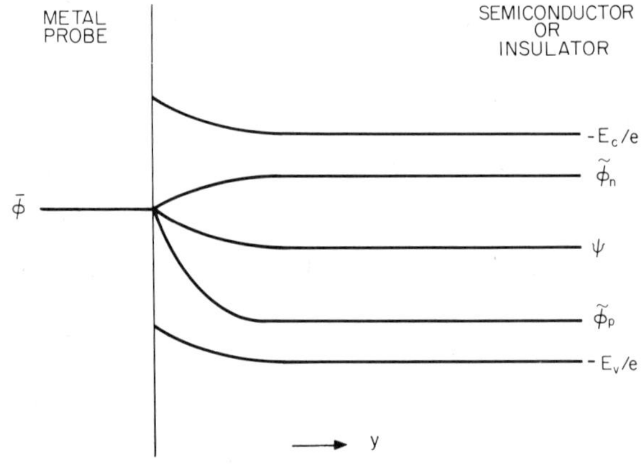

FIG. 61. Schematic electron-potential diagram at the interface of a metal probe and a semiconductor or insulator surface. (After Mayer *et al.*[19])

[85a]The assumption that ϕ_n and ϕ_p merge smoothly together without any discontinuity at the boundary makes possible the simple theoretical arguments in this chapter. However, theoretical considerations indicate that the extremely small diffusion length for holes in the metal would make any discontinuity negligibly small.

It is reasonable to assume that the currents flowing to the probe are radial sufficiently near the probe. Restricting the following to a one-dimensional argument, the y components of $\mathbf{J}_n$ and $\mathbf{J}_p$ along the y axis contain only the currents flowing to the probe, and can be written as

$$J_n^y = -e\mu_n n \, d\phi_n/dy, \qquad J_p^y = -e\mu_p p \, d\phi_p/dy.$$

Because of the assumption that the probe draws no current, it follows from the above that $J^y = 0$ and $J_n^y = -J_p^y = J_1(y)$ everywhere on the y axis. Multiplying these equations by $\tilde{n}/n$ and $\tilde{p}/p$, respectively, and integrating along the y axis from $y = 0$ at the contact to $y = d$ (where d is large enough that $\phi_n = \tilde{\phi}_n$ and $\phi_p = \tilde{\phi}_p$) in the interior of the semiconductor, one obtains

$$\int_0^d \tilde{n}\frac{J_n^y}{n}\,dy = -e\mu_n\tilde{n}(\tilde{\phi}_n - \bar{\phi}),$$

$$\int_0^d \tilde{p}\frac{J_p^y}{p}\,dy = -e\mu_p\tilde{p}(\tilde{\phi}_p - \bar{\phi}). \tag{143}$$

Adding the two expressions and dividing the sum by $e(\tilde{n}\mu_n + \tilde{p}\mu_p)$, one obtains

$$\bar{\phi} - \frac{\mu_n\tilde{n}\tilde{\phi}_n + \mu_p\tilde{p}\tilde{\phi}_p}{\tilde{n}\mu_n + \tilde{p}\mu_p} = \Delta, \tag{144}$$

where

$$\Delta \equiv \frac{1}{e(\tilde{n}\mu_n + \tilde{p}\mu_p)}\int_0^d \left(\frac{\tilde{n}J_n^y}{n} + \frac{\tilde{p}J_p^y}{p}\right)dy. \tag{145}$$

In order to interpret the probe voltage $\bar{\phi}$ in terms of material parameters, it is necessary to estimate the value of Δ. This will require the use of several assumptions. By using the relations $J^y = 0$ and $\tilde{n} \approx \tilde{p}$, Eq. (145) may be written as

$$\Delta = \frac{1}{e(\mu_n + \mu_p)}\int_0^d \frac{[n(y) - p(y)]J_1(y)}{n(y)p(y)}\,dy. \tag{146}$$

Thus, if the net charge $n - p$ is sufficiently small compared with the product np, Δ can be small compared with $\bar{\phi} - \tilde{\phi}_n$ or $\bar{\phi} - \tilde{\phi}_p$, and we can approximate from Eq. (144) that

$$\bar{\phi} \approx \frac{\mu_n\tilde{\phi}_n + \mu_p\tilde{\phi}_p}{\mu_n + \mu_p} = \frac{b\tilde{\phi}_n + \tilde{\phi}_p}{b + 1}, \tag{147}$$

where $b = \mu_n/\mu_p$. As can be seen from Eq. (144), Δ must be negligible in the case of an extrinsic semiconductor, where the majority-carrier density is much greater than the minority-carrier density, if the probe voltage is to measure the quasi-Fermi level of the majority carriers as suggested by Macdonald.[32] This suggests that it might be reasonable to assume that Δ is negligible for all cases, and that the probe potential can be interpreted as the mobility weighted mean of the quasi-Fermi levels.

The remainder of this appendix presents, for the case where $\tilde{n} \approx \tilde{p}$, strong experimental evidence supporting the validity of Eq. (147).

23. *p–i–n* STRUCTURE

The distributions of the carrier concentration and potential in a forward-biased *p–i–n* junction are shown in Fig. 62. In the intrinsic region, consideration of space-charge neutrality requires that $n(x) \approx p(x) > n_i$. The injected carrier densities may exceed the intrinsic carrier concentration by a factor of from 1.5 to 10^4. The carrier distributions are characterized by an exponential decrease from the *n–i* and *p–i* junctions that may be approximated by a factor $\exp(x/L_a)$ as shown in Section 4.

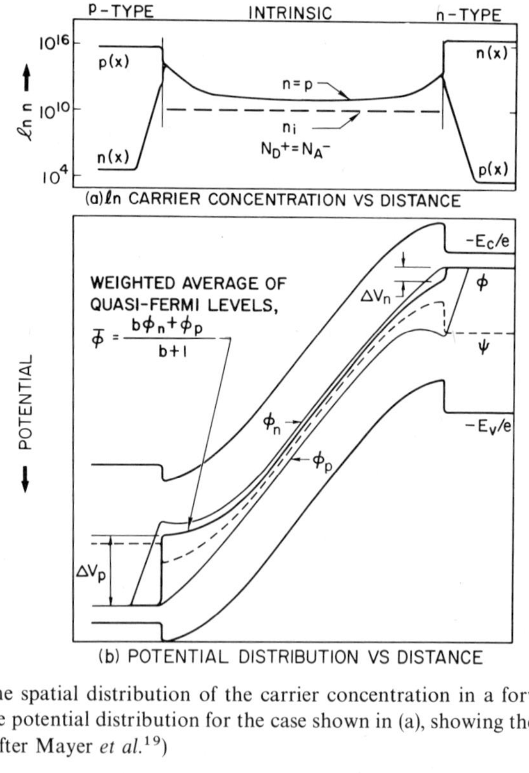

FIG. 62. (a) The spatial distribution of the carrier concentration in a forward-biased *p–i–n* structure. (b) The potential distribution for the case shown in (a), showing the influence of $\bar{\phi}$ on ΔV_n and ΔV_p. (After Mayer *et al.*[19])

These variations in carrier concentration are reflected in the variation of the quasi-Fermi levels ϕ_n and ϕ_p shown in Fig. 62. The weighted average of the quasi-Fermi levels $\bar{\phi}$ has the same general shape of ϕ_n or ϕ_p, except at the junctions, where there is a relatively steep step. The width of the step is determined by the Debye length $X_0 = (\beta\varepsilon/en)^{1/2}$. Note that the value of n to be used in the formula for the Debye length is that given by the carrier concentration at the junction $\tilde{n}_1$. Thus, the width of the step will decrease with increasing injection level. Applying the definition of the quasi-Fermi levels, Eqs. (140) and (147), to the bulk,

$$\bar{\phi} - \psi = -\beta\frac{b-1}{b+1}\ln\frac{\tilde{n}_1}{n_1}. \tag{148}$$

The relationship between the probe potential and carrier concentration in the bulk can be found from Eq. (142) as follows. Since Eq. (142) is general, it can be applied to the current in the bulk. With the quantities in the bulk indicated by tildes, Eq. (142) can be written as

$$\tilde{\mathbf{J}} = -e\mu_p\tilde{p}\,\nabla\tilde{\phi}_p - e\mu_n\tilde{n}\,\nabla\tilde{\phi}_n. \tag{149}$$

It should be noted that $\tilde{\mathbf{J}}$ is the current flowing in the bulk of the sample, and is different from J_y used to indicate the current flowing toward the metal probe. When $\tilde{\mathbf{J}}$ flows in the x direction, and when $\tilde{n}(x) = \tilde{p}(x)$, one can simplify Eq. (149) with the help of Eq. (148) as $\tilde{\mathbf{J}} = -e\mu_p(b+1)\tilde{n}(x)\mathscr{E}^{\dagger}$, where $\mathscr{E}^{\dagger} = d\bar{\phi}(x)/dx$.

This shows that by measuring $\bar{\phi}(x)$ and $\tilde{\mathbf{J}}$, one can derive the carrier concentration $\tilde{n}(x)$ from

$$\tilde{n}(x) = |J[e\mu_p(b+1)\mathscr{E}^{\dagger}]^{-1}|. \tag{150}$$

The following very simple test can be used to determine whether the probe measures the weighted mean of the quasi-Fermi levels $\bar{\phi}$ or the electrostatic potential ψ. Assuming that Boltzmann statistics apply across the junction, the voltage step that the probe would measure if it measured ψ rather than $\bar{\phi}$ is given by

$$\begin{aligned}\Delta V_{p,n}(\text{EP}) &= \phi_p{}^* - \psi = \psi - \phi_n{}^* \\ &= \beta\ln(\tilde{n}_1/n_i)\end{aligned} \tag{151}$$

for both the p–i and n–i junctions. Here, $\phi_p{}^*$ and $\phi_n{}^*$ are the quasi-Fermi levels for the majority carriers in the appropriate regions and $\tilde{n}_1$ is measured at the appropriate junction. On the other hand, when the probe measures $\bar{\phi}$ rather than ψ, the step ΔV_p at the p–i junction is given by [using Eqs. (148)

and (151)]

$$\Delta V_p(\mathrm{QFL}) = \phi_p{}^* - \bar{\phi}$$
$$= \frac{2b}{b+1}\beta \ln \frac{\tilde{n}_1}{n_i} \tag{152}$$

and the step ΔV_n at the n–i junction is given by

$$\Delta V_n(\mathrm{QFL}) = \bar{\phi} - \phi_n{}^*$$
$$= \frac{2}{b+1}\beta \ln \frac{\tilde{n}_1}{n_i}. \tag{153}$$

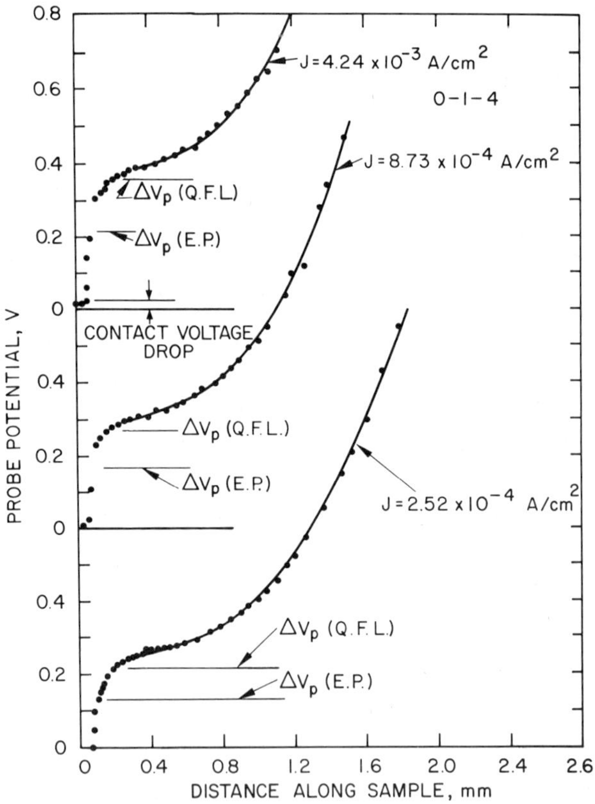

FIG. 63. Detail of the measured probe potential near the p–i junction versus distance along a silicon p–i–n sample biased in the insulator regime for three values of the current density. (After Mayer *et al.*[19])

24. Potential-Step Measurements

The potential distributions near the *n–i* and *p–i* junctions were determined[19] for the *p–i–n* junctions discussed in Section 8. The distribution near the *p–i* junction of 0-1-4 is shown in Fig. 63 for three values of the sample current. The width of the potential step decreased with increased carrier injection levels, as found in all samples and as predicted by the decrease in the Debye length mentioned above. The values of ΔV_p(QFL) and ΔV_p(EP) were calculated from the values of *n* in Fig. 17 (Section 8) extrapolated to the junction, and from Eqs. (152) and (151), respectively. Figure 64 shows the potential distribution near the *n–i* junction in sample 0-1-1. The voltage step is clearly outside experimental error. The values of ΔV_n(QFL) and ΔV_n(EP) calculated from the extrapolated values of the carrier concentration and the *n–i* junction are shown in Fig. 64.

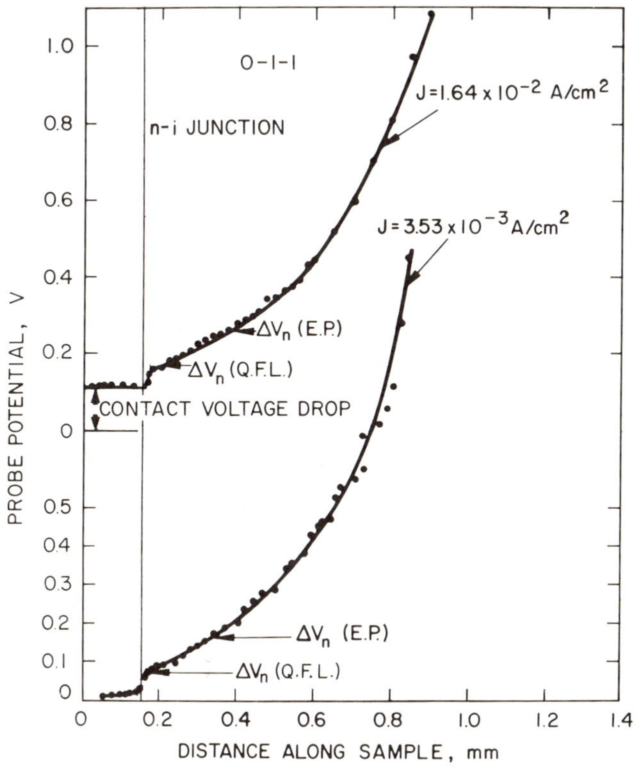

FIG. 64. Detail of the measured probe potential near the *n–i* junction versus distance along a silicon *p–i–n* sample biased in the insulator regime for two values of the current density. (After Mayer *et al.*[19])

In six samples, the carrier distribution followed the exponential decrease over a sufficient range of concentration to permit extrapolation to the junctions. The values of ΔV could be determined from the value of the measured probe potential at the "knee" of the potential distribution with an uncertainty of ± 0.02 V. The values of the carrier concentration at the n–i and p–i junction are shown in Fig. 65 as a function of the potential step ΔV. The values of ΔV calculated using the weighted average of the quasi-Fermi levels and using the change in electrostatic potential are denoted by broken lines in Fig. 65. The experimental points are in close agreement with the values of ΔV calculated from the quasi-Fermi level.

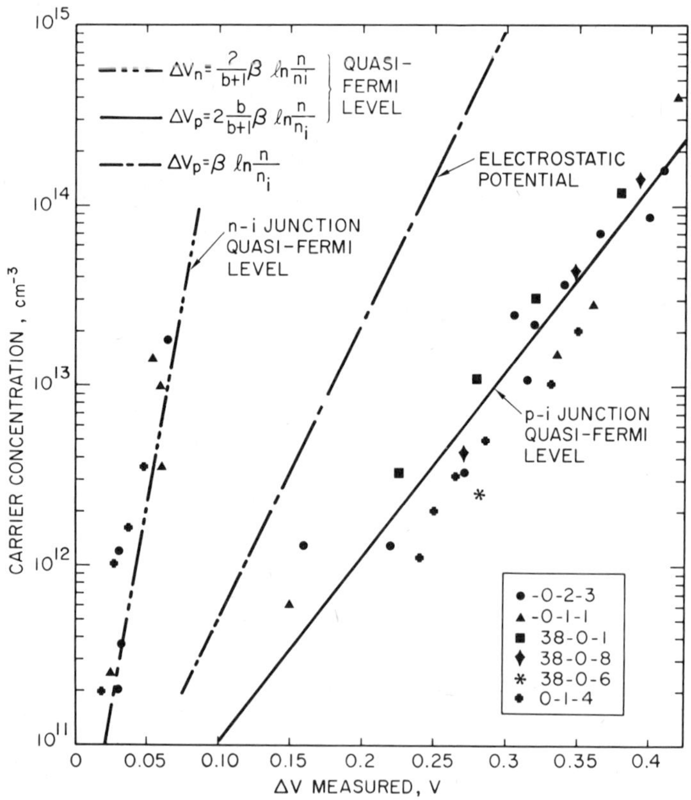

FIG. 65. Plot of the measured values of the carrier concentration at the p–i and n–i junctions versus the potential drop ΔV at the same junction (from data similar to those in Figs. 63 and 64). The straight lines represent, from left to right, the theoretical relations in Eqs. (153), (151), and (152). (After Mayer *et al.*[19])

It should be noted here that for these samples, the maximum value of $d\tilde{n}/dx$ was small enough to be neglected in the current equation, so that

$$\frac{d\bar{\phi}}{dx} = \frac{d\psi}{dx} - \beta\frac{b-1}{b+1}\frac{1}{\tilde{n}}\frac{d\tilde{n}}{dx} \approx \frac{d\psi}{dx}$$

from Eq. (148); Eq. (150) gives the same value for $\tilde{n}(x)$ whether the probe measured ψ or $\bar{\phi}$. Since the ΔV are direct experimental quantities and since the calculation of the $\tilde{n}_1$ from direct experimental quantities does not depend on the assumption to be checked, the experimental relation between ΔV and $\tilde{n}_1$ can be used to establish the validity of a theoretical relation and thus the validity of the assumption on which it is based.

Thus, the close agreement of the experimental data with the theoretical relations given in Eqs. (152) and (153) and the violent disagreement with the theoretical relations given in Eq. (151) strongly confirm that the probe measures $\bar{\phi}$ and not ψ for the case $n \approx p$.

List of Symbols[85b]

a	Half-width of slab, see Section 6
b	μ_n/μ_p
B	Susceptance of sample
B_0, B_1	Coefficients in the small-signal expansion of B (Section 19 only)
c_n, c_p	$\langle v_{th}\sigma_n\rangle, \langle v_{th}\sigma_p\rangle$
C_0	Geometrical capacitance
D_a	Ambipolar diffusion constant
e	Electronic charge
$\mathscr{E}$	Electric field
$\mathscr{E}^\dagger$	Measured electric field
$\mathscr{E}_0{}^\dagger$	Maximum value of $\mathscr{E}^\dagger$, see Eq. (40)
$\mathscr{E}^*$	$\mathscr{E}^\dagger/E$ (normalization constant)
$\mathscr{E}_i, \mathscr{E}_f$	Initial, final steady-state values of $\mathscr{E}$
$\mathscr{E}_1$	Value of $\mathscr{E}$ at $t = t_1$ (Section 18 only)
$\mathscr{E}_0, \mathscr{E}_1$	Coefficients in small-signal expansion of $\mathscr{E}$ (Section 19 only)
E	Reduced value of $\mathscr{E}^\dagger$, see Section 5c
E_0	Maximum value of E
E_F	Fermi level
E_R	Recombination-center energy level
f	Frequency
Δf	Frequency bandwidth of noise measurement, see Section 20.
G	Conductance of sample
G_0, G_1	Coefficients in the small-signal expansion of G
$\langle i^2\rangle$	Mean-square fluctuation in I, see Eq. (138)
I	Total current
I_0, I_1	Coefficients in the small-signal expansion of I (Section 19 only)
I_{eq}	Equivalent noise current, see Section 20.
J	Total current density
J_n, J_p	Electron, hole current densities
J_{diff}	Diffusion-driven contribution to J
J_{field}	Field-driven contribution to J
$J_{i,f}$	Initial, final steady-state values of J
J_{tr}	Transition current density, see Eq. (29)
J_{SCL}	Steady-state current density for single-carrier, trap-free, space-charge-limited current
J_0	$\varepsilon\mu V^2/L^3$ (Section 21 only)
J_1	Value of J at $t = t_1$ (Section 18 only)
J_0, J_1	Coefficients in small-signal expansion of J (Section 19 only)

[85b]Symbols that are defined and used only in a short space of text are not listed here.

$\mathscr{J}$ Reduced value of J, defined in Section 5c for Part II

$\mathscr{J}$ Reduced value of J, defined in Section 11a for Part IV

k Boltzmann's constant

K $eN_R n_{R,0} c_n/b$

K' Constant, see Eq. (36)

K'' $ebn_{R,0}^2 N_R c_n/p_0$

l_1 Inductance in equivalent circuit for the small-signal impedance (Section 19 only)

L Sample length

L_a Ambipolar diffusion length

L_{eff} Effective length $= L - Q(2L_a)$

L_n, L_p Electron, hole diffusion lengths

L' Length of diffusion wing, see Section 5a

n_0, p_0 Thermal equilibrium electron, hole densities (except in Section 19)

$\delta n, \delta p$ Excess injected electron, hole densities

n, p Total electron, hole densities

n_0, n_1 Coefficients in small-signal expansion of n (Section 19 only)

n_1, p_1 Electron, hole densities for $E_F = E_R$ (except in Sections 18, 19)

n_1 Value of n at $t = t_1$ (Section 18 only)

n_2, p_2 Constants, see Eq. (52)

n_3, p_3 Constants, see Eq. (52)

n_i Intrinsic value of n_0, p_0 (except in Section 18)

n_i, n_f Initial, final steady-state values of $N(x)$ (Section 18 only)

n_{min} Minimum value of n

n_n, p_n Electron, hole densities in n region

n_p, p_p Electron, hole densities in p region

n_R, p_R Electron, hole densities on recombination centers ($n_R + p_R \equiv N_R$)

$n_{R,0}, p_{R,0}$ Thermal equilibrium values of n_R, p_R

n^* n/η, normalization constant

N_A Density of acceptor levels

N_D Density of donor levels

N_R Density of recombination centers

δp_R Change in the density of holes occupying recombination centers from thermal equilibrium

$\mathscr{P}_i$ Constant, see Section 5b

Q Constant in definition of L_{eff} (except in Part VI)

Q $|B_1|/G_1$, small-signal figure of merit (Part VI only)

r Steady-state recombination rate density (except in Section 19)

r, r_1 Resistances in equivalent circuit for the small-signal impedance (Section 19 only)

r_n, r_p Electron, hole recombination rate densities

R_0 V_0/I_0 (Section 19 only)

s Surface recombination velocity

t Time

t_1 Defined as $\Theta \ll t_1 \ll \tau$ (Section 18 only)

t_1 Transit time of leading edge of injected space charge (Section 21 only)

t_r $L^2/\mu V$, transit time in uniform field (Section 21 only)

T Absolute temperature, °K

u $\delta p/\delta n$

u_0 n_3/p_3, thermal equilibrium value of u

v $u - u_0$

v_{th} Thermal velocity of carriers

V Applied voltage, measured potential

V_1 Limiting voltage for pure diffusion-dominated case, see Eq. (27a) (except in Sections 18 and 19)

V_1 Value of V at $t = t_1$ (Section 18 only)

V_0, V_1 Coefficients in small-signal expansion of V (Section 19 only)

V_B Breakdown voltage, see Section 16

V_i, V_f Initial, final steady-state values of V

V_J Voltage drop across junction

$V_{J,0}, V_{J,1}$ Coefficients in the small-signal expansion of V_J (Section 19 only)

V_{TFL} Voltage at which the SCLC trap-filled-limit transition takes place

$\mathscr{V}$ Reduced value of V, see Section 11a

x Position along direction of current flow

y x/L_a, reduced value of x (except in Sections 6, 22, and 23)

y	Position transverse to direction of current flow (Sections 6, 22, and 23 only)
y_2	$x/2L_a$
y'	Reduced value of x, see Eq. (13)
y'_L	Value of y' at boundary
Y	Admittance of sample
Y_0, Y_1	Coefficients in the small-signal expansion of Y (Section 19 only)
z	$(x/L)^{1/2}$
Z	Impedance of sample
Z_0, Z_1	Coefficients in the small-signal expansion of Z (Section 19 only)
α_0	Constant, see Section 6
β	kT/e (except in Part IV)
β	c_n/c_p (in Part IV only)
γ	Constant, see Eq. (37)
ε	$\kappa\varepsilon_0$, permittivity of sample
η	Reduced value of n, Section 5a
η_0	Minimum value of η
Θ	$\frac{3}{2}C_0R_0$, average high-level dielectric relaxation time
Θ'	High-level local dielectric relaxation time
Θ_i	Intrinsic dielectric relaxation time
$\Theta_{p,0}$	$\varepsilon/e\mu_p p_0$, low-level dielectric relaxation time for p-type material
$\Theta_{\Omega,n}$	$\varepsilon/e\mu_n\|n_0 - p_0\|$, "ohmic" dielectric relaxation time
μ_n, μ_p	Electron, hole mobilities
σ_n, σ_p	Electron, hole capture cross sections
τ	High-level lifetime ($\delta n/r$)
τ_c	Trapping time constant (see Section 21)
τ_{eff}	Steady-state effective value of τ for sample with finite cross section
$\tau_0{}^*$	Decay time constant (lowest-order mode) for sample with finite cross section
τ_1	Low-level lifetime (minority carrier)
τ_n, τ_p	Electron, hole lifetimes
$\tau_{n,0}, \tau_{p,0}$	Thermal equilibrium values of τ_n, τ_p
$\bar{\phi}$	Average of quasi-Fermi levels that gives the measured potential, see Eq. (147)
ϕ_n, ϕ_p	Electron, hole quasi-Fermi levels
ψ	Electrostatic potential
ω	$2\pi f$

CHAPTER 5

The Photoconductor–Metal Contact

W. Ruppel

I. Introduction

Photoconduction in semiconductors occurs due to a change of the free-charge carrier concentration upon irradiation with light or with particles, such as electrons. Although the mobility of the free carriers is also influenced by light,[1] this change may, in general, be neglected with respect to that of the free-carrier concentrations. The total current in a photoconductor of uniform temperature is carried by both electrons and holes. Each of these currents is proportional to the gradient of the electrochemical potential of the electrons or holes, E_{fn} or E_{fp}, respectively. The total current density i is

$$i = ne\mu_n \operatorname{grad} E_{fn} + pe\mu_p \operatorname{grad} E_{fp}, \tag{1}$$

where n and p denote the total (dark + light) carrier concentrations and μ_n and μ_p the free-carrier mobilities. If only one kind of free carrier, e.g., the

[1] R. H. Bube and H. E. MacDonald, *Phys. Rev.* **121**, 473 (1961).

electrons, contributes to the total current, only one of the two terms in Eq. (1), e.g., the first, needs to be considered.

In the volume of a uniform photoconductor, if contact influences may be neglected, the free-carrier concentration, and with it the chemical potential of the electrons, is constant. If the free-electron concentration is written as the product of their generation rate g_n and their lifetime τ_n, the current i of Eq. (1) reduces to

$$i = eg\mu_n\tau_n F, \tag{2}$$

where, under these conditions, the field strength F is equal to the gradient of any of the potentials shown in Fig. 1.

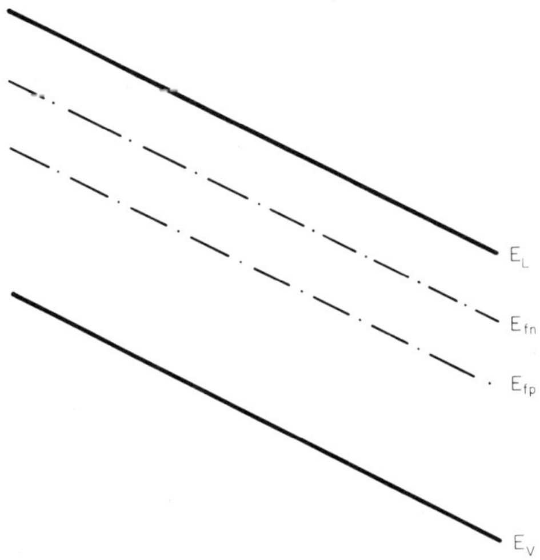

FIG. 1. Electrostatic potentials E_L and E_V and electrochemical potentials E_{fn} and E_{fp} of the free electrons and holes in the volume of an illuminated uniform photoconductor under applied voltage. The abscissa is distance along the specimen. Subscripts L and V refer to conduction and valence bands, respectively.

Equation (2) shows that the quality of a photoconductor is best characterized by its majority-carrier range per unit field $\mu_n\tau_n$. The μ_n accounts for the limitation of this range by scattering, and the τ_n by loss of the free carrier.

The application of Eq. (2) to the evaluation of a photoconductor therefore requires the experimental realization of the potential distribution of a homogeneous photoconductor of Fig. 1. Ryvkin[2] discusses the avoidance of the

[2] S. M. Ryvkin, "Fotoelektricheskie Iavleniia v Poluprovodnikakh." Fizmatgiz, Moscow, 1963 [*English Transl.*: "Photoelectric Effects in Semiconductors," pp. 27–32. Consultants Bureau, New York, 1964].

inhomogeneity of a contact by choosing a sufficiently long sample and by probe measurements. Diemer[3] proposed the use of a ring-shaped photoconductor as the secondary winding of a transformer. Ogawa[4] and Carrelli *et al.*[5] measured the change in conductivity under illumination by a change of the torque on a photoconductor sample in a rotating electric field. Also, the PEM effect in photoconductors may be detected without electrodes applied to the sample.[6]

A metal contact introduces an inhomogeneity into the photoconductor. If we assume the carrier concentration at the photoconductor–metal interface to be fixed to its dark value due to a high interface recombination rate, the potential distribution of an n-type photoconductor in the vicinity of the cathode under applied voltage will look as shown in Fig. 2.

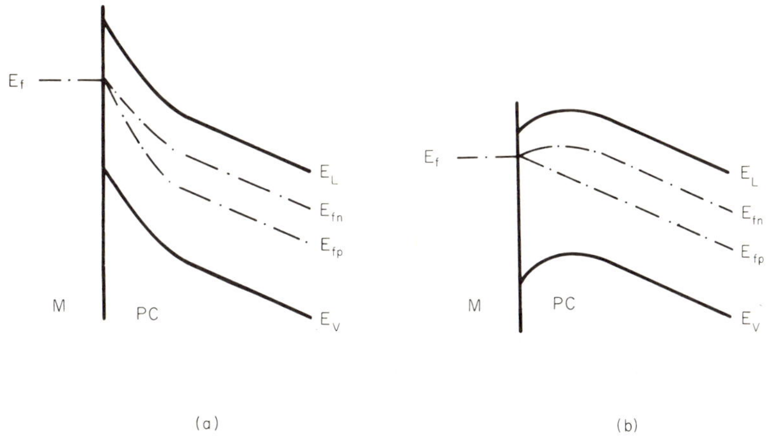

FIG. 2. Potential distribution at the illuminated photoconductor–metal contact under applied voltage. (a) Blocking contact. (b) Ohmic contact.

The contact space-charge layer extends more deeply into the photoconductor for insulating than for highly conducting material. Complications of the simple uniform-field case are expected for insulating large-band-gap photoconductors, with which this chapter will be mainly concerned. The position of E_{fn} and E_{fp} in a space-charge layer at the surface for Ge- and Si-type photoconductors has been discussed by Zuev.[7]

[3] G. Diemer, *Philips Res. Rept.* **18**, 127 (1963).

[4] T. Ogawa, *J. Appl. Phys.* **32**, 584 (1961).

[5] A. Carrelli, F. Fittipaldi, and L. Pauciulo, *J. Phys. Chem. Solids* **28**, 297 (1967).

[6] J. Hlávka, *Phys. Letters* **27A**, 131 (1968).

[7] V. A. Zuev, *Ukr. Fiz. Zh.* **13**, 38 (1968).

First, the photoconductor metal contact will be considered under the application of an externally applied voltage. The impact of carrier injection and extraction by the contacts on the photoconductive gain will be discussed. The experimental realization of injection and extraction by ohmic and blocking contacts will be reviewed for the different types of photoconduction measurements and photoconductor materials. Applications to devices will not be described in detail, nor will the problem of noise be covered. For two accounts of this topic, one detailed and the other more general, the reader is referred to the work of van Vliet[8] and Rose.[9]

Second, the illuminated photoconductor–metal contact without externally applied voltage will be discussed. The potential distribution at the contact is mainly inferred from observations of the photo-emf and of the total sample charge as a function of illumination.

II. The Photoconductor–Metal Contact under Externally Applied Voltage

1. General Survey

While the excess carriers in a photoconductor are, in general, created in the volume of the photoconductor, the resulting photocurrent under applied voltage depends not only on the $\mu\tau$ product of the majority carriers, but also, to a large extent, on the potential distribution in the neighborhood of the contacts. In particular, the contact determines the maximum achievable gain; i.e., the maximum number of elementary charges that are displaced across the two electrodes per incident and absorbed photon. The maximum gain achievable for different contacts will be discussed in Section 2. Measured gains are reviewed in Section 3, and the experimental methods of applying ohmic and blocking contacts to various photoconductive materials are outlined in Section 4. At sufficiently high applied voltage, the gain for operation with ohmic contacts is limited by majority-carrier injection. The injected carriers are mainly bound in localized states. Therefore, the occupancy of the states in the forbidden gap is altered by injection, and, consequently, the spectral response of a photoconductor in the long-wavelength part of its spectrum may be altered. This will be shown in Section 5. Under the action of a blocking contact, carriers are extracted from the photoconductor and the maximum gain is limited to unity because one elementary charge carrier may at maximum be displaced across the distance between the two electrodes. The transit time of a carrier generated near one electrode may be measured, and, from it, the mobility of the charge carrier may be computed.

[8] K. M. van Vliet, *Appl. Opt.* **6**, 1145 (1967).

[9] A. Rose, "Concepts in Photoconductivity and Allied Problems," pp. 97–117. Wiley (Interscience), New York, 1963.

This detection of transient photocurrents under light or electron-pulse operation has become a powerful tool for the determination of mobilities in low-mobility insulating materials (Section 6). Another manifestation of transient photocurrents is in photosensitive dielectrics of capacitors, which will also be briefly reviewed in Section 6.

Some influences of the illuminated metal–photoconductor system on the conduction through the photoconductor are schematically summarized in Table I.

TABLE I

THE PHOTOCONDUCTOR–METAL CONTACT UNDER APPLIED VOLTAGE

Conditions	References		
Ohmic contact (Photocurrent saturation due to carrier injection)	10–14, 16, 18–21, 27, 31–35, 46, 48, 50–69, 100–103	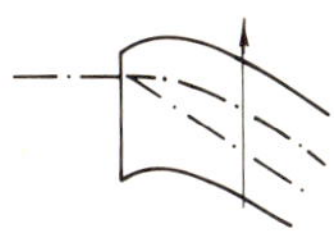	1
Blocking contact (Photocurrent saturation due to carrier extraction)	17, 22–30, 34, 35, 55 66, 71, 72, 76–78	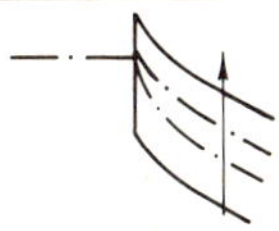	2
Photoemission of majority carriers into the photoconductor	90–96, 132, 138, Chapter 2 in this volume	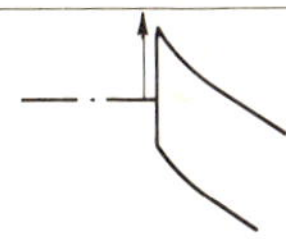	3
Creation of free carrier reservoir by excitation in the photoconductor	49, 111–131	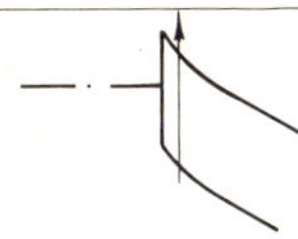	4
Long-wavelength sensitization by space-charge-limited trap filling	97–99, 104–109	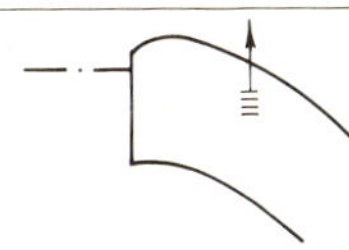	5

2. MAXIMUM PHOTOCONDUCTIVE GAIN

The importance of the contacts for the performance of a photoconductor can, in principle, be best understood by considering the photoconductive gain. First, let the voltage applied to the photoconductor be small enough that the free-carrier concentrations in the photoconductor do not depend on the applied voltage, but only on the incident light intensity. Let it, on the other hand, be large enough that the gradient of the electrostatic potential may be considered as the driving force for the total current. This condition is certainly fulfilled when the applied voltage is large with respect to any photo-emf arising at the contacts, the latter being due to nonbalancing counteracting electrostatic and chemical potential gradients of the electron and the hole gas. If, further, the contribution of only the majority carriers to the photocurrent is considered, the gain may be expressed by either the ratio of the majority-carrier lifetime τ_{maj} to the majority-carrier transit time T_{maj} across the interelectrode distance L, or as the ratio of the majority carrier range S_{maj} to L, where S_{maj} is the displacement of the majority carrier in the electric field during its lifetime τ_{maj}. It should be noted that again the physically relevant parameter is, in any event, the displacement of the charge carrier per unit field, i.e., its $\mu\tau$ product. If relations are considered that concern only a single carrier type, the range relations may be reduced to relations expressed in terms of either times or distances.

Although the carrier transit time between the electrodes is inversely proportional to the applied voltage, the gain does not increase indefinitely when the applied voltage is increased. The maximum achievable gain is closely connected with the injection or extraction of both majority and minority carriers which occur at a sufficiently elevated voltage. This connection was first noted and elaborated in detail by Rose,[10] Stöckmann,[11] and Redington,[12] and forms an essential part of many monographs on photoconductivity.[9,13–15] The relations between the gain and carrier injection and extraction apply to arbitrary excitation rates[16] and to arbitrary trap depths of the photoconductor.[17]

In the following, the essential physical features leading to the gain limitation and therefore to saturation of the photocurrent will be outlined. The results are summarized in Table II.

Let the voltage be raised until the minority-carrier transit time T_{min} becomes short with respect to the minority-carrier lifetime τ_{min}. This condition is equivalent to L being small with respect to the minority-carrier range $S_{min} = \mu_{min}\tau_{min}F$. The total conductivity is assumed to be large enough that no space charge is formed in the photoconductor due to injection or extrac-

[10] A. Rose, *Helv. Phys. Acta* **30**, 242 (1957).

[11] F. Stöckmann, *Z. Physik* **147**, 544 (1957).

[12] R. W. Redington, *J. Appl. Phys.* **29**, 189 (1958).

[13] R. H. Bube, "Photoconductivity of Solids." Wiley, New York, 1960.

[14] F. Stöckmann, *in* "Halbleiterprobleme" (F. Sauter, ed.), Vol. 6, p. 279. Vieweg, Braunschweig, 1961.

[15] L. Heijne, *in* "Festkörperprobleme" (O. Madelung, ed.), Vol. 6, p. 127. Vieweg, Braunschweig, 1967.

[16] F. Stöckmann, *Z. Physik* **180**, 184 (1964).

[17] H. Kiess, *J. Phys. Chem. Solids* **28**, 1473 (1967).

TABLE II

THE PHOTOCONDUCTIVE GAIN IN DIFFERENT INJECTION RANGES

Voltage range	Conditions	Gain
$T_{maj} \gg \tau_{RC}$, $T_{min} \gg \tau_{min}$	No injection	τ_{maj}/T_{maj}
$T_{maj} \gg \tau_{RC}$, $T_{min} \ll \tau_{min}$	Injection of minority carriers	μ_{maj}/μ_{min}
$T_{maj} \ll \tau_{RC}^0$, $T_{min} \gg \tau_{min}$	Injection of majority carriers	τ_{maj}/τ_{RC}

tion of majority carriers. This implies that T_{maj} is large with respect to the dielectric relaxation time τ_{RC} of the photoconductor, or, in terms of characteristic lengths, that L is large with respect to $\mu_{maj}\tau_{RC}F$.

The minority-carrier lifetime τ_{min} being limited for both an ohmic and a blocking contact to the minority carrier transit time T_{min}, the gain contribution of majority-carrier charge that may be displaced across the space-charge-free photoconductor during a minority-carrier transit equals the ratio of the two mobilities μ_{maj}/μ_{min}. The ratio of the mobilities represents the maximum gain achievable under the given conditions if the contribution of the minority carriers to the photocurrent is neglected. It corresponds to the gain obtained by minority-carrier injection into the base of a bipolar transistor.

If the photoconductor is sufficiently insulating and the minority-carrier range is short, injection or extraction sets in for majority rather than for minority carriers. This condition may be characterized by T_{maj} being short with respect to the dielectric relaxation time τ_{RC}^0 of the low-voltage, space-charge-free regime, or by the condition that the space-charge layer which emerges from the majority-carrier-emitting contact extends across the entire photoconductor. In case of an ohmic majority-carrier-emitting contact, the space charge has the same sign as the majority carriers. For a blocking contact, it is of the opposite sign. Since T_{maj} is equal to τ_{RC} in the space-charge-limited voltage regime, the maximum gain is limited to τ_{maj}/τ_{RC}. This result applies to both majority-carrier injection and extraction. In the case of majority-carrier injection, the gain may greatly exceed unity. It was pointed out by Rose and Lampert[18] that gains exceeding this ratio by a factor $M > 1$ are to be expected if the traps in which part of the injected space charge is localized are not identical with the traps that determine the rise and decay rate of the photocurrent upon a change in illumination. For particular sets of

[18] A. Rose and M. A. Lampert, *Phys. Rev.* **113**, 1227 (1959).

traps and recombination centers, the maximum gain was also numerically computed by Harth[19] and Harth and Dommaschk.[20]

In the case of majority-carrier extraction due to a blocking majority-carrier-emitting contact, τ_{maj} is limited to T_{maj}, and, therefore, to τ_{RC}. The maximum gain is unity.

In addition to determining the photoconductive gain, it was pointed out by Lampert and Rose[9,21] that the contact to a photoconductor may also determine the response time of the photoconductor. When the illumination is changed, the new steady-state photocurrent will require a different height of the contact potential barrier, and therefore a charge relaxation at the contact, which may become the response-limiting process, in contrast to the response of the volume of the photoconductor.

3. Photoconductive Gain—Experimental

Photocurrents corresponding to unity gain, called "primary photocurrents," were identified by Gudden, Pohl, and co-workers in a number of substances, such as diamond and ZnS,[22] the alkali halides,[23] and the Ag and Tl halides.[24,25] Primary photocurrents in CdS with metal contacts were discussed by van Heerden.[26] Heijne[27] observed primary photocurrents carried by holes in p-PbO layers sandwiched between a conducting SnO_2 glass coating and an evaporated Ag electrode. For layers converted to n-type but using the same electrodes, an electron photocurrent with a gain greater than unity appeared. The observation that contacts blocking to one type of carriers may be ohmic for the opposite carrier type is important for the achievement of double injection, and will be discussed in Section 4.

Photocurrents with unity gain are also observed with nonmetallic contacts. An example is the photoconductive layer in a Vidicon-type television-camera tube, where the contact is formed by an electron beam, or the electrophotographic layer, where the contact is formed by energetically deep-lying ions which are applied to the layer by a corona discharge. It is remarkable that such photoconductors whose $\mu\tau$ product differs by several orders of magni-

[19] W. Harth, *Z. Physik* **184**, 198 (1965).

[20] W. Harth and W. Dommaschk, *Z. Naturforsch.* **20a**, 1313 (1965).

[21] M. A. Lampert and A. Rose, *Phys. Rev.* **113**, 1236 (1959).

[22] B. Gudden and R. W. Pohl, *Z. Physik* **16**, 170 (1923); **30**, 14 (1924).

[23] R. W. Pohl, *Physik. Z.* **39**, 36 (1938).

[24] K. Hecht, *Z. Physik* **77**, 235 (1932).

[25] W. Lehfeldt, *Nachr. Ges. Wiss. Göttingen, Math.-physik. Kl., Fachgruppe II* **1**, 171 (1935).

[26] P. J. van Heerden, *Phys. Rev.* **106**, 468 (1957).

[27] L. Heijne, *J. Phys. Chem. Solids* **22**, 207 (1962).

tude as CdS,[28] ZnO,[29] and amorphous Se[30] all exhibit the same order of sensitivity when operated by photoinduced discharge in the Electrofax or xerography process. The requirement for the gain of unity, that the transit time of the majority carriers be shorter than their lifetime, is fulfilled for the thin layers of all these materials.

If the majority-carrier-supplying contact is ohmic and the majority-carrier lifetime is longer than the transit time, the gain exceeds unity, and the corresponding photocurrents are called "secondary photocurrents." Gains of the order of 10^4 were observed by Lappe[31] for evaporated CdS layers under electron excitation, and M-factors up to $M = 500$ were reported for CdS single crystals with melted In contacts by Bube and Barton.[32]

The saturation of photocurrent corresponding to a limitation of the gain by extraction of minority carriers due to a blocking minority-carrier-supplying contact was observed by Kiess[33] for Cd_4GeS_6. In this case, as in the case where there is minority-carrier injection under space-charge-free conditions because the majority-carrier transit time is shorter than the dielectric relaxation time, the observed gain was limited to the ratio μ_{maj}/μ_{min}. The depletion of majority carriers concomitant with the extraction of minority carriers was demonstrated by showing that there is an increase of the sample resistance measured transversely to the photocurrent path with the onset of saturation.

The character of a contact as ohmic or blocking with respect to a charge carrier, and the corresponding current–voltage characteristic of the photocurrent, may change with varying illumination intensity of the photoconductor or with varying applied voltage. In particular, increasing light intensity at constant voltage may cause a contact to change from ohmic to blocking. This happens if, at low light levels, the majority-carrier concentration in the volume is smaller than that at the surface, but, for high light levels, becomes larger than the illumination-independent concentration at the surface. The transition from majority-carrier accumulation to majority-carrier depletion is connected with a change of sign of the total crystal charge, as will be discussed in Section 9 for the case of the photoconductor without applied voltage. For constant applied voltage, this transition from an ohmic to a blocking contact has been demonstrated for CdS with In electrodes by the transition from negative to positive crystal charge with increasing illumination[34] (Fig. 3).

[28] M. Smith and A. J. Behringer, *J. Appl. Phys.* **36**, 3475 (1965).

[29] H. J. Gerritsen, W. Ruppel, and A. Rose, *Helv. Phys. Acta* **30**, 504 (1957).

[30] H. T. Li and P. J. Regensburger, *J. Appl. Phys.* **34**, 1730 (1963).

[31] F. Lappe, *Z. Physik* **154**, 267 (1959).

[32] R. H. Bube and L. A. Barton, *RCA Rev.* **20**, 564 (1959).

[33] H. Kiess, *J. Phys. Chem. Solids* **28**, 1465 (1967).

[34] W. Ruppel, *J. Phys. Chem. Solids* **22**, 199 (1962).

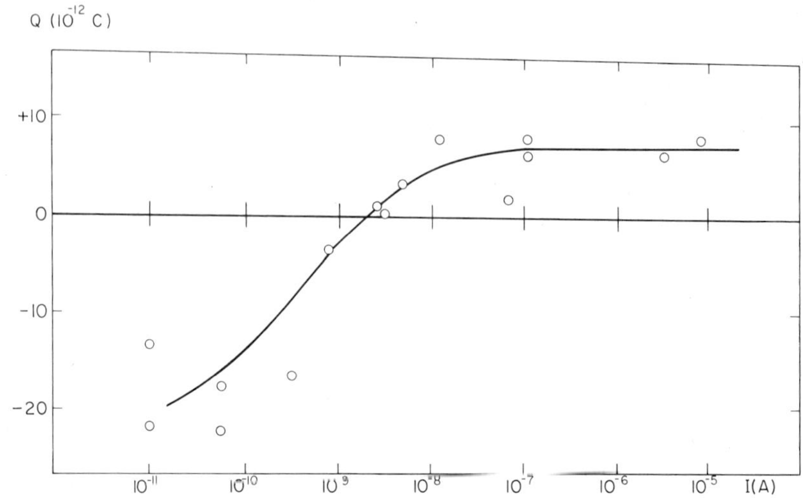

FIG. 3. Total charge of a CdS crystal versus photocurrent through the crystal at 100 V, indium electrodes. (After Ruppel.[34])

The transition of photocurrents from more-than-linear as a function of applied voltage to less-than-linear with increasing illumination as a consequence of majority-carrier exhaustion at the contact has also been observed for other photoconductors; e.g., crystalline Se.[35] Figure 4 shows the potential distribution for a crystalline Se layer with gold electrodes for four different illumination intensities and different voltages. The upper left picture indicates, by the curvature of the potential distribution at high voltages, space-charge-limited hole injection into the *p*-Se layer in the dark. The lower pictures show negative space charge at high light intensity, due to the conversion of the anode from ohmic to blocking and the corresponding extraction of holes from the Se.

The measurement of the potential distribution represents generally a sensitive method by which those current-saturation mechanisms in photoconductors that are not due to contact-controlled gain limitation may be distinguished from those mechanisms discussed here. The potential distribution that accompanies current saturation by traveling-wave amplification of acoustic waves in piezoelectric photoconductors, such as CdS, when the drift velocity of the free-charge carriers exceeds the phase velocity of the acoustic wave, first observed by Smith,[36] was measured by In probes distributed uniformly along the sample[37] or by detecting the linear electrooptical effect

[35] M. Polke, *Phys. Status Solidi* **5**, 279 (1964).

[36] R. W. Smith, *Phys. Rev. Letters* **9**, 87 (1962).

[37] J. H. MacFee and P. K. Tien, *J. Appl. Phys.* **37**, 2754 (1966).

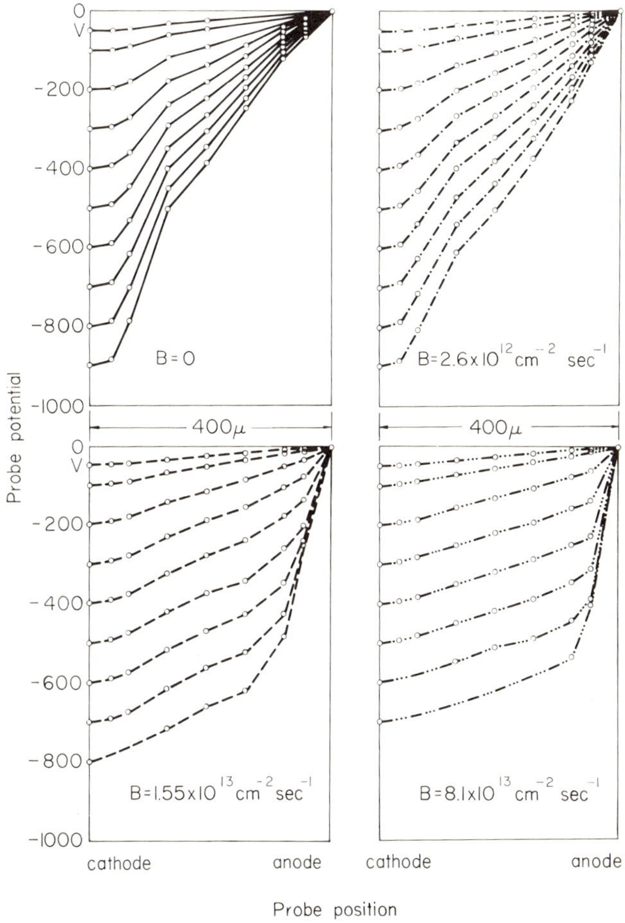

FIG. 4. Potential distribution in a crystalline selenium layer with gold electrodes for different light intensities *B* and different applied voltages. The potential distribution was determined by probe measurements. (After Polke.[35])

using a He–Ne laser.[38] Kiess[39] also observed, in CdS, a photocurrent that was saturated as a function of the applied voltage. The measurement of the potential distribution showed, however, that the saturation could neither be ascribed to contact influences nor to the acoustoelectric effect. Rather, the

[38] H. J. Fossum and A. Rannestad, *J. Appl. Phys.* **38**, 5177 (1967).

[39] H. Kiess, *Phys. Status Solidi* **4**, 107 (1964).

crystal divided into a high- and a low-field domain corresponding to a low- and a high-conductivity region. The relative width of these two domains depends on the applied voltage.[40] This type of current saturation is due to negative differential conductivity in the photoconductors, which comes about from field-enhanced recombination (field quenching).[41] This effect can be used to determine, quite accurately, metal–photoconductor work functions at high current densities.[41a,41b]

4. Electrode Materials

a. *Inorganic Photoconductors*

In narrow-band-gap photoconductors, in which the photocurrent is only a small perturbation of the dark current because of the high dark conductivity, the contact to the photoconductor poses no particular problems that would go beyond those of rectifier and transistor technology. In large-band-gap photoconductors, among the binary III–V and II–VI compounds or ternary compounds, the performance of a photoconductor depends critically on the contact properties. In the following, the discussion will be limited to large-band-gap photoconductors.

(1) *Ohmic Contacts.* Concerning III–V compounds, the application of ohmic contacts to GaAs in semiconducting devices was discussed by Libov *et al.*[42] These authors review solubility, melting point, vapor pressure, electrical and thermal conductivity, and thermal expansion of metals with respect to the formation of a majority-carrier accumulation layer. Metal combinations for the formation of ohmic contacts on *n*- and *p*-type GaAs by chemical deposition are reported by Goldberg *et al.*[43] Ohmic and blocking alloyed contacts to GaP are described by Ignatkina *et al.*[44] Zinc as an acceptor material and Te as a donor are applied by fusion in combination with other metals. The ohmic and blocking action of the contacts is demonstrated by current–voltage characteristics and electroluminescent light output.

[40] H. Kiess and F. Stöckmann, *Phys. Status Solidi* **4**, 117 (1964).

[41] K. W. Böer, *Phys. Rev.* **ω39**, A1949 (1965).

[41] K. W. Böer, *Phys. Rev.* **139**, A1949 (1965).

[41a] K. W. Böer, G. A. Dussel, and P. Voss, *Phys. Rev.* **179**, 703 (1969).

[41b] R. J. Stirn, K. W. Böer, G. A. Dussel, and P. Voss, *Proc. 3rd Intern. Conf. on Photoconductivity* (to appear as Supplement, *J. Phys. Chem. Solids*).

[42] L. D. Libov, S. S. Meskin, D. N. Nasledov, V. E. Sedov, and B. V. Tsarenkov, *Pribory i Tekhn. Eksperim.* No. 4, 14 (1965) [*English Transl.*: *Instr. Exptl. Tech.* (*USSR*) No. 4, 746 (1966)].

[43] Yu. A. Goldberg, D. N. Nasledov, and B. V. Tsarenkov, *Pribory i Tekhn. Eksperim.* No. 4, 189 (1966) [*English Transl.*: *Instr. Exptl. Tech.* (*USSR*) No. 4, 969 (1966)].

[44] R. S. Ignatkina, L. D. Libov, and S. S. Meskin, *Pribory i Tekhn. Eksperim.* No. 3, 242 (1965) [*English Transl.*: *Instr. Exptl. Tech.* (*USSR*) No. 3, 726 (1965)].

For large-band-gap II–VI compounds, it was shown by Mead[45] that, due to the largely heteropolar binding character of these compounds, the potential distribution in the photoconductor in contact with a metal is determined by the metal work function in conjunction with the photoconductor work function and electron affinity, and not by metal–photoconductor interface states. Consequently, low-work-function metals are expected to form majority-carrier accumulation layers, and therefore ohmic contacts to n-type photoconductors.

It was shown in the preceding section that in order to achieve high-gain photoconduction, for the majority-carrier injection from an ohmic contact, a low-resistance region in the photoconductor adjacent to the metal is required. It should be mentioned that, according to Stöckmann,[46] a low contact resistance, and, therefore, space-charge-limited majority-carrier injection, may be obtained for blocking contacts also. This is the case if, above a certain field at the contact, the current increases rapidly as a function of field strength, as, for instance, in the case of Schottky emission or tunneling. Thus, according to Shivonen and Boyd,[47] a low contact resistance to CdS may be formed with almost any metal contact by tunneling through the surface barrier with an intense current pulse.

The concept of using a low-work-function metal as an ohmic contact to an n-type photoconductor led Smith[48] to use In and Ga as ohmic contacts to CdS. Gallium and, particularly, In have been used successfully as ohmic contacts not only to CdS crystals for the observation of steady-state[32] and of transient[49] photocurrents in CdS, but for almost any large-band-gap n-type photoconductor. Indium is also observed to act as a reliable ohmic contact in CdS thin-film transistor devices,[50,51] even when the CdS is deposited by a chemical spraying technique.[52,53] It was pointed out by Diemer[54] that In acts as an ohmic contact possibly also by diffusion of In metal into CdS, which occurs at moderate temperatures. Indium gives rise to shallow donors in CdS, and the increased electron concentration might, at least partly, be due to a higher donor concentration at the surface than in the volume of the CdS. The effect of various application techniques of In to CdS, such as soldering

[45] C. A. Mead, *Solid-State Electron.* **9**, 1023 (1966).
[46] F. Stöckmann, *Phys. Status Solidi* **3**, 221 (1963).
[47] Y. T. Shivonen and D. R. Boyd, *J. Appl. Phys.* **29**, 1143 (1958).
[48] R. W. Smith, *Phys. Rev.* **97**, 1525 (1955).
[49] W. E. Spear and J. Mort, *Proc. Phys. Soc.* (*London*) **81**, 130 (1963).
[50] P. K. Weimer, *Proc. IRE* **50**, 1402 (1962).
[51] R. Zuleeg, *Solid-State Electron.* **6**, 193 (1963).
[52] K. Heime, *Solid-State Electron.* **10**, 732 (1967).
[53] N. Pavaskar and C. Menezes, *Japan. J. Appl. Phys.* **7**, 743 (1968).
[54] G. Diemer, *Physica* **26**, 889 (1960).

or evaporation, and the influence of application temperature on the injection properties of electrons and holes into CdS were studied in detail by Marlor and Woods.[55] Gershun *et al.*[56] point out that the contact properties depend on the crystallographic CdS crystal face onto which In is applied.

As an ohmic metal contact which is stable up to higher temperatures, Böer and Lubitz[57] proposed Al. Aluminum was observed by Pizzarello[58] to be ohmic also to CdS thin films. The formation of Al_2O_3 due to oxygen adsorbed at the CdS surface is avoided, according to Böer and Hall,[59] by depositing onto CdS a sequence of metal layers, preferably Ti–Al–Pt. The first layer is supposed to remove adsorbed gases from the CdS surface by chemical reaction, the second layer forms the desired ohmic contact, and the third layer acts as protective layer. Excellent ohmic performance has been observed up to 350°C.

Various attempts to establish the desired contact potential difference between CdS and the contact metal have been made by raising the CdS work function through cleaning of the CdS surface. The exposure of the CdS surface to a glow discharge as proposed by Butler and Muscheid[60] and Fassbender[61] can, however, greatly damage the CdS surface and make it conductive by creating a Cd excess. The influence of the treatment of the CdS prior to the deposition of a metal electrode is discussed by Greiner *et al.*[62] The contact resistance of In and In–Ga contacts to CdS was measured by Johnson and Darsey[63] for etched crystals and crystals with highly resistive surface layers. They conclude that a linear I–V characteristic is a necessary, but not sufficient, condition for neglecting the contact resistance to CdS, and that only etched samples exhibit low contact resistance.

Indium is observed to make a good ohmic contact also to CdSe,[32] even to chemically deposited, heat-treated CdSe layers.[64] Ohmic contacts are further observed for In to ZnS after brief heating to 600°C.[65] Further examples of photoconductors on which ohmic In contacts were observed are melt-

[55] G. A. Marlor and J. Woods, *Proc. Phys. Soc.* (*London*) **81**, 1013 (1963).

[56] A. S. Gershun, L. A. Sysoev, and B. L. Timan, *Fiz. Tverd. Tela* **8**, 1633 (1966) [*English Transl.: Soviet Phys.–Solid State* **8**, 1302 (1966)].

[57] K. W. Böer and K. Lubitz, *Z. Naturforsch.* **17a**, 397 (1962).

[58] F. A. Pizzarello, *J. Appl. Phys.* **38**, 1752 (1967).

[59] K. W. Böer and R. B. Hall, *J. Appl. Phys.* **37**, 4739 (1966).

[60] W. M. Butler and W. Muscheid, *Ann. Physik* **14**, 215 (1954); **15**, 82 (1954).

[61] J. Fassbender, *Z. Physik* **145**, 301 (1956)

[62] R. A. Greiner, R. F. Miller and C. Retherford, *J. Appl. Phys.* **28**, 1358 (1957).

[63] R. T. Johnson, Jr. and D. M. Darsey, *Solid-State Electron.* **11**, 1015 (1968).

[64] H. Okimura and Y. Sakai, *Japan. J. Appl. Phys.* **7**, 731 (1968).

[65] G. F. Alfrey and I. Cooke, *Proc. Phys. Soc.* (*London*) **B70**, 1096 (1957).

grown PbI_2[66] and ternary sulfides[67] to which In amalgam was applied. Among these compounds, $ZnIn_2S_4$ with ohmic contacts is a particularly sensitive photoconductor,[68] in which, furthermore, a negative differential resistance was observed for ohmic contacts upon illumination with a high monochromatic light intensity.[69] For a recent review of the electric and photoelectric properties of ternary compounds of the type $A^{II}B^{IV}C_2{}^{V}$, the reader is referred to the work of Borshevskii *et al.*[70]

(2) *Blocking Contacts.* For devising a blocking metal contact to largely heteropolar photoconductors, the work-function concept may again serve as a rough guide, at least to the extent as to exclude low-work-function metals. For the observation of primary photocurrents, van Heerden[26] used Au contacts, although Au is also occasionally reported as not being blocking to CdS. Sign and magnitude of the photovoltaic signal detected at a Au electrode, not only in contact with CdS, but also with other photoconductors, e.g., PbI_2,[66] point to a majority-carrier-depletion space-charge layer in the photoconductor adjacent to the Au electrode. Semitransparent Au layers are used with good success as blocking contacts to a variety of low-mobility photoconductors, as will be described in more detail in Section 6. Also, colloidal graphite pasted on the photoconductor is observed to make good blocking contact not only to CdS,[55] but also to the low-mobility photoconductors discussed in Section 6.

Excellent blocking action is reported for evaporated Te on CdS single crystals[71] and on CdS thin films.[72] The CdS–Te blocking contact is applied to the fabrication of diode-logic matrices[73] and the solid-state Vidicon.[74] Dutton and Muller[75] have shown that upon deposition of Te on CdS thin films, a CdTe layer of about 100 Å thickness is formed at the Te–CdS interface, giving rise to a CdTe–CdS heterojunction. The existence of the interfacial CdTe layer was demonstrated by X-ray diffraction and by the observation that the threshold photon energy of the photocurrent coincides with the band gap of CdTe.

[66] G. D. Currie, J. Mudar, and O. Risgin, *Appl. Opt.* **6**, 1137 (1967).

[67] J. A. Beun, R. Nitsche, and M. Lichtensteiger, *Physica* **26**, 647 (1960).

[68] R. Nitsche and W. J. Merz, *Helv. Phys. Acta* **35**, 274 (1962).

[69] W. Ludwig and G. Voigt, *Phys. Status Solidi* **24**, K161 (1967).

[70] A. S. Borshevskii, N. A. Goryunova, F. P. Kesamanly, and D. N. Nasledov, *Phys. Status Solidi* **21**, 9 (1967).

[71] W. Ruppel and R. W. Smith, *RCA Rev.* **20**, 702 (1959)

[72] J. Dresner and F. V. Shallcross, *Solid-State Electron.* **5**, 205 (1962).

[73] H. Fuchs and K. Heime, "Microminiaturization," p. 335. Oldenbourg, München, 1966.

[74] P. K. Weimer, H. Borkan, G. Sadasiv, L. Meray-Horvath, and F. V. Shallcross, *Proc. IEEE* **52**, 1479 (1962).

[75] R. W. Dutton and R. S. Muller, *Solid-State Electron.* **11**, 749 (1968).

Good blocking contacts to CdS thin films are further provided by vapor deposition of heterojunctions, e.g., by direct deposition of CdTe, Al_2O_3, or SiO_x. The rectification properties of these heterojunctions can be accounted for by contact field emission.[76]

In order to study high-field effects in photoconductors by transient or ac photocurrents, blocking electrodes may also be formed by any metal separated from the photoconductor by an insulating spacer like mica or a Mylar sheet, as used, e.g., by Many[77] and Onuki and Miyake.[78]

Strongly blocking contacts to photoconductors are not only formed by metals or solid heterojunctions, but may be provided also by the ions of a corona discharge,[29] as used in the electrophotographic process, or by an electrolyte.[79,80]

(3) *Double-Injecting Contacts.* Extreme blocking action for one type of carrier favors the injection of the opposite charge carrier. Thus, Marlor and Woods[55] observed hole injection into CdS from electron-blocking contacts, such as colloidal dispersions of Ag or graphite, or from evaporated layers of Cu, Ag, and Au, as shown by the appearance of recombination radiation and a negative differential resistance region in the *I–V* characteristic. Hole injection from Ag-paste electrodes into CdS at fields $F > 10^3$ V/cm and under illumination in order to produce a high field at the anode was also observed by Many.[77] A negative differential resistance region, according to the theory of Lampert,[81] was found by Litton and Reynolds[82] using selected CdS crystals with an In cathode and a Ag anode; by Rushby and Woods[83] using CdS with an In cathode and a Au anode; and by Anderson and Mitchell,[84] these authors injecting across a *p*-CdS layer produced by P-ion implantation. Under illumination, the negative differential resistance disappears and the photoconductor is converted from the low-current into the high-current state. This conversion from an insulator region into a semiconductor region is analogous to that observed for double injection in *n–i–p* structures of GaAs, Si, and Ge.[85] Double injection has also been reported by Rohde[86]

[76] R. S. Muller and R. Zuleeg, *J. Appl. Phys.* **35**, 1550 (1964).

[77] A. Many, *J. Phys. Chem. Solids* **26**, 575 (1965).

[78] M. Onuki and T. Miyake, *J. Phys. Soc. Japan* **18**, 151 (1963).

[79] R. Williams, *Phys. Rev.* **123**, 1645 (1961); **125**, 850 (1962).

[80] A. Many, *J. Phys. Chem. Solids* **26**, 587 (1965).

[81] M. A. Lampert, *Phys. Rev.* **125**, 126 (1962).

[82] C. W. Litton and D. C. Reynolds, *Phys. Rev.* **133**, A536 (1964).

[83] A. N. Rushby and J. Woods, *Brit. J. Appl. Phys.* **17**, 1187 (1966).

[84] W. W. Anderson and J. T. Mitchell, *Appl. Phys. Letters* **12**, 334 (1968).

[85] N. Holonyak, Jr., S. W. Ing, Jr., R. C. Thomas, and S. F. Bevacqua, *Phys. Rev. Letters* **8** (1962).

[86] H. J. Rohde, *Phys. Status Solidi* **23**, 277 (1967).

for CdS from an In cathode and Au, Te, and Cu_2S anodes, and by Ludwig and Zeise[87] for ZnS from two In electrodes. Double injection triggered by light, analogous to that observed by Many,[77] was observed by Vishchakas *et al.*[88] for CdSe whiskers by an enhancement of the space-charge-limited current. The current increase was explained by an effective reduction of the transit distance of the carriers.

The electrolyte contact being particularly blocking for one kind of carrier, also facilitates the injection of the opposite carrier. Van Ruyven and Williams[89] have found the potential barrier between fuming sulfuric acid and a ZnS single crystal high enough as to permit the injection of holes into the ZnS, as detected by the emission of recombination radiation.

b. *Molecular and Organic Photoconductors*

For the discussion of contacts to organic and molecular photoconductors, it is again indispensable to consider electrolytic contacts as well as metallic contacts. CuI electrolytic contacts were used as ohmic contacts for holes to anthracene as a photoconductor by Kallmann and Pope[90] and by Boroffka,[91] who observed an increase of hole generation at the CuI–anthracene interface upon illumination. Also, Ce^{4+} in solution was observed to be ohmic for holes to anthracene.[90] The observed saturation current was equal for the dark current and the photocurrent, and was therefore due to a limited supply of holes by this anode. The theory of ohmic contacts to organic photoconductors in the form of aqueous solutions, especially the iodine–iodide solution, was treated by Mark and Helfrich.[92] Recently, Litvinenko and Fridkin[93] have discussed the change of the concentration of free I_2 for the iodide contact upon illumination. The level of hole injection is raised by the photochemical dissociation of HI and also by the dissociation of the I_2 molecule. The authors observe a corresponding transition from linear current–voltage characteristics in the dark to space-charge-limited currents proportional to V^2 at high illumination levels.

Metals are also observed to emit holes into anthracene,[94] as described in more detail in Chapter 2 of this volume, dealing with injection by internal

W. Ludwig and V. Zeise, *Phys. Status Solidi* **7**, 143 (1964).

Yu. K. Vishchakas, A. A. Smilga, and G. B. Yushka, *Fiz. Tekhn. Poluprov.* **1**, 1733 (1967) [*English Transl.*: *Soviet Phys.–Semicond.* **1**, 1437 (1968)].

L. J. Van Ruyven and F. E. Williams, *Phys. Rev. Letters* **16**, 889 (1966).

H. Kallmann and M. Pope, *J. Chem. Phys.* **32**, 300 (1960).

I. Boroffka, *Z. Physik*, **160**, 93 (1960).

Mark and W. Helfrich, *J. Appl. Phys.* **33**, 205 (1962).

Yu. Litvinenko and V. M. Fridkin, *Fiz. Tverd. Tela* **9**, 3615 (1967) [*English Transl.*: *Soviet Phys.–Solid State* **9**, 2845 (1968)].

Williams and J. Dresner, *J. Chem. Phys.* **46**, 2133 (1967).

photoemission. Also, electron emission from metals into anthracene was detected.[95,96] Mehl and Funk[95] observed space-charge-limited electron injection from a K–Na eutectic. According to Baessler and Vaubel,[96] the photocurrent with Na and Cs cathodes is due to excitation of electrons from the metal cathode and from surface states of the anthracene into two separate conduction bands of the anthracene. This conclusion is based upon the comparison of the results obtained with Ca, Mg, and Ce cathodes. Mehl,[97] on the other hand, interprets the photocurrent as being due to the interaction of light with trapped space charge by releasing either trapped carriers or triplet and singlet excitons in anthracene, since a space-charge-limited current is also observed without illumination. This would be a particular example of sensitization of a photoconductor by injection of carriers and release of the trapped space charge by light, for which more examples will be given in the next section. An analogous interpretation of the photocurrent in anthracene with a Ce^{4+}-solution anode or with an evaporated CuI anode and a Ag cathode by release of injected holes from traps is given by Jansen *et al.*[98] and by Gugeshashvili *et al.*[99]

Phthalocyanine with ohmic Ag-paste contacts was observed to show space-charge-limited conduction in the dark and a linear *I–V* characteristic under illumination, indicating a photogenerated carrier concentration greatly exceeding the dark concentration.[100]

Ohmic contacts to molecular crystals that lead to majority-carrier injection in the dark may also be provided by electrolytes, as shown by Mehl *et al.*[101,102] for sulfur and by Lohmann for iodine.[103] The formation of free-carrier reservoirs at the contacts for the measurement of photocurrent pulses in order to determine the carrier mobilities in these materials will be discussed in Section 6.

5. Injection Sensitization

The operation of a photoconductor with ohmic contacts for majority carriers in a voltage range in which the majority-carrier transit time between the electrodes becomes equal to the dielectric relaxation time leads to space-charge-limited injection of majority carriers, as was pointed out in Section 2.

[95] W. Mehl and B. Funk, *Phys. Letters* **25A**, 364 (1967).

[96] H. Baessler and G. Vaubel, *Solid State Commun.* **6**, 97, 631 (1968).

[97] W. Mehl, *Solid State Commun.* **6**, 549 (1968).

[98] P. Jansen, W. Helfrich, and N. Riehl, *Phys. Status Solidi* **7**, 851 (1964).

[99] M. I. Gugeshashvili, I. A. Eligulashvili, G. A. Nakashidze, L. D. Rozenshtein, and V. V. Chavchanidze, *Fiz. Tekhn. Poluprov.* **2**, 144 (1968) [*English Transl.*: *Soviet Phys.–Semicond.* **2**, 125 (1968)].

[100] G. H. Heilmeier and G. Warfield, *J. Chem. Phys.* **38**, 163 (1963).

[101] W. Mehl, F. Lohmann, J. L. Brebner, and E. Mooser, *Phys. Letters* **23**, 34 (1966).

[102] F. Lohmann and W. Mehl, *J. Phys. Chem. Solids* **28**, 1317 (1967).

[103] F. Lohmann, *J. Phys. Chem. Solids* **29**, 1693 (1968).

The injected majority-carrier space charge is usually located in traps, thereby changing the occupancy of the traps. If, under illumination, these trapped carriers are released, the photoconductor appears as being sensitized in its long-wavelength spectral-response range. This possibility of sensitization by space-charge-limited trap filling was analyzed by Helfrich.[104] The light-intensity dependence of the photocurrent is related to the voltage dependence of the space-charge-limited dark current by the energy distribution of the traps. As pointed out in the last section, these results were applied to anthracene.[98]

Sensitization by space-charge-limited trap filling was also observed by Driedonks and Zijlstra[105] for CdS with In electrodes and by Zibuts *et al.*[106] for CdS and Cu-doped Si. Long time effects could also be explained by a rearrangement of the trap occupancy under space-charge-limited current-flow conditions.[107]

According to Izvozchikov,[108] the same phenomenon accounts for the observation of the photoconductivity measured by pulses in PbO. For longer pulses, a higher photocurrent maximum is found, because traps were filled by majority-carrier injection from the ohmic contact and were emptied by light.

Negative photoconductivity may occur in the injection range. Bräunlich[109] explained the observation of a negative photoconductivity in CdS by double injection and, in particular, by an injection-determined occupancy of two classes of recombination centers such as was used by Stöckmann[110] to account for negative photoconductivity in Ge.

6. Transient Photocurrents

a. Determination of the Drift Mobility

Transient photocurrents have become an important tool in the investigation of low-mobility semiconductors. The drift mobility of either type of charge carrier may be determined by measuring the transit time of the charge carrier across the sample. A necessary requirement for this method to operate is a carrier lifetime not short with respect to the transit time, or, expressed in characteristic distances, the carrier range not short with respect

[104] W. Helfrich, *Phys. Status Solidi* **7**, 863 (1964).

[105] F. Driedonks and R. J. J. Zijlstra, *Phys. Letters* **23**, 527 (1966).

[106] Yu. A. Zibuts, L. G. Paritskii, and S. M. Ryvkin, *Fiz. Tekhn. Poluprov.* **1**, 724 (1967) [*English Transl.*: *Soviet Phys.–Semicond.* **1**, 602 (1967)].

[107] Zh. G. Dokholyan, Yu. A. Zibuts, L. G. Paritskii, and A. I. Rozental, *Fiz. Tekhn. Poluprov.* **2**, 137 (1968) [*English Transl.*: *Soviet Phys.–Semicond.* **2**, 118 (1968)].

[108] V. A. Izvozchikov, *Fiz. Tekhn. Poluprov.* **2**, 254 (1968) [*English Transl.*: *Soviet Phys.–Semicond.* **2**, 209 (1968)].

[109] P. Bräunlich, *Phys. Status Solidi* **21**, 383 (1967).

[110] F. Stöckmann, *Z. Physik* **143**, 348 (1955).

to the interelectrode distance. By this method, carrier drift mobilities were determined in CdS[49], amorphous Se,[111–117] monoclinic Se,[118] As_4S_4,[119] iodine,[120–123] sulphur,[124–131] KCl,[132] anthracene,[133–137] and liquid *n*-hexane.[138]

Light, strongly absorbed in a surface layer of the sample close to the contact, X rays, or electron beams create a reservoir of free carriers in the photoconductor at the contact, or free carriers are emitted by the metal electrode under illumination. The metal electrode itself forms a blocking contact to the photoconductor. Either this blocking contact is a potential barrier in the photoconductor, or one may be provided by separating the photoconductor sample from the metal electrode by means of an insulating spacer. For the short photocurrent pulses, the insulating spacer and the photoconducting sample act as a capacitive voltage divider.

The photogenerated free-carrier reservoir adjacent to the metal electrode forms an ohmic contact to the illuminated portion of the photoconductor, and the current pulse through the sample is space-charge-limited for insula-

[111] W. E. Spear, *Proc. Phys. Soc.* (*London*) **B70**, 669 (1957).
[112] W. E. Spear, *Proc. Phys. Soc.* (*London*) **76**, 826 (1960).
[113] J. L. Hartke, *Phys. Rev.* **125**, 1177 (1962).
[114] Yu. K. Vishchakas, G. B. Yushka, A. D. Petravichus, and A. Yu. Matulenis, *Fiz. Tverd. Tela* **8**, 1616 (1966) [*English Transl.*: *Soviet Phys.–Solid State* **8**, 1283 (1966)].
[115] R. M. Blakney and H. P. Grunwald, *Phys. Rev.* **159**, 664 (1967).
[116] M. D. Tabak, *Trans. AIME* **239**, 330 (1967).
[117] M. D. Tabak and P. J. Wachter, Jr., *Phys. Rev.* **173**, 899 (1968).
[118] W. E. Spear, *Chem. Solids* **21**, 110 (1961).
[119] G. B. Street and W. D. Gill, *Phys. Status Solidi* **18**, 601 (1966).
[120] A. Many, M. Simhony, S. Z. Weisz, and J. Levinson, *J. Phys. Chem. Solids* **22**, 285 (1961).
[121] A. Many, S. Z. Weisz, and M. Simhony, *Phys. Rev.* **126**, 1989 (1962).
[122] A. Many, M. Simhony, S. Z. Weisz, and Y. Teucher, *J. Phys. Chem. Solids* **25**, 721 (1964).
[123] M. Simhony and J. Gorelik, *J. Phys. Chem. Solids* **26**, 2077 (1965).
[124] A. R. Adams and W. E. Spear, *J. Phys. Chem. Solids* **25**, 1113 (1964).
[125] A. R. Adams, D. J. Gibbons, and W. E. Spear, *Solid State Commun.* **2**, 387 (1964).
[126] C. A. Mead, *Phys. Letters* **11**, 212 (1964).
[127] K. K. Thornber and C. A. Mead, *J. Phys. Chem. Solids* **26**, 1489 (1965).
[128] A. Many, M. Simhony, and Y. Grushkevitz, *J. Phys. Chem. Solids* **26**, 1925 (1965).
[129] W. E. Spear and A. R. Adams, *J. Phys. Chem. Solids* **27**, 281 (1966).
[130] D. J. Gibbons and W. E. Spear, *J. Phys. Chem. Solids* **27**, 1917 (1966).
[131] W. D. Gill, G. B. Street, and R. E. MacDonald, *J. Phys. Chem. Solids* **28**, 1517 (1967).
[132] H. Hirth and U. Tödheide-Haupt, *Phys. Status Solidi* **31**, 425 (1969).
[133] R. G. Kepler, *Phys. Rev.* **119**, 1226 (1960).
[134] O. H. LeBlanc, Jr., *J. Chem. Phys.* **33**, 626 (1960).
[135] M. Silver, M. Swicord, R. C. Jarnagin, A. Many, S. Z. Weisz, and M. Simhony, *J. Phys. Chem. Solids* **23**, 419 (1962).
[136] C. Bogus, *Z. Physik* **184**, 211 (1965).
[137] Y. Watanabe, N. Saito, and Y. Inuishi, *Japan. J. Appl. Phys.* **7**, 854 (1968).
[138] O. H. LeBlanc, Jr., *J. Chem. Phys.* **30**, 1443 (1959).

ting samples. For low light levels, the photocarrier reservoir at the electrode may be exhausted, and the current becomes space-charge-free.[120] A strongly illuminated part of a photoconductor forming an ohmic contact to an unilluminated part of the same sample has also been demonstrated by steady-state space-charge-limited current flow across the unilluminated part in CdS.[139]

Particular difficulties are encountered when either of the metal electrodes forms an ohmic contact to the sample because of dark injection of majority carriers. Spear and Mort[49] did not observe any transient photocurrent signal with two In electrodes to CdS. The signal was masked by heavy injection following the excitation pulse. Gold which is at least partially blocking, also led to negative results. The reason was apparently polarization of the sample, due to the buildup of trapped space charge. This would tend to decrease the applied field throughout most of the sample. The difficulties were resolved by using "mixed" electrodes, namely, a blocking electrode at the start of the carrier and an ohmic collector electrode for the type of carrier making the transit. In this case, during the "off" period of the applied field, when both electrodes are at ground potential, a reverse injection through the collector electrode reduces the buildup of polarizing space charge similarly to self-neutralization observed by Freeman *et al.*[140] For the observation of electron transit in CdS,[49] the best structure was a Pyrex film on the CdS as insulating spacer carrying an evaporated Au layer as cathode and an In layer on the CdS as anode. For the observation of hole transit, an Ag-paste cathode on CdS was used.

Complications by injection at the unilluminated contact were also encountered for sulfur[131] and realgar, As_4S_4.[119] The occurrence of hole injection at the unilluminated anode upon illumination of the cathode was verified for As_4S_4 by the suppression of the hole injection when the metal anode was separated from the As_4S_4 sample by a Mylar sheet. In general, for the illuminated contact to sulfur, an evaporated semitransparent Au layer or a conducting glass or quartz window pressed or glued onto the sample surface is used. Surface damaging of the sulfur has been found to have no influence on the results.[131] A simple unilluminated collector electrode, colloidal graphite, was used.

In order to prevent undesired injection at the unilluminated contact, for the measurements on amorphous Se, the electrodes were separated by a 2000-Å-thick, insulating ZnS layer[113] or plastic film[117] from the photoconductor, while, for the illuminated contact, evaporated Au or a conducting SnO_2-coated glass electrode was used.

[139] R. Fischer, F. Stöckmann, and J. Stuke, *Phys. Status Solidi* **17**, 335 (1966).

[140] J. R. Freeman, H. P. Kallmann, and M. Silver, *Rev. Mod. Phys.* **33**, 553 (1961).

b. Photocapacitors

Transient photocurrents occur in an ac-operated photoconductor which forms part of the dielectric of a capacitor. The capacitor acts as a light detector if, by illumination of the photoconductor, the capacitance is changed. Photodielectric and photoconductive capacitance changes of a capacitor filled with a photoconductor as dielectric were discussed in detail by Reuber.[141]

In particular, the photoconductor layer may form part of a two-layer system, e.g., CdS with $BaTiO_3$ or silicone plastic,[142] CdS or CdSe films on glass or SiO_2,[143] or phthalocyanine as the photoconductive layer.[144]

In these cases, the entire photoconductive layer changes its conductivity under illumination. On the other hand, the capacitance may be modulated by a change of a photoconductor space-charge barrier-layer capacitance adjacent to either a different dielectric[145] or metal electrode.[146]

Lee and Henisch[145] describe the photocapacitive response of an *n*-type Si–collodion contact under applied bias voltage. They observe two opposite effects of the light on the barrier-layer capacitance of the inversion layer in the Si adjacent to the collodion: a small increase of the barrier-layer thickness due to a flattening of the bands, and a corresponding decrease of the capacitance on one hand, but also a strong opposite effect upon illumination of the electrode edges of the capacitor. The latter effect is ascribed to the creation of a high hole concentration diffusing into the barrier-layer region, thereby partly neutralizing the space charge and decreasing the space-charge-layer thickness.

The change of the capacitance upon illumination of an *n*-type GaAs space-charge barrier layer in contact with a Au electrode without applied bias voltage was reported by Yamamoto and Ota.[146] The effect is applied to light detection using frequency modulation and the control of parametric oscillation by the modulated light beam of a GaAs diode.

[141] C. Reuber, *in* "Festkörperprobleme" (O. Madelung, ed.), Vol. 8, p. 175. Vieweg, Braunschweig, 1968.

[142] Y. T. Shivonen, D. R. Boyd, and E. L. Kitts, *Proc. IEEE* **53**, 378 (1965).

[143] A. N. Zyuganov and S. V. Svechnikov, *Radiotekhn. i Elektron.* **10**, 2264 (1965) [*English Transl.*: *Radio Eng. Electron.* (*USSR*) **10**, 1940 (1965)].

[144] Yu. A. Vidadi, L. D. Rozenshtein, and E. A. Chistyakov, *Fiz. Tekhn. Poluprov.* **1**, 1257 (1967) [*English Transl.*: *Soviet Phys.–Semicond.* **1**, 1049 (1968)].

[145] S. Lee and H. K. Henisch, *Solid-State Electron.* **11**, 301 (1968).

[146] T. Yamamoto and Y. Ota, *Proc. IEEE* **54**, 691 (1966); *Solid-State Electron.* **11**, 219 (1968).

III. The Photoconductor–Metal Contact without Externally Applied Voltage

7. General Survey

Carrier injection into a photoconductor and extraction from a photoconductor across a metal contact under applied voltage are determined by the relation of the potential drop across the space-charge region at the metal–photoconductor contact and the total potential drop across the photoconductor. A majority-carrier accumulation layer is the most important realization of a majority-carrier-injecting contact, and a majority-carrier depletion layer that of an extracting contact. Schottky emission over a contact potential barrier or tunneling through the barrier were, however, examples for the possibility of space-charge-limited majority-carrier-injection processes despite majority-carrier depletion at the contact. The observation of the current–voltage characteristic under applied voltage yields, therefore, no direct experimental information on the potential distribution at the metal–photoconductor contact. This information is obtained without externally applied voltage from the photovoltaic response of the inhomogeneity provided by the contact, and from the determination of the total charge which the photoconductor acquires in equilibrium with its contacts.

When the photoconductor is illuminated under open-circuit conditions, the electrostatic potentials for a one-dimensional geometry $E_c(x)$ and $E_v(x)$ of the electrons and holes as well as their respective electrochemical potentials $E_{fn}(x)$ and $E_{fp}(x)$ are determined by Eq. (1) of Section 1 with $i = 0$.[147] If the free-carrier-concentration distributions $n(x)$ and $p(x)$ differ by only distance-independent terms from a Boltzmann distribution, the absolute amount of the photo-emf V_{ph} at the n-type photoconductor–metal contact upon illumination of the space-charge layer in the n-type photoconductor is given by[148]

$$V_{ph} = \frac{kT}{e} \ln\left(1 + \frac{\mu_p \Delta p}{\mu_n n_0}\right) \tag{3}$$

for illumination intensities not approaching saturation of V_{ph}. The sign of V_{ph} is opposite for a depletion and an accumulation layer. The free-carrier concentrations at the photoconductor–metal interface are assumed to be light-independent. Here, Δp denotes the photogenerated hole concentration in the space-charge layer of the photoconductor; n_0, in the case of a majority-carrier depletion layer, denotes the dark electron concentration at the photoconductor-metal interface, and, in the case of a majority-carrier accumulation layer, is the dark electron concentration in the volume of the photo-

[147] J. Tauc, "Photo and Thermoelectric Effects in Semiconductors." Pergamon Press, New York, 1962.

[148] W. Ruppel, *Phys. Status Solidi* **5**, 657, 667 (1964).

conductor. The majority-carrier volume dark concentration being much larger than that at the interface for a depletion layer, it is not surprising that photovoltages indicating by their sign an accumulation layer have only rarely been reported.[149–151] In Fig. 5, $E_c(x)$, $E_v(x)$, $E_{fn}(x)$, and $E_{fp}(x)$ are shown for an illuminated depletion and accumulation layer of a photoconductor adjacent to a metal. It may be noted that only in the case of a depletion layer does the main drop of the majority-carrier electrochemical potential $E_{fn}(x)$ occur across the illuminated space-charge layer. This is why the photo-emf for this type of contact can be derived from one-carrier rectification considerations although, as Eq. (1) shows, the occurrence of a steady-state photo-emf requires two kinds of mobile carriers. In the case of the majority-carrier accumulation layer, the main drop of $E_{fn}(x)$ occurs in the unilluminated region in the photoconductor, in which the majority-carrier concentration decays within a Debye length and the minority-carrier concentration within a diffusion length to their thermal equilibrium values.

The sign of the space-charge layer is determined by the direction of the observed photo-emf. The mobilities being known, e.g., from transient photocurrent measurements (Section 6), the minority-carrier lifetime may be inferred if the minority-carrier generation rate and the dark majority-carrier concentration are known.

The main problem in the interpretation of photovoltaic measurements by minority-carrier generation lies in the fact that an alternative possibility

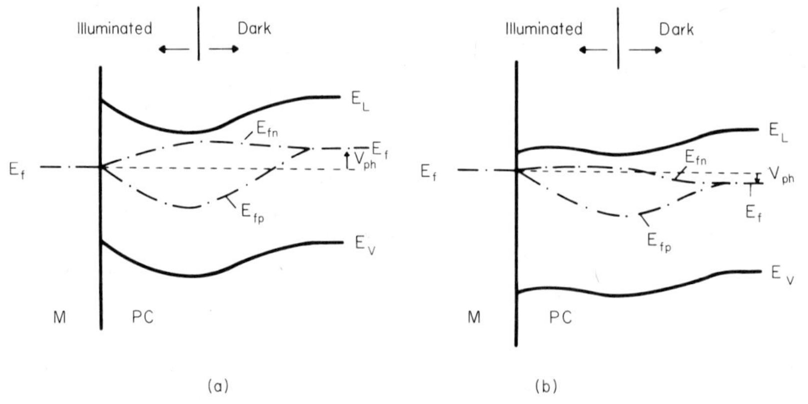

FIG. 5. Potential distribution and steady-state photovoltage V_{ph} at the illuminated photoconductor–metal contact without applied voltage. (a) Blocking contact. (b) Ohmic contact.

[149] W. Ruppel, *Helv. Phys. Acta* **34**, 790 (1961).
[150] O. Dehoust, *Z. Angew. Phys.* **14**, 404 (1962).
[151] G. O. Müller, *Phys. Status Solidi* **3**, 523 (1963).

to the generation of free minority carriers by the illumination of the photoconductor–metal contact is the photoemission of majority carriers from the metal into the photoconductor, thereby setting up a photovoltage in an analogous way to the emission of electrons from a metal in a vacuum photocell. Both the free minority-carrier generation and the photoemission may occur at smaller photon energies than that of the photoconductor band gap. The processes are schematically indicated in Table III, and will be discussed in the following section. The situation is particularly complicated in the case of an illuminated CdS–Cu contact, which has become of foremost technical importance for the construction of an effective large-area solar cell.

Apart from photovoltaic measurements, the potential distribution in the photoconductor adjacent to the metal contact may be determined by the detection of the total charge on the photoconductor after removal of the metal electrodes. This method is applicable to measurements both with and without applied voltage and has yielded information not only about the space-charge layer in the photoconductor at the metal contact, but also

TABLE III

THE PHOTOCONDUCTOR–METAL CONTACT WITHOUT APPLIED VOLTAGE

Conditions	References	
Photovoltage by two-step minority-carrier excitation for photon energies less than band gap	156, 158–161	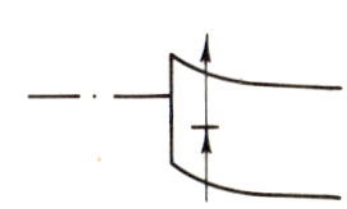 1
Photovoltage by photoemission for photon energies less than band gap	45, Chapter 2 in this volume	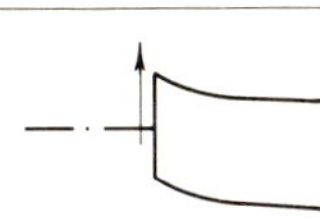 2
Photovoltage by minority-carrier creation in impurity band	163–174	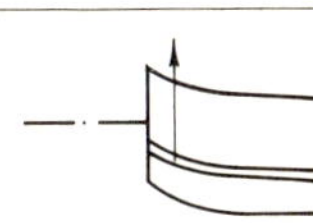 3
Photovoltage by excitation in heterojunction	152–155	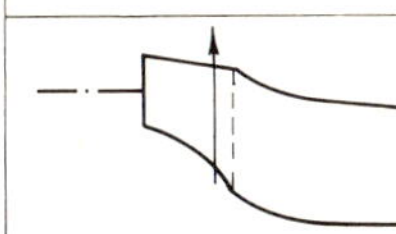 4
Total crystal charge at different illuminations (with and without applied voltages)	162, 175–180	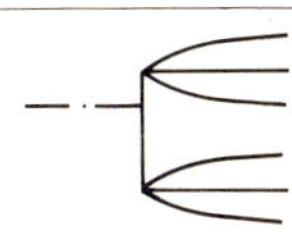 5

about the relaxation of this space charge in the photoconductor after removal of the metal electrode.

8. Steady-State Photovoltage

While a steady-state photovoltage at a photoconductor–metal contact under illumination with photons of energy larger than the band gap may be interpreted in a straightforward way by free-pair generation in the photoconductor space-charge layer, the occurrence of a steady-state photovoltage for photon energies smaller than the band gap presents a particularly sensitive effect for the investigation of the inhomogeneity provided by the contact.

A photovoltage at photon energies smaller than the band gap was first observed in CdS crystals and CdS powder with Cu, Ag, Au, and Pt contacts by Reynolds *et al.*[152] In order to account for the generation of mobile minority carriers by light, these authors[153] assumed the existence of an impurity band in the CdS formed by the 3d levels of the Cu diffused into the CdS during the activation process of the CdS. This point of view is essentially shared by Fabricius[154] and Grimmeiss and Memming,[155] but extensive recent investigations[155a] of the CdS–Cu contact point to the conclusion that a Cu_2S–CdS heterojunction is formed at this contact and must be considered as the seat of the photovoltage.

Not considering further the formation of impurity bands in the photoconductor, essentially three different mechanisms may account for the observation of a steady-state photovoltage at the photoconductor–metal contact upon illumination with photon energies smaller than the band gap:

1. Photoemission of electrons or holes from the metal into the photoconductor.
2. Generation of mobile minority carriers in the photoconductor space charge layer by a two-step excitation process via centers in the forbidden gap.
3. Band-to-band excitation in a heterojunction formed as an intermediate layer between the metal and the photoconductor.

A photoemission current is not driven by the gradient of the electrochemical potential, and, consequently, Eq. (1) does not apply. A steady-state photovoltage by photoemission therefore does not require the presence of both kinds of mobile charge carriers; rather, the photoemission of one kind of carrier alone from the metal is sufficient to establish a steady-

[152] D. C. Reynolds, G. Leies, L. L. Antes, and R. E. Marburger, *Phys. Rev.* **96**, 533 (1954).

[153] D. C. Reynolds and S. J. Czyzak, *Phys. Rev.* **96**, 1705 (1954).

[154] E. D. Fabricius, *J. Appl. Phys.* **33**, 1597 (1962).

[155] H. G. Grimmeiss and R. Memming, *J. Appl. Phys.* **33**, 2217, 3596 (1962).

[155a] L. R. Shiozawa, G. A. Sullivan, and F. Augustine, *Record IEEE Photovoltaic Specialists Conf., Pasadena, California, November 1968*, IEEE Cat. No. 68 C 63-ED.

state photovoltage. For a discussion of photoemission, the reader is referred to Chapter 2 in this volume. A comprehensive review of photovoltage-current data interpreted by photoemission is also given by Mead.[45]

The generation of mobile minority carriers in the space charge layer of high-resistivity CdS crystals in contact with Au was demonstrated by relating the enhancement of the photovoltage upon illumination with low-photon-energy light to the quenching of the photoconduction by the same light under applied voltage.[156] The generation of free minority carriers is known to be a necessary requirement for the observation of photoconduction quenching.[157] Free minority carriers created by low-energy photons, therefore, quench a photocurrent produced by short-wavelength excitation and an externally applied voltage, but enhance a photovoltage excited by illumination in the fundamental absorption range. This mechanism of the photovoltage generation being established, minority-carrier lifetimes were determined by this method for CdS[158] and ZnS.[159] A photovoltaic response of a CdS barrier layer due to excitation of impurities at long-wavelength illumination is also claimed by DucCuong and Blair.[160] These authors observed an enhancement of the spectral response by additional illumination with bandgap light. They explain it by free-minority-carrier generation under the long-wavelength illumination. Similarly, free-minority-carrier generation was demonstrated by Komashchenko and Fedorus[161] by the thermal quenching of the photoconduction in CdSe crystals under applied voltage for the contact of CdSe with Ag, Au, and Pt. For evaporated Cu both on CdSe crystals and films,[162] a different mechanism, the formation of a heterojunction, was claimed.

The heterojunction concept for the contact to Cd chalcogenides was first invoked by Cusano[163] for the p-Cu_2Te and n-CdTe junction and by Keating[164] for the p-Cu_2S and n-CdS junction. Keating demonstrated the injection of holes into CdS from Cu_2S,[164] and compared the magnitude of the open-circuit photovoltage with a large-signal theory of the photovoltage at abrupt heterojunctions.[165] Cu_2S applied chemically or by vapor deposition to CdS apparently forms a heterojunction, whether applied to

[156] W. Palz and W. Ruppel, *Phys. Status Solidi* **6**, K164 (1964); **15**, 649 (1966).

[157] A. Rose, *RCA Rev.* **12**, 362 (1951).

[158] W. Palz and W. Ruppel, *Phys. Status Solidi* **15**, 665 (1964).

[159] H. Friedrich, L. A. Sysoev, and K. Thiessen, *Phys. Status Solidi* **26**, 107 (1968).

[160] N. DucCuong and J. Blair, *J. Appl. Phys.* **37**, 1660 (1966).

[161] V. N. Komashchenko and G. A. Fedorus, *Fiz. Tekhn. Poluprov.* **1**, 495 (1967) [*English Transl.: Soviet Phys.–Semicond.* **1**, 411 (1967)].

[162] V. N. Komashchenko and G. A. Fedorus, *Ukr. Fiz. Zh.* **13**, 688 (1968).

[163] D. A. Cusano, *Solid-State Electron.* **6**, 217 (1963); *Rev. Phys. Appl.* **1**, 195 (1966).

[164] P. N. Keating, *J. Phys. Chem. Solids* **24**, 1101 (1963).

[165] P. N. Keating, *J. Appl. Phys.* **36**, 564 (1965).

CdS single crystals,[166] evaporated layers[167–169] or layers deposited by a chemical spray technique.[170] The chemical reactions occurring at the interface upon immersion of the CdS layer into an aqueous CuCl solution and leading to the formation of the heterojunction were studied by Hill and Keramidas.[171] Singer and Faeth[172] observed the formation of a Cu_2S single crystal on a CdS single crystal in a saturated CuCl solution, and, by considering the growth kinetics, they established the lattice relationship between the CdS and Cu_2S crystals. According to Pavelets' and Fedorus,[173] even the evaporation of Cu onto CdS films without further heat treatment leads to a formation of a heterojunction of p-$Cu_{2-x}S$ with n-CdS if a photovoltage is observed. In low-resistivity CdS, the formation of a p-$Cu_{2-x}S$ layer requires the deposition of a thin sulfur film prior to that of copper because there may not be enough sulfur available in the CdS in this case to react with the Cu.

The problem of finding the potential distribution at the Cu–CdS contact reduces to that of a heterojunction between degenerate p-$Cu_{2-x}S$ and n-CdS. There is no doubt that due to the diffusion of Cu acceptors into the CdS, an electron depletion layer is formed in the n-CdS at one side of the heterojunction. Measurements of the spectral response and of the current–voltage characteristic of the heterojunction lead to a potential distribution in the dark as sketched in Fig. 6. The potential at the very transition between the Cu_2S and the CdS is not drawn because it does not seem to be altogether clear at the present time. As Mytton[168] points out, it is likely that a potential barrier exists for electrons in this region more than a few angstroms thick. This barrier would seriously impede the electron flow from the Cu_2S to the CdS under illumination.

For a description of the preparation of CdS thin-film solar cells and their applications, the reader is referred to Shirland.[174]

9. Total Charge

Integration of the Poisson equation over the entire crystal shows that the total charge of a crystal is determined by the electric field strength at its

[166] B. Selle, W. Ludwig, and R. Mach, *Phys. Status Solidi* **24**, K149 (1967).

[167] M. Balkanski and B. Choné, *Rev. Phys. Appl.* **1**, 179 (1966).

[168] R. J. Mytton, *Brit. J. Appl. Phys.* [2], **1**, 721 (1968).

[169] I. V. Egorova, *Fiz. Tekhn. Poluprov.* **2**, 319 (1968) [*English Transl.*: *Soviet Phys.–Semicond.* **2**, 266 (1968)].

[170] R. R. Chamberlin and J. S. Skarman, *Solid-State Electron.* **9**, 819 (1966).

[171] E. R. Hill and B. G. Keramidas, *Rev. Phys. Appl.* **1**, 189 (1966).

[172] J. Singer and P. A. Faeth, *Appl. Phys. Letters* **11**, 130 (1967).

[173] S. Yu. Pavelets' and G. A. Fedorus, *Ukr. Fiz. Zh.* **11**, 687 (1966).

[174] F. A. Shirland, *Advan. Energy Conversion* **6**, 201 (1966).

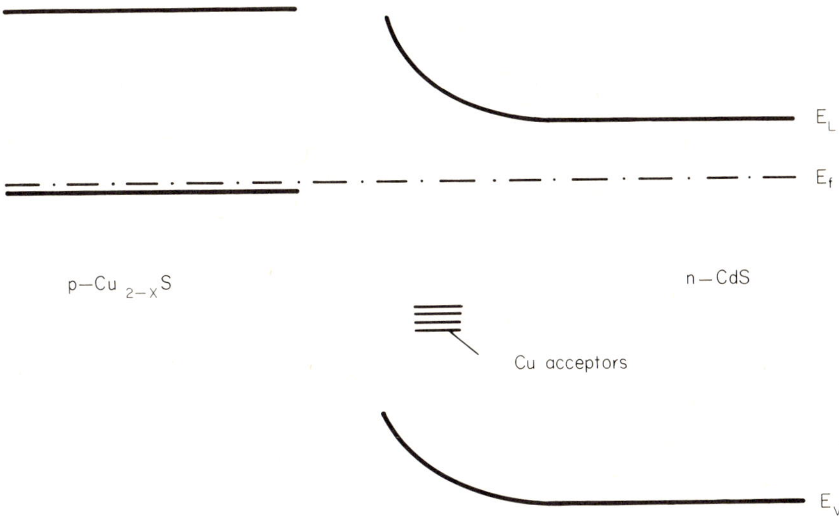

FIG. 6. Potential distribution at a Cu_2S–CdS heterojunction in the dark. The center part, not yet definitely established, is not drawn. (After Selle *et al.*[166] and Mytton.[168])

boundaries. Therefore, the charge of a photoconductor is closely related to its contacts. Smith and Rose[175] detected the majority-carrier excess charge with applied voltage under space-charge-limited majority-carrier injection conditions by removing the metal electrodes from the photoconductor and dropping it onto the pan of an electrometer. The same method was also used later by other authors using different electrode materials[176,177] and voltage and illumination ranges in which the carrier reservoir at the injecting contact was exhausted and the charge had changed sign.[34] On the other hand, the total crystal charge may also be detected by measuring the charge that leaves a crystal after removal of the applied voltage. By this effect, which was mentioned earlier in connection with the determination of the drift mobility by transient currents (Section 6) and discussed by Spear and Mort,[49] Gershun *et al.*[178] established that the electrons injected from an In contact into the CdS are localized in a layer less than 10^{-5} cm thick adjacent to the In electrode.

The crystal charge without applied voltage may be detected more simply than under applied voltage by removing only one electrode from the photo-

[175] R. W. Smith and A. Rose, *Phys. Rev.* **97**, 1531 (1955).

[176] K. W. Böer and U. Kümmel, *Z. Naturforsch.* **12a**, 667 (1957); *Ann. Physik* **2**, 217 (1958).

[177] K. W. Böer, U. Kümmel, and K. E. Schroeter, *Z. Physik* **167**, 403 (1962).

[178] A. S. Gershun, L. A. Sysoev, and B. L. Timan, *Fiz. Tverd. Tela* **8**, 3712 (1966) [*English Transl.: Soviet Phys.–Solid State* **8**, 2982 (1967)].

conductor connected to an electrometer.[179] If the metal electrode is removed rapidly with respect to the dielectric relaxation time of the space charge in the photoconductor and if it is removed far enough that no field lines of the space charge in the photoconductor end on it, the space charge in the photoconductor is quantitatively shown by the electrometer. Measurements of this type were carried out on CdS with various contacts under different illumination conditions. Since, according to Mead,[45] the potential distribution at the metal–CdS contact is not determined by surface states, but by the work functions of metal and CdS, the gradient of the electrostatic potential in the CdS at the CdS–metal interface is a function of the metal work function. As a consequence, the charge that is accumulated in the CdS crystal also depends

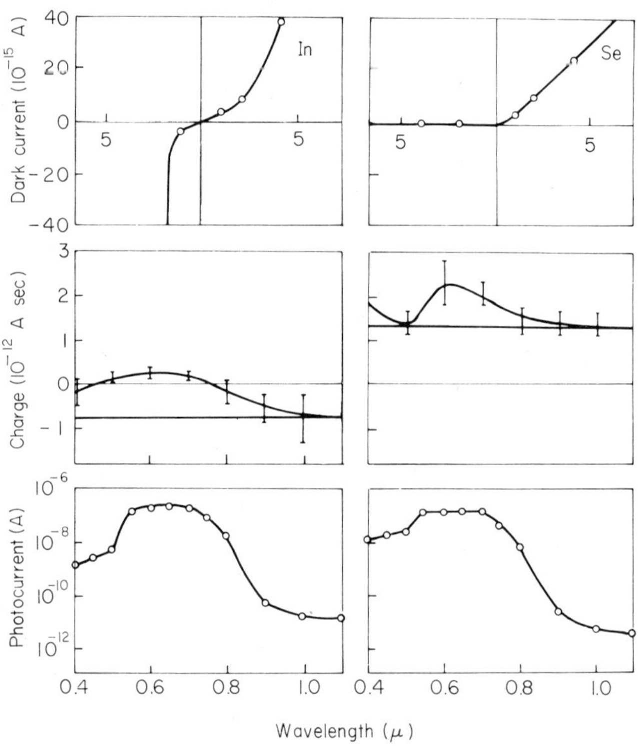

FIG. 7. Upper diagrams: Dark current versus voltage for CdS crystal with contact as indicated. Middle diagrams: Charge on CdS crystal in the dark (horizontal bar) and under illumination as a function of wavelength. Contacts as upper diagram. Lower diagrams: Photocurrent versus wavelength under 10 V externally applied and the same illumination as for the charge measurement. (After Ruppel and Schlaile.[179])

[179] W. Ruppel and H. G. Schlaile, *Phys. Status Solidi* **15**, 675 (1966).

on the metal work function. If it is smaller than the work function of the CdS crystal, the CdS should become negatively charged in contact with the metal, and vice versa. Under illumination with weakly absorbed light, the majority-carrier concentration increased more in the volume than at the interface, because of fast recombination at the interface. It follows that the crystal charge becomes more positive with increasing illumination. For an *n*-type photoconductor with ohmic contacts, such as CdS–In, the total charge is expected to change from negative to positive upon strong illumination.

These expectations are borne out by the observations.[179] Figure 7 shows, in the upper diagrams, the dark current–voltage characteristics of a CdS–In and a CdS–crystalline Se contact. While the In contact forms an electron accumulation layer in the CdS, the *p*-Se forms a rectifying heterojunction with CdS. In the middle diagrams, the charge on the CdS upon removal of the In or Se, respectively, is shown. As expected, the charge is negative in the dark in the case of electron accumulation in contact with In, and positive in case of electron depletion in contact with Se. Under illumination, the charge becomes more positive in any event, and changes sign in case of the contact with In. The lowest diagrams show the close correspondence of the photoconduction with the wavelength dependence of the charge on the illuminated crystal. It confirms the contention that the change under illumination of the crystal charge localized in the contact region of the CdS is indeed due to an increase in the free-electron concentration in the volume of the CdS.

It is an old experience that charge is accumulated on insulators by repeated contact with metals. By repeated contact of photoconductive additively colored KCl with metals, it was shown[180] that the charge ultimately obtained on the insulator is given both by the amount of charge transferred between metal and insulator upon each contact and by the amount of charge that relaxes due to the volume conductivity of the insulator, which, in the case, of a photoconductor, depends on the illumination.

[180] G. Weingärtner, U. Tödheide-Haupt, and W. Ruppel, *Z. Angew. Phys.* **25**, 257 (1968).

Author Index

Numbers in parentheses are footnote numbers and are inserted to enable the reader to locate those cross references where the author's name does not appear at the point of reference in the text.

N

O

P

Subject Index

D

T